Statistical Analysis with Excel™ FOR DUMMIES®

Joseph Schmuller, PhD

WILEY

Wiley Publishing, Inc.

Statistical Analysis with Excel™ For Dummies®

Published by
Wiley Publishing, Inc.
111 River Street
Hoboken, NJ 07030-5774

Library of Congress Control Number: 2005920688

ISBN: 0-7645-7594-5

Manufactured in the United States of America

10 9 8 7 6 5 4 3 2

1B/RX/QS/QV/IN

WILEY

About the Author

A veteran of over twenty years in Information Technology, **Joseph Schmuller** is a Technical Architect at Blue Cross-Blue Shield of Florida. He is the author of several books on computing, including the three editions of *Teach Yourself UML in 24 Hours* (SAMS), and he has written numerous articles on advanced technology. From 1991 through 1997, he was Editor-in-Chief of *PC AI* magazine.

He is a former member of the American Statistical Association, and he has taught statistics at the undergraduate and graduate levels. He holds a B.S. from Brooklyn College, an M.A. from the University of Missouri-Kansas City, and a Ph.D. from the University of Wisconsin. He and his family live in Jacksonville, Florida, where he is an Adjunct Professor at the University of North Florida.

Dedication

To Jesse Edward Sprague, my best friend in the whole world

Author's Acknowledgments

One thing I have to tell you about writing a *For Dummies* book — it's an incredible amount of fun. You get to air out your ideas in a friendly, conversational way and you get a chance to throw in some humor, too.

Best of all, you get to work with a wonderfully supportive team. Acquisitions editor Tom Heine and project editor Sarah Hellert of Wiley Publishing, Inc. have been encouraging, cooperative, and patient from beginning to end. Technical editor Dennis Short of Purdue University helped make the book you're reading as technically bulletproof as possible. Copy editor Kim Heusel tightened up the wording. Any errors that remain are in the author's court. Tom, Sarah, Kim, and Dennis have my deepest thanks.

The Wiley Publishing team includes many others whose contributions were crucial. Production coordinator Adrienne Martinez had the Herculean task of coordinating text, images, equations, scheduling, and more. Graphics technicians Karl Brandt, Lauren Goddard, Lynsey Osborn, and Mary Virgin are among those who ensured that the graphics you see are top-notch. Carrie Foster and other page layout technicians laid out the book. All these fine professionals have my respect and my appreciation.

My sincerest thanks to my agent, David Fugate of Waterside Productions, for getting me involved in another great project.

A long overdue thank-you to mentors in college and graduate school who helped shape my statistical knowledge: Mitch Grossberg (Brooklyn College), Mort Goldman, Al Hillix, Larry Simkins, and Jerry Sheridan (University of Missouri-Kansas City), and Cliff Gillman and John Theios (University of Wisconsin-Madison).

I thank my mother and my brother David for their love and support and for always being there for me, and Kathryn for so much more than I can say.

Finally, a special note of thanks to my friend Brad, who suggested this whole thing in the first place!

Publisher's Acknowledgments

We're proud of this book; please send us your comments through our online registration form located at www.dummies.com/register/.

Some of the people who helped bring this book to market include the following:

Acquisitions, Editorial, and Media Development

Project Editor: Sarah Hellert

Acquisitions Editor: Tom Heine

Copy Editor: Kim Heusel

Technical Editor: Dennis R. Short, PhD

Editorial Manager: Robyn Siesky

Media Development Manager: Laura VanWinkle

Media Development Supervisor: Richard Graves

Editorial Assistant: Adrienne D. Porter

Cartoons: Rich Tennant (www.the5thwave.com)

Composition Services

Project Coordinator: Adrienne Martinez

Layout and Graphics: Lauren Goddard, Denny Hager, Joyce Haughey, Stephanie D. Jumper, Barry Offringa, Lynsey Osborn, Jacque Roth, Julie Trippetti

Proofreaders: Laura Albert, Leeann Harney, Carl Pierce, TECHBOOKS Production Services

Indexer: TECHBOOKS Production Services

Special Help
Rebecca Huehls

Publishing and Editorial for Technology Publishing

Richard Swadley, Vice President and Executive Group Publisher

Barry Pruett, Vice President and Publisher, Visual/Web Graphics

Andy Cummings, Vice President and Publisher, Technology Dummies

Mary Bednarek, Executive Acquisitions Director, Technology Dummies

Mary C. Corder, Editorial Director, Technology Dummies

Publishing for Consumer Dummies

Diane Graves Steele, Vice President and Publisher

Joyce Pepple, Acquisitions Director

Composition Services

Gerry Fahey, Vice President of Production Services

Debbie Stailey, Director of Composition Services

Contents at a Glance

Table of Contents

Introduction

*W*hat? Yet another statistics book? Well . . . this is a statistics book, alright, but in my humble (and thoroughly biased) opinion, it's not *just* another statistics book.

What? Yet another Excel book? Same thoroughly biased opinion — it's not just another Excel book.

So here's the deal. Many statistics books teach you the concepts but don't give you a way to apply them. That often leads to a lack of understanding. With Excel, you have a ready-made package for applying statistics concepts.

Looking at it from the opposite direction, many Excel books show you Excel's capabilities, but don't tell you about the concepts behind them. Before I tell you about an Excel statistical tool, I give you the statistical foundation it's based on. That way, you understand the tool when you use it — and you use it more effectively.

I didn't want to write a book that's just "select this menu" and "click this button." Some of that is necessary, of course, in any book that shows you how to use a software package. My goal was to go way beyond that.

I also didn't want to write a statistics "cookbook": When-faced-with-problem-#310-use-statistical-procedure-#214. My goal was to go way beyond that, too.

Bottom line: This book isn't just about statistics or just about Excel — it sits firmly at the intersection of the two. In the course of telling you about statistics, I cover every Excel statistical feature. (Well . . . *almost*. I left one out. It's called Fourier Analysis. All the necessary math to understand it would take a whole book, and you might never use this tool, anyway.)

About This Book

Although statistics involves a logical progression of concepts, I organized this book so you can open it up in any chapter and start reading. The idea is for you to find what you're looking for in a hurry and use it immediately — whether it's a statistical concept or an Excel tool.

On the other hand, cover-to-cover is okay if you're so inclined. If you're a statistics newbie and you have to use Excel for statistical analysis, I recommend you begin at the beginning — even if you know Excel pretty well.

What You Can Safely Skip

Any reference book throws a lot of information at you, and this one is no exception. I intended it all to be useful, but I didn't aim it all at the same level. So if you're not deeply into the subject matter, you can avoid paragraphs marked with the Technical Stuff icon.

Every so often, you'll run into sidebars. They provide information that elaborates on a topic, but they're not part of the main path. If you're in a hurry, you can breeze past them.

Because I wrote this book so you can open it up anywhere and start using it, step-by-step instructions appear throughout. Many of the procedures I describe have steps in common. After you go through some of the procedures, you can probably skip the first few steps when you come to a procedure you haven't been through before.

Foolish Assumptions

This is not an introductory book on Excel or on Windows, so I'm assuming:

- ✔ You know how to work with Windows.

  I don't go through the details of pointing, clicking, how to select a menu, and so forth.

- ✔ You have Excel installed on your computer and you can work along with the examples.

  I don't take you through the steps of Excel installation. Incidentally, I use Excel 2003. If you're using Excel 97 or Excel 2000, that's okay. The statistical functionality is the same. Some of the screen shots in the book will look a little different from what appears on your computer, however. In a section in Chapter 2, I address earlier versions.

- ✔ You've worked with Excel before, and you understand the essentials of worksheets and formulas.

  If you don't know much about Excel, consider looking into Greg Harvey's excellent Excel books in the *For Dummies* series.

How This Book Is Organized

I organized this book into five parts.

Part 1: Statistics and Excel: A Marriage Made in Heaven

In Part I, I provide a general introduction to statistics and to Excel's statistical capabilities. I discuss important statistical concepts and describe useful Excel techniques. If it's been a long time since your last course in statistics or if you never had a statistics course at all, start here. If you haven't worked with Excel's built-in functions (of any kind) definitely start here.

Part II: Describing Data

Part of statistics is to take sets of numbers and summarize them in meaningful ways. Here's where you find out how to do that. We all know about averages and how to compute them. But that's not the whole story. In this part, I tell you about additional statistics that fill in the gaps, and I show you how to use Excel to work with those statistics. I also introduce Excel graphics in this part.

Part III: Drawing Conclusions from Data

Part III addresses the fundamental aim of statistical analysis: To go beyond the data and help decision-makers make decisions. Usually, the data are measurements of a sample taken from a large population. The goal is to use these data to figure out what's going on in the population.

This opens a wide range of questions: What does an average mean? What does the difference between two averages mean? Are two things associated? These are only a few of the questions I address in Part III, and I discuss the Excel functions and tools that help you answer them.

Part IV: Working with Probability

Probability is the basis for statistical analysis and decision-making. In Part IV, I tell you all about it. I show you how to apply probability, particularly in the area of modeling. Excel provides a rich set of built-in capabilities that help you understand and apply probability. Here's where you find them.

Part V: The Part of Tens

Part V meets two objectives. First, I get to stand on the soapbox and rant about statistical peeves and about helpful hints. The peeves and hints total up to ten. Also, I discuss ten (okay, eleven) Excel things I couldn't fit in any other chapter. They come from all over the world of statistics. If it's Excel and statistical, and if you can't find it anywhere else in the book, you'll find it here.

Pretty handy, the Part of Tens.

In addition to performing calculations, Excel serves another purpose: Record-keeping. Although it's not a dedicated database, Excel does offer some database functions. Some of them are statistical in nature. I introduce Excel database functions in the appendix, along with pivot tables that allow you to turn your database inside out and look at your data in different ways.

Icons Used in This Book

As is the case with all *For Dummies* books, icons appear all over. Each one is a little picture in the margin that lets you know something special about the paragraph it's next to.

This icon points out a hint or a shortcut that helps you in your work and makes you an all-around better human being.

This one points out timeless wisdom to take with you long after you finish this book, grasshopper.

Pay attention to this icon. It's a reminder to avoid something that might gum up the works for you.

As I mentioned in "What You Can Safely Skip," this icon indicates material you can blow past if statistics and Excel aren't your passion.

Where to Go from Here

You can start the book anywhere, but here are a few hints. Want to learn the foundations of statistics? Turn the page. Introduce yourself to Excel's statistical features? That's Chapter 2. Want to start with graphics? Hit Chapter 3. For anything else, find it in the Table of Contents or in the index and go to it.

If you have half as much fun reading and using this book as I had writing it, you'll have a blast.

Part I

Statistics and Excel: A Marriage Made in Heaven

The 5th Wave   By Rich Tennant

©RICHTENNANT

"Get ready, Mona — here come the stats."

In this part . . .

Part I deals with the foundations of statistics and with the statistics-related tasks that Excel can perform. On the statistics side, this part introduces samples and populations, hypothesis testing, the two types of errors in decision-making, independent and dependent variables, and probability. It's a brief introduction to all the statistical concepts I explore in the rest of the book. On the Excel side, I focus on cell referencing, and on how to use worksheet functions, array functions, and data analysis tools. My objective is to get you thinking about statistics conceptually, and about Excel as a statistical analysis tool.

Chapter 1

Evaluating Data in the Real World

· ·

In This Chapter

▶ Introducing statistical concepts

▶ Generalizing from samples to populations

▶ Getting into probability

▶ Making decisions

▶ Understanding important Excel fundamentals

· ·

*T*he field of statistics is all about decision-making — decision-making based on groups of numbers. Statisticians constantly ask questions: What do the numbers tell us? What are the trends? What predictions can we make?

To answer these questions, statisticians have developed an impressive array of analytical tools. These tools help us to make sense of the mountains of data that are out there waiting for us to delve into, and to understand the numbers we generate in the course of our own work.

The Statistical (and Related) Notions You Just Have to Know

Because intensive calculation is often part and parcel of the statistician's toolset, many people have the misconception that statistics is about number crunching. Number crunching is just one small part of the path to sound decisions, however.

By shouldering the number-crunching load, software increases our speed of traveling down that path. Some software packages are specialized for statistical analysis and contain many of the tools that statisticians use. Although not marketed specifically as a statistical package, Excel provides a number of these tools, which is the reason I wrote this book.

I said that number crunching is a small part of the path to sound decisions. The most important part is the concepts statisticians work with, and that's what I'll talk about for most of the rest of this chapter.

After that, I'll tell you about some important Excel fundamentals.

Samples and populations

On election night, TV commentators routinely predict the outcome of elections before the polls close. Most of the time they're right. How do they do that?

The trick is to interview a sample of voters after they cast their ballots. Assuming the voters tell the truth about whom they voted for, and assuming the sample truly represents the population, network analysts use the sample data to generalize to the population of voters.

This is the job of a statistician — to use the findings from a sample to make a decision about the population from which the sample comes. But sometimes those decisions don't turn out the way the numbers predicted. Flawed pre-election polling led to the memorable picture of President Harry Truman holding up a copy of the *Chicago Daily Tribune* with the famous, but wrong, headline "Dewey Defeats Truman" after the 1948 election. Part of the statistician's job is to express how much confidence he or she has in the decision. Another election-related example speaks to the idea of confidence in a decision. Pre-election polls (again, assuming a representative sample of voters) tell you the percentage of sampled voters who prefer each candidate. The polling organization adds how accurate they believe the polls are. When you hear a newscaster say something like "accurate to within 3 percent," you're hearing a judgment about confidence.

Here's another example. Suppose you've been assigned to find the average reading speed of all fifth-grade children in the U.S., but you haven't got the time or the money to test them all. What would you do?

Your best bet is to take a sample of fifth-graders, measure their reading speeds (in words per minute), and calculate the average of the reading speeds in the sample. You can then use the sample average as an estimate of the population average.

Estimating the population average is one kind of *inference* that statisticians make from sample data. I discuss inference in more detail in the section "Inferential Statistics."

Now for some terminology you have to know: Characteristics of a population (like the population average) are called *parameters*, and characteristics of a sample (like the sample average) are called *statistics*. When you confine your field of view to samples, your statistics are *descriptive*. When you broaden your horizons and concern yourself with populations, your statistics are *inferential*.

Now for a notation convention you have to know: Statisticians use Greek letters (μ, σ, ρ) to stand for parameters, and English letters ($\bar{x}$, s, r) to stand for statistics. Figure 1-1 summarizes the relationship between populations and samples, and parameters and statistics.

Figure 1-1: The relationship between populations, samples, parameters, and statistics.

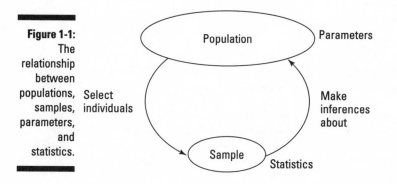

Variables: Dependent and independent

Simply put, a *variable* is something that can take on more than one value. (Something that can have only one value is called a *constant*.) Some variables you might be familiar with are today's temperature, the Dow Jones Industrial Average, your age, and the value of the dollar against the euro.

Statisticians care about two kinds of variables, *independent* and *dependent*. Each kind of variable crops up in any study or experiment, and statisticians assess the relationship between them.

For example, imagine a new way of teaching reading that's intended to increase the reading speed of fifth-graders. Before putting this new method into schools, it would be a good idea to test it. To do that, a researcher would randomly assign a sample of fifth-grade students to one of two groups: One group receives instruction via the new method; the other receives instruction via traditional methods. Before and after both groups receive instruction, the researcher measures the reading speeds of all the children in this study. What happens next? I'll get to that in the upcoming section entitled "Inferential Statistics: Testing Hypotheses."

For now, understand that the independent variable here is Method of Instruction. The two possible values of this variable are New and Traditional. The dependent variable is reading speed.

In general, the idea is to try to find out if changes in the independent variable are associated with changes in the dependent variable.

In the examples that appear throughout the book, I'll show you how to use Excel to calculate various characteristics of groups of scores. I'd like you to bear in mind that each time I show you a group of scores, I'm really talking about the values of a dependent variable.

Types of data

Data come in four kinds. When you work with a variable, the way you work with it depends on what kind of data it is.

The first variety is called *nominal* data. If a number is a piece of nominal data, it's just a name. Its value doesn't signify anything. A good example is the number on an athlete's jersey. It's just a way of identifying the athlete and distinguishing him or her from teammates. The number doesn't indicate the athlete's level of skill.

Next comes *ordinal* data. Ordinal data are all about order, and numbers begin to take on meaning over and above just being identifiers. A higher number indicates the presence of more of a particular attribute than a lower number. One example is Moh's Scale. Used since 1822, it's a scale whose values are 1 through 10. Mineralogists use this scale to rate the hardness of substances. Diamond, rated at 10, is the hardest. Talc, rated at 1, is the softest. A substance that has a given rating can scratch any substance that has a lower rating.

What's missing from Moh's Scale (and from all ordinal data) is the idea of equal intervals and equal differences. The difference between a hardness of 10 and a hardness of 8 is not the same as the difference between a hardness of 6 and a hardness of 4.

Interval data provides equal differences. Fahrenheit temperatures provide an example of interval data. The difference between 60 degrees and 70 degrees is the same as the difference between 80 degrees and 90 degrees.

Here's something that might surprise you about Fahrenheit temperatures: A temperature of 100 degrees is not twice as hot as a temperature of 50 degrees. For ratio statements (twice as much as, half as much as) to be valid, zero has to mean the complete absence of the attribute you're measuring. A temperature of 0 degrees F doesn't mean the absence of heat — it's just an arbitrary point on the Fahrenheit scale.

The last data type, *ratio* data, includes a meaningful zero point. For temperatures, the Kelvin scale gives us ratio data. One hundred degrees Kelvin is twice as hot as 50 degrees Kelvin. This is because the Kelvin zero point is *absolute zero*, where all molecular motion (the basis of heat) stops. Another example is a ruler. Eight inches is twice as long as 4 inches. A length of zero means a complete absence of length.

Any of these types can form the basis for an independent variable or a dependent variable. The analytical tools you use depend on the type of data you're dealing with.

A little probability

When statisticians make decisions, they express their confidence about those decisions in terms of probability. They can never be certain about what they decide. They can only tell you how probable their conclusions are.

So what is probability? The best way to attack this is with a few examples. If you toss a coin, what's the probability that it comes up heads? Intuitively, you know that if the coin is fair, you have a 50-50 chance of heads and a 50-50 chance of tails. In terms of the kinds of numbers associated with probability, that's 1/2.

How about rolling a die (one member of a pair of dice)? What's the probability that you roll a 3? Hmmm . . . a die has six faces and one of them is 3, so that ought to be 1/6, right? Right.

Here's one more. You have a standard deck of playing cards. You select one card at random. What's the probability that it's a club? Well, a deck of cards has four suits, so that answer is 1/4.

I think you're getting the picture. If you want to know the probability that an event occurs, figure out how many ways that event can happen and divide by the total number of events that can happen. In each of the three examples, the event we were interested in (head, 3, or club) only happens one way.

Things can get a bit more complicated. When you toss a die, what's the probability that you'll roll a 3 or a 4? Now you're talking about two ways the event you're interested in can occur, so that's $(1 + 1)/6 = 2/6 = 1/3$. What about the probability of rolling an even number? That has to be 2, 4, or 6, and the probability is $(1 + 1 + 1)/6 = 3/6 = 1/2$.

On to another kind of probability question. Suppose you roll a die and toss a coin at the same time. What's the probability you roll a 3 and the coin comes up heads? Consider all the possible events that could occur when you roll a die and toss a coin at the same time. Your outcome could be a head and 1-6, or a tail and 1-6. That's a total of 12 possibilities. The head-and-3 combination can only happen one way. So, the answer is 1/12.

In general the formula for the probability that a particular event occurs is

$$\text{Pr(event)} = \frac{\text{Number of ways the event can occur}}{\text{Total number of possible events}}$$

I began this section by saying that statisticians express their confidence about their decisions in terms of probability, which is really why I brought up this topic in the first place. This line of thinking leads us to *conditional* probability — the probability that an event occurs given that some other event occurs. For example, suppose I roll a die, take a look at it (so that you can't see it), and I tell you that I've rolled an even number. What's the probability that I've rolled a 2? Ordinarily, the probability of a 2 is 1/6, but I've narrowed the field. I've eliminated the three odd numbers (1, 3, and 5) as possibilities. In this case, only the three even numbers (2, 4, and 6) are possible, so now the probability of rolling a 2 is 1/3.

Exactly how does conditional probability play into statistical analysis? Read on.

Inferential Statistics: Testing Hypotheses

In advance of doing a study, a statistician draws up a tentative explanation — a *hypothesis* — as to why the data might come out a certain way. After the study is complete and the sample data are all tabulated, he or she faces the essential decision a statistician has to make — whether or not to reject the hypothesis.

That decision is wrapped in a conditional probability question: What's the probability of obtaining the data, given that this hypothesis is correct? Statistical analysis provides tools to calculate the probability. If the probability turns out to be low, the statistician rejects the hypothesis.

Here's an example. Suppose you're interested in whether or not a particular coin is fair — whether it has an equal chance of coming up heads or tails. To study this issue, you'd take the coin and toss it a number of times — say 100. These 100 tosses make up your sample data. Starting from the hypothesis that the coin is fair, you'd expect that the data in your sample of 100 tosses would show 50 heads and 50 tails.

If it turns out to be 99 heads and one tail, you'd undoubtedly reject the fair coin hypothesis. Why? The conditional probability of getting 99 heads and one tail given a fair coin is very low. Wait a second. The coin could still be fair and you just happened to get a 99-1 split, right? Absolutely. In fact, you never really know. You have to gather the sample data (the results from 100 tosses) and make a decision. Your decision might be right, or it might not.

Juries face this all the time. They have to decide among competing hypotheses that explain the evidence in a trial. (Think of the evidence as data.) One hypothesis is that the defendant is guilty. The other is that the defendant is not guilty. Jury members have to consider the evidence and, in effect, answer a conditional probability question: What's the probability of the evidence given that the defendant is not guilty? The answer to this question determines the verdict.

Null and alternative hypotheses

Consider once again that coin-tossing study I just mentioned. The sample data are the results from the 100 tosses. Before tossing the coin, you might start with the hypothesis that the coin is a fair one, so that you expect an equal number of heads and tails. This starting point is called the *null hypothesis*. The statistical notation for the null hypothesis is H_0. According to this hypothesis, any heads-tails split in the data is consistent with a fair coin. Think of it as the idea that nothing in the results of the study is out of the ordinary.

An alternative hypothesis is possible — that the coin isn't a fair one, and it's loaded to produce an unequal number of heads and tails. This hypothesis says that any heads-tails split is consistent with an unfair coin. The alternative hypothesis is called, believe it or not, the *alternative hypothesis*. The statistical notation for the alternative hypothesis is H_1.

With the hypotheses in place, toss the coin 100 times and note the number of heads and tails. If the results are something like 90 heads and 10 tails, it's a good idea to reject H_0. If the results are around 50 heads and 50 tails, don't reject H_0.

Similar ideas apply to the reading-speed example I gave earlier. One sample of children receives reading instruction under a new method designed to increase reading speed, the other learns via a traditional method. Measure the children's reading speeds before and after instruction, and tabulate the improvement for each child. The null hypothesis, H_0, is that one method isn't different from the other. If the improvements are greater with the new method than with the traditional method — so much greater that it's unlikely that the methods aren't different from one another — reject H_0. If they're not, don't reject H_0.

Notice that I *didn't* say "accept H$_0$." The way the logic works, you *never* accept a hypothesis. You either reject H$_0$ or don't reject H$_0$.

Notice also that in the coin-tossing example I said around 50 heads and 50 tails. What does around mean? Also, I said if it's 90-10, reject H$_0$. What about 85-15? 80-20? 70-30? Exactly how much different from 50-50 does the split have to be for you to reject H$_0$? In the reading-speed example, how much greater does the improvement have to be to reject H$_0$?

I won't answer these questions now. Statisticians have formulated decision rules for situations like this, and we'll explore those rules throughout the book.

Two types of error

Whenever you evaluate the data from a study and decide to reject H$_0$ or to not reject H$_0$, you can never be absolutely sure. You never really know what the true state of the world is. In the context of the coin-tossing example, that means you never know for certain if the coin is fair or not. All you can do is make a decision based on the sample data you gather. If you want to be certain about the coin, you'd have to have the data for the entire population of tosses — which means you'd have to keep tossing the coin until the end of time.

Because you're never certain about your decisions, it's possible to make an error regardless of what you decide. As I mentioned before, the coin could be fair and you just happen to get 99 heads in 100 tosses. That's not likely, and that's why you reject H$_0$. It's also possible that the coin is biased, and yet you just happen to toss 50 heads in 100 tosses. Again, that's not likely and you don't reject H$_0$ in that case.

Although not likely, those errors are possible. They lurk in every study that involves inferential statistics. Statisticians have named them *Type I* and *Type II*.

If you reject H$_0$ and you shouldn't, that's a Type I error. In the coin example, that's rejecting the hypothesis that the coin is fair, when in reality it is a fair coin.

If you don't reject H$_0$ and you should have, that's a Type II error. That happens if you don't reject the hypothesis that the coin is fair, and in reality it's biased.

How do you know if you've made either type of error? You don't — at least not right after you make your decision to reject or not to reject H$_0$. (If it's possible to know, you wouldn't make the error in the first place!) All you can do is gather more data and see if the additional data are consistent with your decision.

If you think of H_0 as a tendency to maintain the status quo and not interpret anything as being out of the ordinary (no matter how it looks), a Type II error means you missed out on something big. Looked at in that way, Type II errors form the basis of many historical ironies.

Here's what I mean: In the 1950s, a particular TV show gave talented young entertainers a few minutes to perform on stage and a chance to compete for a prize. The audience voted to determine the winner. The producers held auditions around the country to find people for the show. Many years after the show went off the air, the producer was interviewed. The interviewer asked him if he had ever turned down anyone at an audition that he shouldn't have.

"Well," said the producer, "once a young singer auditioned for us and he seemed really odd."

"In what way?" asked the interviewer.

"In a couple of ways," said the producer. "He sang really loud, gyrated his body and his legs when he played the guitar, and he had these long sideburns. We figured this kid would never make it in show business, so we thanked him for showing up, but we sent him on his way."

"Wait a minute, are you telling me you turned down . . ."

"That's right. We actually said 'no' . . . to Elvis Presley!"

Now *that's* a Type II error.

Some Things About Excel You Absolutely Have to Know

Although I'm assuming you're not new to Excel, I think it's wise to take a little time and space up front to discuss a few Excel fundamentals that figure prominently in the number-crunching part of statistical work. Knowing these fundamentals helps you work efficiently with Excel formulas.

Autofilling cells

The first important fundamental is *autofill*, Excel's capability for repeating a calculation throughout a worksheet. Insert a formula into a cell, and you can drag that formula into adjoining cells.

Figure 1-2 is a worksheet of expenditures for R&D in science and engineering at colleges and universities for the years shown. The data, taken from a U.S. National Science Foundation report, are in millions of dollars. I left a column on the right for the total for each field, and a row at the bottom for the total for each year.

If I want to create a formula to calculate the first row total (for Physical Sciences), one way (among several) is to enter

= D2 + E2 + F2 + G2

into cell H2. (A formula always begins with "=".) Press Enter and the total appears in H2.

Now, to put that formula into cells H3 through H10, the trick is to position the cursor on the lower-right corner of H2 until the cursor changes to a "+". Hold down the left mouse button, and drag the mouse through the cells. That "+" is called the cell's *fill handle*.

When you finish dragging, release the mouse button and the row totals appear. This saves huge amounts of time, because you don't have to reenter the formula eight times.

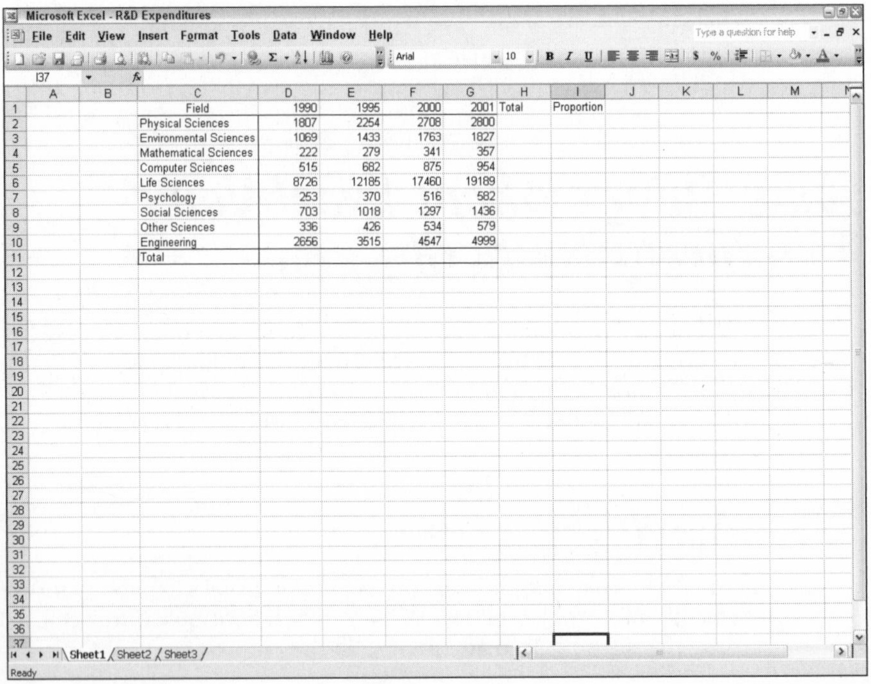

	C	D	E	F	G	H	I
1	Field	1990	1995	2000	2001	Total	Proportion
2	Physical Sciences	1807	2254	2708	2800		
3	Environmental Sciences	1069	1433	1763	1827		
4	Mathematical Sciences	222	279	341	357		
5	Computer Sciences	515	682	875	954		
6	Life Sciences	8726	12185	17460	19189		
7	Psychology	253	370	516	582		
8	Social Sciences	703	1018	1297	1436		
9	Other Sciences	336	426	534	579		
10	Engineering	2656	3515	4547	4999		
11	Total						

Figure 1-2: Expenditures for R&D in science and engineering.

Same thing with the column totals. One way to create the formula that sums up the numbers in the first column (1990) is to enter

=D2 + D3 + D4 + D5 + D6 + D7 + D8 + D9 + D10

into cell D11. Position the cursor on D11's fill handle, drag through row 11 and release in column H, and you autofill the totals into E11 through H11.

Dragging isn't the only way to do it. The other way is to select the row or column of cells you want to autofill (including the one that contains the formula), and click the Edit menu. From the Edit menu, select Fill, and then choose Series from the popup menu. This opens the Series dialog box (see Figure 1-3). In this dialog box, click the Autofill radio button, click OK, and you've accomplished the same thing as dragging and dropping.

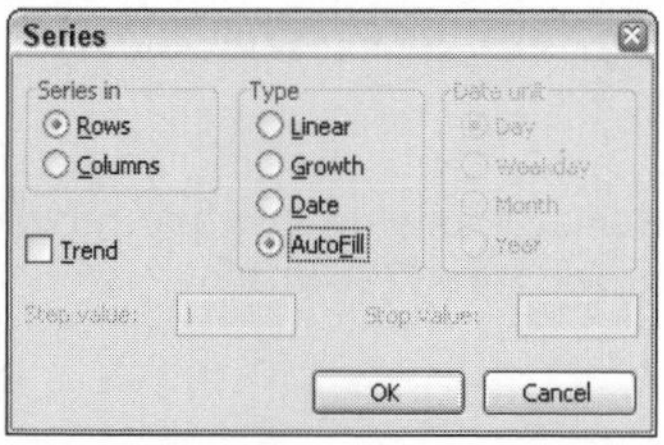

Figure 1-3: The Series dialog box.

I bring all this up because statistical analysis often involves repeating a formula from cell to cell. The formulas are usually more complex than the ones in this section, and you might have to repeat them many times, so it pays to know how to autofill.

Referencing cells

The second important fundamental is the way Excel references worksheet cells. Consider again the worksheet in Figure 1-2. Each autofilled formula is slightly different from the original. This, remember, is the formula in cell H2:

= D2 + E2 + F2 + G2

After autofill, the formula in H3 is

= D3 + E3 + F3 + G3

and the formula in H4 is . . . well, you get the picture.

This is perfectly appropriate. I want the total in each row, so Excel adjusts the formula accordingly as it automatically inserts it into each cell. This is called *relative referencing* — the reference (the cell label) gets adjusted relative to where it is in the worksheet. Here, the formula directs Excel to total up the numbers in the cells in the four columns immediately to the left.

Now for another possibility: Suppose I want to know each row total's proportion of the grand total (the number in H11). That should be straightforward, right? Create a formula for I2, and then autofill cells I3 through I10.

Similar to the earlier example, I start by entering this formula into I2:

= H2/H11

Press Enter and the proportion appears in I2. Position the cursor on the fill handle, drag through column I, release in I10, and . . . D'oh!!! Figure 1-4 shows the unhappy result — the extremely ugly #/DIV0! in I3 through I10. What's the story?

The story is this: Unless you tell it not to, Excel uses relative referencing when you autofill. So, the formula inserted into I3 is not

=H3/H11

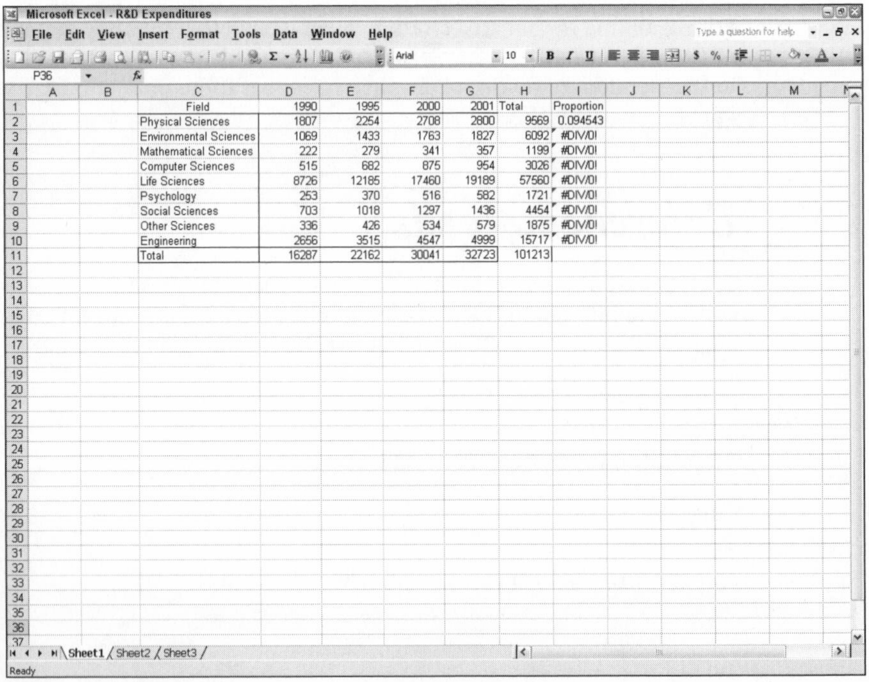

Field	1990	1995	2000	2001	Total	Proportion
Physical Sciences	1807	2254	2708	2800	9569	0.094543
Environmental Sciences	1069	1433	1763	1827	6092	#DIV/0!
Mathematical Sciences	222	279	341	357	1199	#DIV/0!
Computer Sciences	515	682	875	954	3026	#DIV/0!
Life Sciences	8726	12185	17460	19189	57560	#DIV/0!
Psychology	253	370	516	582	1721	#DIV/0!
Social Sciences	703	1018	1297	1436	4454	#DIV/0!
Other Sciences	336	426	534	579	1875	#DIV/0!
Engineering	2656	3515	4547	4999	15717	#DIV/0!
Total	16287	22162	30041	32723	101213	

Figure 1-4: Whoops! Incorrect autofill!

instead, it's

=H3/H12.

Why does H11 become H12? Relative referencing assumes that the formula means divide the number in the cell by whatever number is nine cells south of here in the same column. Because H12 has nothing in it, the formula is telling Excel to divide by zero, which is a no-no.

The idea is to tell Excel to divide all the numbers by the number in H11, not by whatever number is nine cells south of here. To do this, you work with *absolute referencing*. You show absolute referencing by adding dollar signs ($) to the cell ID. The correct formula for I2 is

= H2/$H$11.

This tells Excel not to adjust the column and not to adjust the row when you autofill. Figure 1-5 shows the worksheet with the proportions. In the figure, cell I10 is selected. Note the formula in the Formula Bar — the elongated white box next to the button labeled f_x. (I discuss the Formula Bar and that button in Chapter 2.)

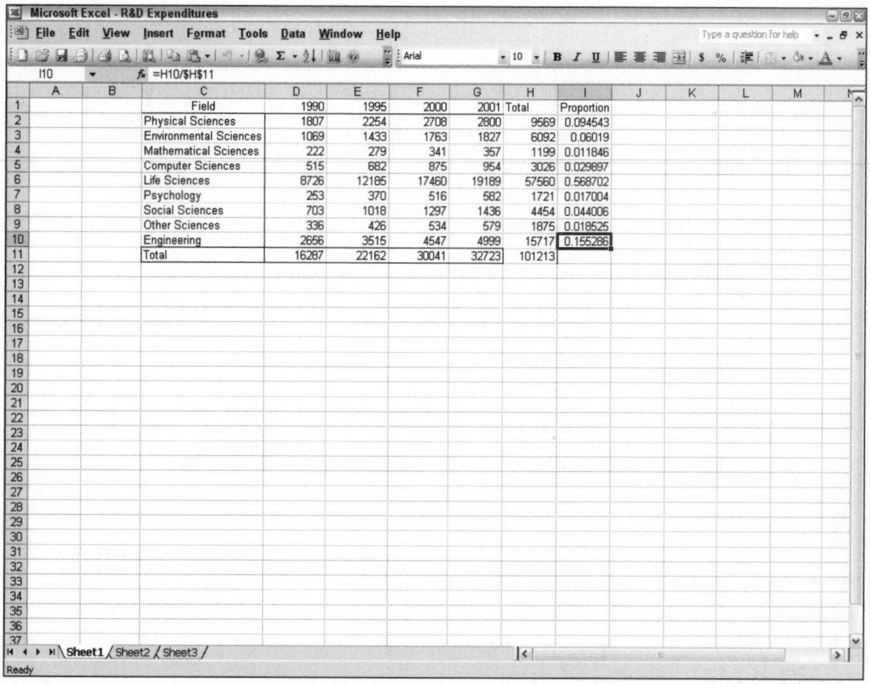

Figure 1-5: Autofill based on absolute referencing.

To convert a relative reference into absolute reference format, select the cell address (or addresses) you want to convert, and press F4. F4 serves as a toggle that goes between relative reference (H11, for example), absolute reference for both the row and column in the address ($H$11), absolute reference for the row-part only (H$11), and absolute reference for the column-part only ($H11).

Chapter 2

Understanding Excel's Statistical Capabilities

*I*n this chapter, I introduce you to Excel's statistical functions and data analysis tools. If you've used Excel, and I'm assuming you have, you're aware of Excel's extensive functionality, of which statistical capabilities are a subset. You can enter a piece of data into each worksheet cell, instruct Excel to carry calculations on data that reside in a set of cells, or use one of Excel's worksheet functions to work on data. Each worksheet function is a built-in formula that saves you the trouble of having to direct Excel to perform a sequence of calculations. Formulas, of course, are the business end of Excel. The data analysis tools go beyond the formulas. Each tool provides a set of informative results.

Getting Started

Many of Excel's statistical features are built into its worksheet functions. Excel provides a button that allows you to easily access these functions. This button is called the *Insert Function* button, and it's labeled with the symbol f_x. In Excel 2003, it's on the Formula Bar. (For Excel 97 and Excel 2000, see the section "Those oldies but goodies.") Figure 2-1 shows the location of the Insert Function button and the Formula Bar.

Insert Function Button

Name Bar   Formula Bar

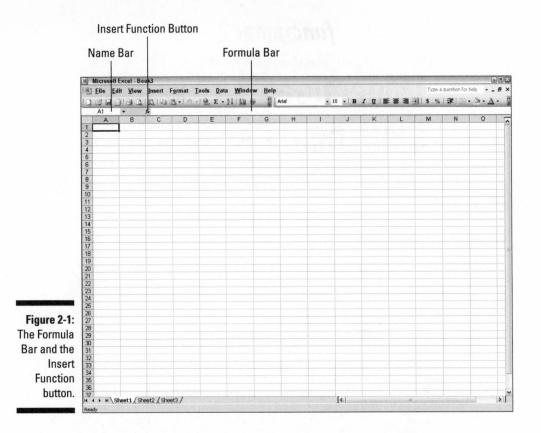

Figure 2-1:
The Formula
Bar and the
Insert
Function
button.

The Formula Bar is like a clone of a cell you select: Information entered into the Formula Bar goes into the selected cell, and vice versa. The Insert Function button, well, inserts a function into both places, and saves you from having to remember the specific details of every worksheet function.

Figure 2-1 also shows the Name Bar, which is something like a running record of what you do in the worksheet. Select a cell, and the cell's address appears in the Name Bar. Click the Insert Function button and the name of the function you selected most recently appears in the Name Bar.

In addition to the formulas, Excel provides a number of data analysis tools you access through the Tools menu.

Setting Up for Statistics

In this section, I show you how to use the worksheet functions and the analysis tools.

Worksheet functions

In general, you use a worksheet function by following these steps:

1. **Type your data into a data array and select a cell for the result.**

2. **Click the Insert Function button (the one labeled f_x) to open the Insert Function dialog box. You can also open the Insert Function dialog box by selecting the Insert menu and then selecting Function.**

3. **In the Insert Function dialog box, select the worksheet formula you want to work with, and click OK to open the Function Arguments dialog box for that function.**

4. **In the Function Arguments dialog box, type the appropriate arguments for the function.**

5. **Click OK to put the result into the selected cell.**

To give you an example, I explore a function that typifies how Excel's worksheet functions work. This function, SUM, adds up the numbers in cells you specify and returns the sum in still another cell that you specify.

Here, step by step, is how to use SUM.

1. **Type your numbers into an array of cells and select a cell for the result.**

   In this example, I've entered 45, 33, 18, 37, 32, 46, and 39 into cells C2 through C8, and selected C9 to hold the sum.

2. **Click the Insert Function button (the button with f_x on it) to open the Insert Function dialog box.**

3. **In the Insert Function dialog box, select SUM.**

   You might have to scroll through the box labeled Select a function to find SUM. Use the upper box (labeled Search for a function) or the middle box (labeled Or Select a category) to help narrow the list in the lower box (Select a function).

4. **Click OK to open the Function Arguments dialog box.**

   Argument is a term from mathematics. It has nothing to do with debates, fights, or confrontations. In mathematics, an argument is a value on which a function does its work.

   Excel guesses that you want to sum the numbers in cells C2 through C8 and identifies that array in the Number1 box. The dialog box shows the sum of the numbers in the array. In this example, the sum is 250. (See Figure 2-2).

5. **Click OK to put the sum into the selected cell.**

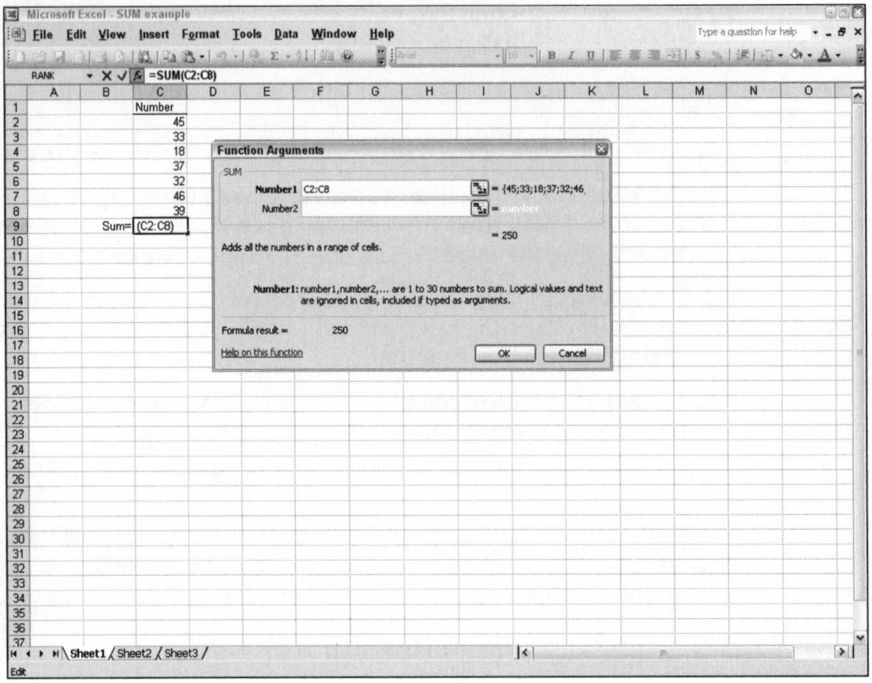

Figure 2-2:
Using SUM.

Note a couple of points. First, as Figure 2-2 shows, the Formula Bar holds

=SUM(C2:C8)

This formula indicates that the value in the selected cell equals the sum of the numbers in cells C2 through C8.

TIP

After you get familiar with a worksheet function and its arguments, you can bypass the Insert Function button and type the function directly into the Formula Bar.

Another noteworthy point is the set of boxes in the Function Arguments dialog box in Figure 2-2. In the figure, you see just two boxes, Number1 and Number2. The data array appears in Number1. So what's Number2 for?

The Number2 box allows you to include an additional argument in the sum. And it doesn't end there. Click in the Number 2 box and the Number 3 box appears. Click in the Number 3 box, and the Number 4 box appears . . . and on and on. The limit is 30 boxes, with each box corresponding to an argument. A value can be another array of cells anywhere in the worksheet, a number, an arithmetic expression that evaluates to a number, or a cell ID. As you type in values, the SUM dialog box shows the updated sum. Clicking OK puts the updated sum into the selected cell.

You won't find this multiargument capability on every worksheet function. Some are designed to work with just one argument. For the ones that do work with multiple arguments, however, you can incorporate data that reside all over the worksheet. Figure 2-3 shows a worksheet with a Function Arguments dialog box that includes data from two arrays of cells, two arithmetic expressions, and one cell. Notice the format of the function in the Formula Bar: A comma separates successive arguments.

I showed you SUM because it's a typical worksheet function, and because it carries out a familiar operation. Although adding numbers together is an integral part of statistical number crunching, SUM is not in the Statistical category of worksheet functions.

If you select a cell in the same column as your data and just below the last data cell, Excel correctly guesses the data array you want to work on. Excel doesn't always guess what you want to do, however. Sometimes when Excel does guess, its guess is incorrect. When either of those things happens, it's up to you to enter the appropriate values into the Function Arguments dialog box.

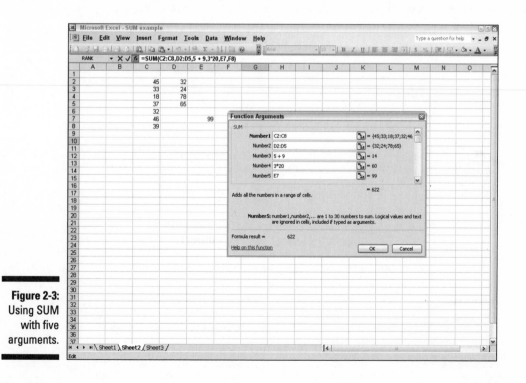

Figure 2-3:
Using SUM
with five
arguments.

Those oldies but goodies

I'm using Excel 2003. Older versions, like Excel 97 and Excel 2000, are a bit different. Although those differences aren't showstoppers, I address them in this section.

To use a worksheet function in Excel 97, you go through a sequence of steps similar to the ones I just described. The Name Box and the Formula Bar are slightly different from their newer counterparts, and the dialog boxes are different, too. What about the all-important Insert Function button? It's there, but it's in a different place and it has a different name.

Figure 2-4 shows the Name Box and the Formula Bar for Excel 97, along with the standard toolbar. Just as in Excel 2003, the Name Box is on the left and the Formula Bar is on the right. The button on the left of the Formula Bar has an equals sign (=), and it's called the *Edit Function* button. The button labeled f_x is on the standard toolbar (not the Formula Bar), and it's called the *Paste Function* button in Excel 97.

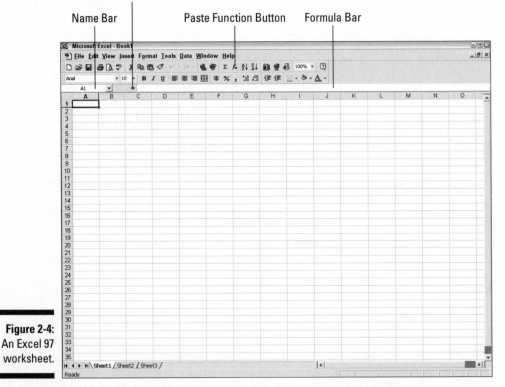

Figure 2-4:
An Excel 97 worksheet.

Here are the steps for using SUM in Excel 97:

1. **Type the numbers and select a cell to hold the sum.**

   I've entered 45, 33, 18, 37, 32, 46, and 39 into C2 through C8, and selected C9.

2. **Click the Paste Function button to open the Paste Function dialog box.**

   Figure 2-5 shows the Paste Function dialog box.

Figure 2-5:
The Excel 97
Paste
Function
dialog box.

Paste Function	
Function category:	Function name:
Most Recently Used	SQRTPI
All	SUBTOTAL
Financial	SUM
Date & Time	SUMIF
Math & Trig	SUMPRODUCT
Statistical	SUMSQ
Lookup & Reference	SUMX2MY2
Database	SUMX2PY2
Text	SUMXMY2
Logical	TAN
Information	TANH

SUM(number1,number2,...)

Adds all the numbers in a range of cells.

[OK]   [Cancel]

3. **Click SUM.**

   From the list on the left (labeled Function Category) select All. In the list on the right (Function Name) scroll down and select SUM.

4. **Click OK to open the SUM dialog box.**

   Figure 2-6 shows the SUM dialog box. Excel puts the data array into the Number1 box and shows the sum (250).

5. **Click OK to put the sum into the selected cell.**

You can also use the Name Box to select a function. With the numbers entered into an array and a cell selected for the result, click the Edit Function button (the one with the = sign), and then click the Name Box's down arrow. This opens a drop-down list of recently used worksheet functions. If the one you're looking for isn't on that list, select More Functions (the last choice on the drop-down list). That opens the Paste Function dialog box and you can take it from there.

In addition to the different name and location of the f_x button, this example shows a couple of other important differences between Excel 97 and Excel 2003. Those differences center around the names of the dialog boxes. First, Excel 2003's Insert Function dialog box is Excel 97's Paste Function dialog box. Also, selecting a function in Excel 97 opens a dialog named for the selected function. In this example, that's the SUM dialog box. Selecting a function in Excel 2003 always opens the Function Arguments dialog box, which displays the name of the selected function and an appropriate layout for entering arguments.

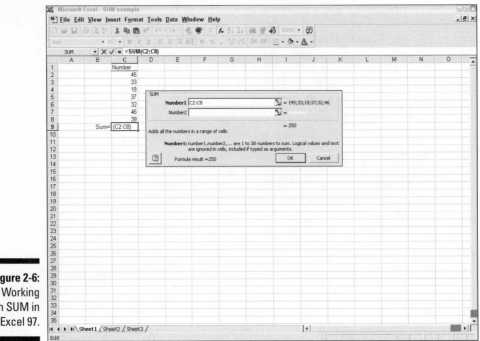

Figure 2-6:
Working
with SUM in
Excel 97.

Earlier versions of Excel have some issues with statistical functions. Excel 2003 addresses those issues. For details, visit `support.microsoft.com` and search for "Excel statistical functions."

Array functions

Most of Excel's built-in functions are formulas that calculate a single value (like a sum) and put that value into a worksheet cell. Excel (regardless of the version) has another type of function. It's called an *array function* because it calculates multiple values and puts those values into an array of cells, rather than into a single cell.

FREQUENCY is a good example of an array function. Its job is to summarize a group of scores by showing how the scores fall into a set of intervals you specify. For example, given these scores:

77, 45, 44, 61, 52, 53, 68, 55

and these intervals:

50, 60, 70, 80

FREQUENCY shows how many are less than or equal to 50 (two in this example), how many are greater than 50 and less than or equal to 60 (that would be three), and so on. The number of scores in each interval is called a *frequency*. A table of the intervals and the frequencies is called a *frequency distribution*.

Here's an example of how to use FREQUENCY:

1. **Type the scores into an array of cells.**

   Figure 2-7 shows a group of scores in cells B2 through B16.

2. **Type the intervals into an array.**

   I've put the intervals in C2 through C9.

3. **Select an array for the frequencies.**

   I've put Frequency as the label at the top of column D, so I select D2 through D10 for the resulting frequencies. Why the extra cell? FREQUENCY returns a vertical array that has one more cell than the frequencies array.

4. **Click the Insert Function button to open the Insert Function dialog box.**

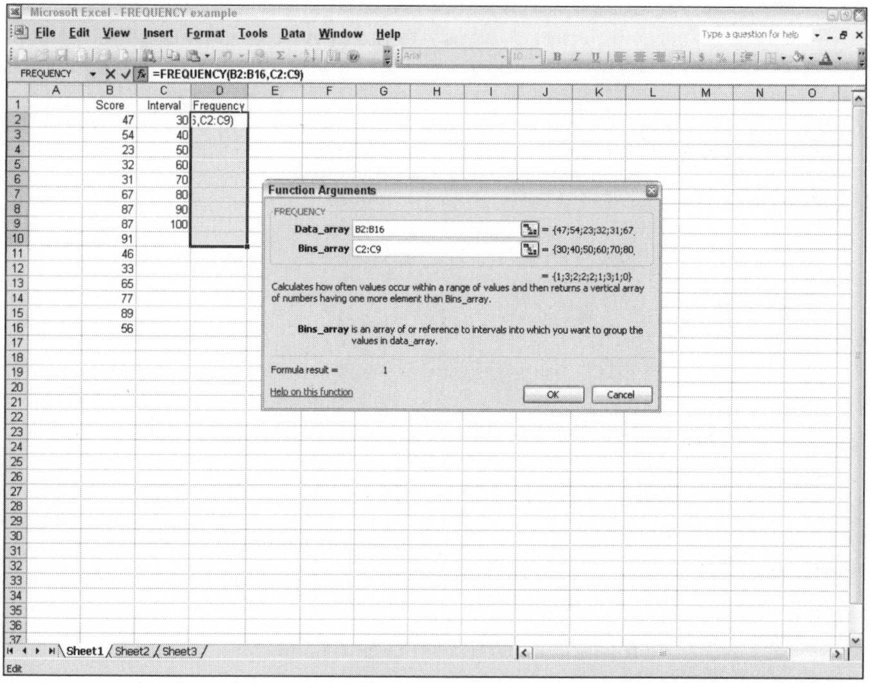

Figure 2-7:
Working with FREQUENCY.

5. **Select FREQUENCY and click OK to open the Function Arguments dialog box.**

   In the Insert Function dialog box, click the down arrow in the middle box (labeled Or Select a Category) and select Statistical from the drop-down list. Scroll down in the bottom box (labeled Select a Function) and select FREQUENCY.

6. **Identify the data array.**

   In the Data_array box enter the cells that hold the scores. In this example, that's B2:B16. I'm assuming you know Excel well enough to know how to do this in several ways.

7. **Identify the intervals array.**

   FREQUENCY refers to intervals as bins, and holds the intervals in the Bins_array box. For this example, C2:C9 goes into the Bins_array box. After you identify both arrays, the Insert Function dialog box shows the frequencies inside a pair of curly brackets.

8. **Press Ctrl+Shift+Enter to close the Function Arguments dialog box.**

   This is VERY important. The tendency is to click OK to put the results into the worksheet, but that doesn't get the job done when you work with an array function. Always use the keystroke combination Ctrl+Shift+Enter to close the Function Arguments dialog box for an array function.

When you close the Function Arguments dialog box, the frequencies go into the appropriate cells, as Figure 2-8 shows.

	A	B	C	D	E
D2			f_x {=FREQUENCY(B2:B16,C2:C9)}		
1		Score	Interval	Frequency	
2		47	30	1	
3		54	40	3	
4		23	50	2	
5		32	60	2	
6		31	70	2	
7		67	80	1	
8		87	90	3	
9		87	100	1	
10		91		0	
11		46			
12		33			
13		65			
14		77			
15		89			
16		56			

Figure 2-8: The finished frequencies.

Note the formula in the Formula Bar:

{= FREQUENCY(B2:B16,C2:C9)}

The curly brackets are Excel's way of telling you this is an array function.

I'm not one to repeat myself, but in this case I'll make an exception. As I said in Step 9, press Ctrl+Shift+Enter whenever you work with an array function. Keep this in mind because the Arguments Function dialog box doesn't provide any reminders. If you click OK after you enter your arguments into an array function, you'll be very frustrated. Trust me.

Data analysis tools

Excel has a set of sophisticated tools for data analysis. Table 2-1 lists the tools I cover. (The one I don't cover, Fourier Analysis, is extremely technical.) Some of the terms in the table may be unfamiliar to you, but you'll know them by the time you finish this book.

Table 2-1	Excel's Data Analysis Tools
Tool	*What it Does*
Anova: Single Factor	Analysis of variance for two or more samples.
Anova: Two Factor with Replication	Analysis of variance with two independent variables, and multiple observations in each combination of the levels of the variables.
Anova: Two Factor without Replication	Analysis of variance with two independent variables, and one observation in each combination of the levels of the variables.
Correlation	With more than two measurements on a sample of individuals, calculates a matrix of correlation coefficients for all possible pairs of the measurements.
Covariance	With more than two measurements on a sample of individuals, calculates a matrix of covariances for all possible pairs of the measurements.
Descriptive Statistics	Generates a report of central tendency, variability, and other characteristics of values in the selected range of cells.
Exponential Smoothing	In a sequence of values, calculates a prediction based on a preceding set of values, and on a prior prediction for those values.
F-Test Two Sample for Variances	Performs an *F*-test to compare two variances.
Histogram	Tabulates individual and cumulative frequencies for values in the selected range of cells.

(continued)

Table 2-1 *(continued)*

Tool	What it Does
Moving Average	In a sequence of values, calculates a prediction, which is the average of a specified number of preceding values.
Random Number Generation	Provides a specified amount of random numbers generated from one of seven possible distributions.
Rank and Percentile	Creates a table that shows the ordinal rank and the percentage rank of each value in a set of values.
Regression	Creates a report of the regression statistics based on linear regression through a set of data containing one dependent variable and one or more independent variables.
Sampling	Creates a sample from the values in a specified range of cells.
t-Test: Two Sample	Three *t*-test tools test the difference between two means. One assumes equal variances in the two samples. Another assumes unequal variances in the two samples. The third assumes matched samples.
z-Test: Two Sample for Means	Performs a two-sample z-test to compare two means when the variances are known.

In order to use these tools, you first have to load them into Excel. On the Tools menu, click Add-Ins to open the Add-Ins dialog box. Click the check box next to the first choice, Analysis ToolPak, and you're good to go. (See Figure 2-9.)

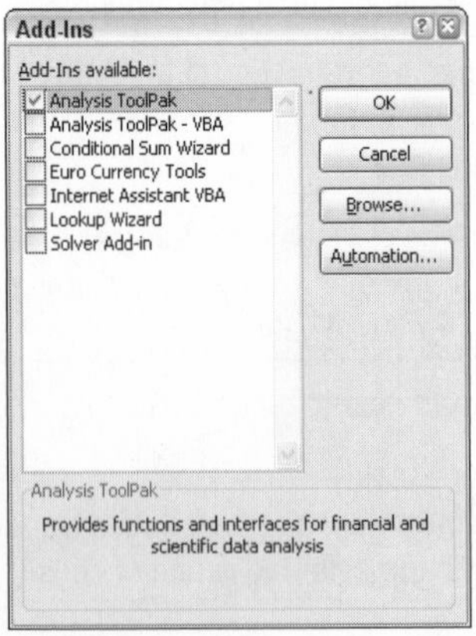

Figure 2-9:
The Add-Ins dialog box.

Adding the Analysis ToolPak adds Data Analysis to the Tools menu. In general, the steps for using a data analysis tool are:

1. **Type your data into an array.**

2. **In the Tools menu, choose Data Analysis to open the Data Analysis dialog box.**

3. **In the Data Analysis dialog box select a data analysis tool.**

4. **Click OK to open the dialog box for the selected tool.**

5. **Identify your input array.**

   In almost all the tools, the dialog box has an area in which you identify the cells that hold the input values for the tool. (The Random Number Generation analysis tool is the exception.) Because each tool has a different purpose, the input areas typically differ slightly from tool to tool.

6. **Identify where you want the output to go.**

   Each analysis tool's dialog box has an area where you select a location for the output: An array of cells on the same page as the input, a new page that the tool creates, or a new workbook.

7. **Click OK to close the dialog box and put the results where you specified in Step 6.**

Here's an example to get you accustomed to using these tools. In this example, I go through the Descriptive Statistics tool. This tool provides a number of statistics that summarize a set of scores. To use this tool, follow these steps:

1. **Type your data into an array.**

   Figure 2-10 shows an array of numbers in cells B2 through B9, with a column header in B1.

2. **From the Tools menu, click Data Analysis to open the Data Analysis dialog box.**

3. **Click Descriptive Statistics to open the Descriptive Statistics dialog box.**

4. **Identify the data array.**

   In the Input Range box, enter the cells that hold the data. For this example, that's B1 through B9. The easiest way to do this is to move the cursor to the upper-leftmost cell (B1), press Shift, and click the lower-rightmost cell (B9). That puts the absolute reference format $B$1:$B$9 into Input Range.

5. **Click the Columns radio button to indicate how the data are organized.** In this example, by columns.

6. **Check the Labels in First Row check box, because the Input Range includes the column heading.**

7. Click the New Worksheet Ply radio button.

This tells Excel to create a new tabbed sheet within the current worksheet, and to send the results to the newly created sheet.

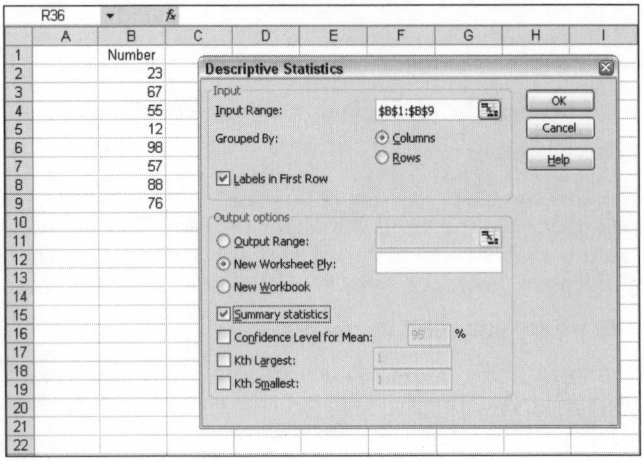

Figure 2-10:
Working
with the
Descriptive
Statistics
analysis
tool.

8. Click the Summary Statistics check box, and leave the others unchecked. Click OK.

The new tabbed sheet (ply) opens, displaying statistics that summarize the data. Figure 2-11 shows the new ply, after I widened Column A.

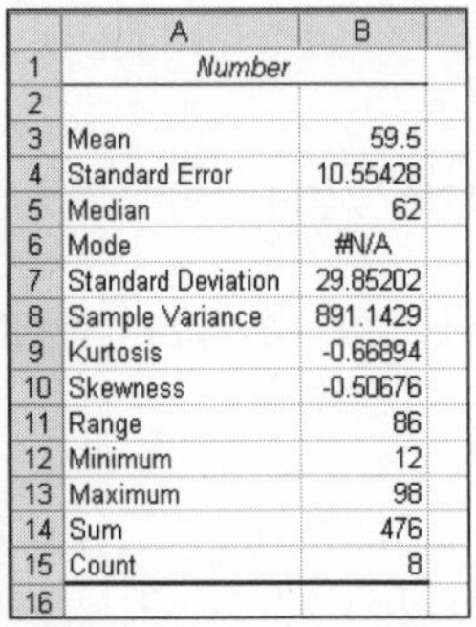

Figure 2-11:
The output
of the
Statistics
Analysis
tool.

For now, I won't tell you the meaning of each individual statistic in the Summary Statistics display. I'll leave that for Chapter 7 when I delve more deeply into descriptive statistics.

Part II
Describing Data

The 5th Wave   By Rich Tennant

"You might want to adjust the value of your 'Nudge' function."

In this part . . .

Here's where you get the details about how to use statistics to summarize and describe data. I begin by showing you how to use the Excel Chart Wizard to produce the kinds of graphs statisticians use. From there, I move on to descriptive statistics — average, variance, standard deviation, and some others. I tell you how to combine a couple of these statistics to standardize scores. Finally, I describe the normal distribution, a very important topic in statistics. Along the way, you find out about Excel functions and data analysis tools that cover all the statistical ideas in this part.

Chapter 3

Show and Tell: Graphing Data

*T*he visual presentation of data is extremely important in statistics. Visual presentation enables you to discern relationships and trends you might not see if you just look at numbers. Visual presentation helps in another way: It's valuable for presenting ideas to groups and making them understand your point of view.

Graphs come in many varieties. In this chapter, I explore the types of graphs you use in statistics and when it's advisable to use them. I also show you how to use Excel to create those graphs.

Why Use Graphs?

Suppose you have to make a pitch to a Congressional committee about commercial space revenues in the early 1990s.

Which would you rather present: The data in Table 3-1 or the graph in Figure 3-1 that shows the same data? (The data, by the way, are from the U.S. Department of Commerce via the Statistical Abstract of the U.S.)

Table 3-1	U.S. Commercial Space Revenues from 1990 to 1994 (In Millions of Dollars)				
Industry	1990	1991	1992	1993	1994
Commercial Satellites Delivered	1,000	1,300	1,300	1,100	1,400
Satellite Services	800	1,200	1,500	1,850	2,330
Satellite Ground Equipment	860	1,300	1,400	1,600	1,970
Commercial Launches	570	380	450	465	580
Remote Sensing Data	155	190	210	250	300
Commercial R&D Infrastructure	0	0	0	30	60
Total	3,385	4,370	4,860	5,295	6,640

US Commercial Space Revenues: 1990–1994

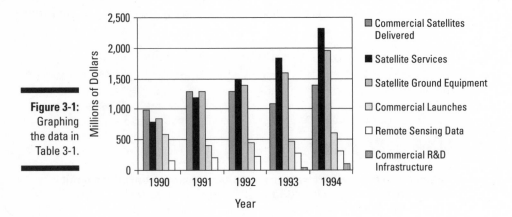

Figure 3-1: Graphing the data in Table 3-1.

Which one makes a bigger and more lasting impact? While the table is certainly informative, you can make a strong argument that the graph gets the point across better and more memorably. Eyes that glaze over when looking at numbers often shine brighter when looking at pictures.

The graph shows you trends you might not see as quickly on the table. (Satellite services rose fastest. Commercial launches, not so much.) Bottom line: Tables are good; graphs are better.

Graphs help bring concepts to life that might otherwise be difficult to understand. In fact, I do that throughout the book. I illustrate points by . . . well . . . illustrating points!

Some Fundamentals

Like the graph in Figure 3-1, most graph formats have a horizontal axis and a vertical axis. The *pie graph*, a format I show you later in this chapter, is an exception. By convention, the horizontal axis is called the *x-axis* and the vertical axis is called the *y-axis*.

Also by convention, what goes on the horizontal axis is called the *independent variable*, and what goes on the vertical axis is called the *dependent variable*. One of Excel's graph formats reverses that convention, and I bring that to your attention when I cover it in the section "Passing the Bar."

Just to give you a heads-up, Excel calls that reversed-axis format a *bar graph*. You might have seen graphs like the one in Figure 3-1 referred to as bar graphs. So have I. Excel calls this a *column graph*, so I use "columns" from here on.

Getting back to "independent" and "dependent," those terms imply that changes in the vertical direction depend (at least partly) on changes in the horizontal direction.

Another fundamental principle of creating a graph: Don't wear out the viewer's eyes! If you put too much into a graph in the way of information or special effects, you defeat the whole purpose of the graph.

For example, in Figure 3-1 I had to make some choices about filling in the columns. Color-coded columns would have been helpful, but the page you're looking at only shows black, white, and shades of gray. Varying shades of gray are a possibility, but they might be too hard to distinguish.

Instead, after black, white, and gray, I used patterning — like the diagonal lines in the Commercial Launches columns. Notice I saved the patterns for the smaller columns. Patterns in the larger ones would have been harder on the eyes. Also, I don't put columns with diagonals in one direction next to columns with diagonals in the other direction — that would have been extremely difficult to look at.

A lot of graph creation comes with experience, and you just have to use your judgment. In this case, my judgment came into play with the horizontal gridlines. In most graphs, I prefer not to have them. Here, they seem to add structure and help the viewer figure out the dollar value associated with each column. But then again, that's just my opinion.

Excel's Graphics Capabilities

As I mentioned a couple of paragraphs ago, the graph in Figure 3-1 is a column graph. It's one of many types of graphs you can create with Excel. Of all the graphics possibilities Excel provides, however, only a few are useful for statistical work. Those are the ones I cover in this chapter.

In addition to the column graph, I show you how to create pie graphs, bar graphs, line graphs, and scatterplots. Excel refers to each one as a *chart* rather than a "graph." In this chapter, I use the two terms interchangeably.

The Chart Wizard

The first thing to do is learn to use Excel's Chart Wizard. (I would have preferred "Graph Wizard," but the designers of Excel apparently forgot to ask me.) I talk about the Wizard here in a general way, and then apply it to specific types of graphs.

 You access the Chart Wizard via the Chart Wizard button on the standard toolbar. The Excel 2003 Chart Wizard button is shown in the margin.

 Here is how the Chart Wizard button appears in earlier versions of Excel.

You start by entering data into an array of worksheet cells. Then, click the Chart Wizard button to open the Wizard.

The Chart Wizard works in four main steps.

1. **Choose a Chart Type.**

   Here's where Excel shows off the range of its graphics capabilities. You can select from an eye-catching array of types and subtypes, as Figure 3-2 shows. If you've selected your data array before you open the Wizard, you're in for a treat: The Wizard grabs your data and takes a guess as to how you've organized it. The Wizard sometimes guesses wrong, however, as I show you in the next section.

   Having trouble choosing a chart type? You can preview how each type handles your data.

   Incidentally, the Chart Type box (on the left) shows you what I meant before in that heads-up I gave you about Column graphs and Bar graphs. The Column graph icon has vertical bars, and the Bar graph icon has horizontal bars.

2. **Tell the Wizard where your data is, and how it's organized.**

   Obviously, this step is extremely important. If you're not careful, the Wizard graphs the wrong numbers and the graph is meaningless. In this step, you tell the Wizard if the data are organized according to rows or columns. You also supply names for different series within your data (like Commercial Launches or Satellite Services) and show the Wizard what to put on the x-axis.

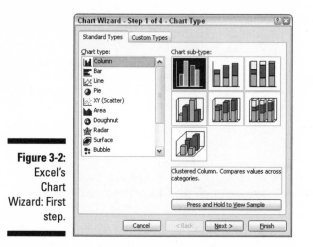

Figure 3-2:
Excel's
Chart
Wizard: First
step.

3. **Choose Chart Options.**

   In this step, you carve out the appearance of the graph. You set the grid-lines, insert titles for the graph and for each axis, and tweak the legend (and, after all, what legend doesn't like to be tweaked?). You can label the data if you like, and you can even insert the data table right into the graph, although I think both those capabilities defeat the purpose.

4. **Find a home for the graph.**

   You can choose to insert the graph onto your worksheet and have it next to your data, you can insert it into another worksheet, or you can create a huge version of it that takes up an entire newly created sheet.

After you finish with the Wizard, you can always go back and change anything you've done.

In the sections that follow, I use the Chart Wizard to create the types of graphs most useful in statistical work.

Becoming a Columnist

In this section, I show you how to use the Chart Wizard to create that spiffy graph in Figure 3-1. Figure 3-3 shows the data in Table 3-1 entered into the cells of a worksheet.

The data are in cells B2:F7. The bottom row (Row G) is not included in the data range, because those cells hold the column totals. The top row is not included because those cells (Row A) hold the years.

Figure 3-3:
Table 3-1
data
entered
into a
worksheet.

	A	B	C	D	E	F
1	Industry	1990	1991	1992	1993	1994
2	Commercial Satellites Delivered	1,000	1,300	1,300	1,100	1,400
3	Satellite Services	800	1,200	1,500	1,850	2,330
4	Satellite Ground Equipment	860	1,300	1,400	1,600	1,970
5	Commercial Launches	570	380	450	465	580
6	Remote Sensing Data	155	190	210	250	300
7	Commercial R&D Infrastructure	0	0	0	30	60
8	Total	3,385	4,370	4,860	5,295	6,640

I start by selecting B2:F7. Here's what happens next:

1. **Click the Chart Wizard button to open the Chart Type dialog box (see Figure 3-2).**

   It opens with Column highlighted in the Chart Type box, and the Clustered Column highlighted in the Chart Sub-Type box. That's the subtype for producing the graph. If you click and hold down the button labeled Press and Hold to View Sample, the Chart Sub-Type box becomes the Sample box and it shows you a preview of the graph (see Figure 3-4).

Figure 3-4:
Previewing
the finished
graph —
based on
a bad
assumption
by the
Wizard.

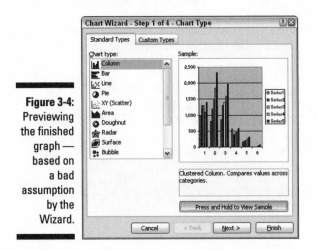

This is an example of what happens when the Wizard makes a bad guess. It looks OK at first because it shows five series — five bars at each point on the horizontal axis. The legend holds five symbols, too. So, what's wrong?

For one thing, the number of categories is six, not five. The problem is that the Wizard assumes the data are organized into columns — that each column is a series of data. That's not the case. Here, the data are organized by rows. Each row represents a different series of data — like Remote Sensing Data or Satellite Equipment.

In the next step, you set the Wizard straight.

2. **Click Next to open the Chart Source Data dialog box.**

   This dialog box opens on the Data Range tab with that erroneous graph showing. The data range is already filled in (if you select the data range before you summon the Wizard), and the Columns option is selected by default. Selecting the Rows option makes everything right, as Figure 3-5 shows.

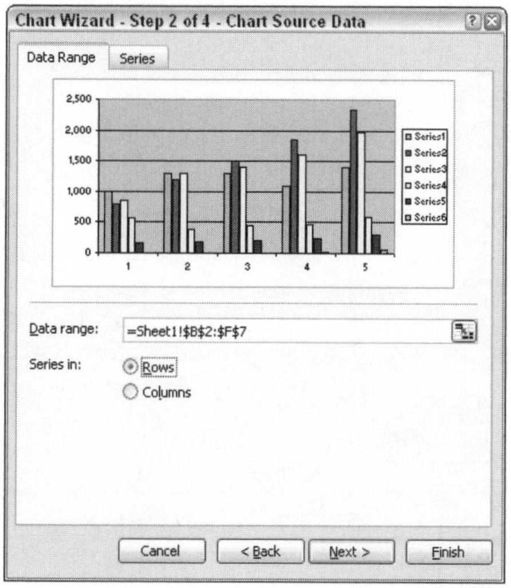

Figure 3-5:
The Chart
Wizard's
second
dialog box,
Chart
Source
Data.

3. **Click the Series tab and name each data series.**

   The Series tab appears in Figure 3-6. To get rid of the default names for each data series, highlight a series in the Series box and type its new name in the Name box. The Values box holds the cell range for that series. In the figure, I've renamed two of the data series. Before moving on, of course, you rename all of them.

   In the Category (X) Axis Labels box, I typed the cell range that holds the years 1990 through 1994. (Newcomers to the Chart Wizard often forget this step.)

   When you enter a cell range into a box in the Wizard, here's the way to do it: Click the box, then click the first cell in the range. Press and hold Shift, and then click the last cell in the range. That enters the range in the proper format for the Wizard.

4. **Click Next to open the Chart Options dialog box.**

   It's time to surround the graph with clever titles, format the background, and generally add to the graph's information value and appeal. The dialog box opens on the Axes tab. I selected the Titles tab to show you how to label the graph and the axes. Figure 3-7 shows that I filled in the Chart Title box, the Category (X) Axis box, and the Value (Y) Axis box. The entries show up immediately on the graph in the dialog box.

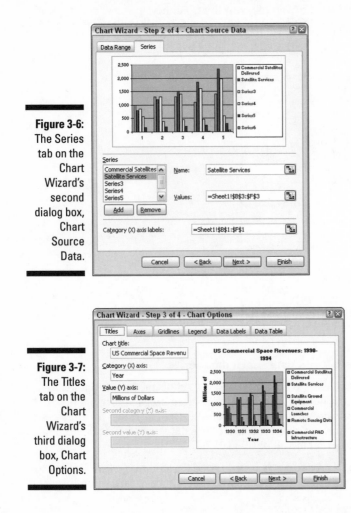

Figure 3-6: The Series tab on the Chart Wizard's second dialog box, Chart Source Data.

Figure 3-7: The Titles tab on the Chart Wizard's third dialog box, Chart Options.

Use the remaining tabs to work with the Axes, Gridlines, Legend, Data Labels, and Data Table. With the Data Labels, you can add the year, the number of dollars, and the value to each column in the graph. Generally, I recommend against all of them, because they add needless clutter to this type of graph.

5. **Click Next to open the Chart Location dialog box.**

   The graph can go into the same worksheet as the data, or into a different worksheet. You can place it onto a new worksheet and have it be the entire page, and you can name the new page's tab. This dialog box (see Figure 3-8) gives you the choice. I selected the As Object In option and placed it on the same page as the data, so you could see what everything looks like together.

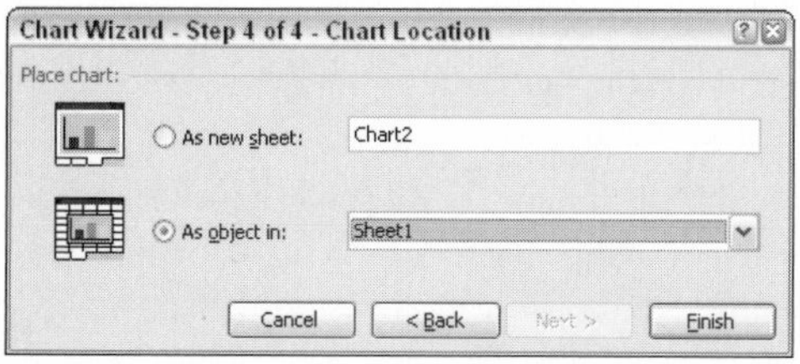

Figure 3-8: The Chart Wizard's fourth dialog box, Chart Location.

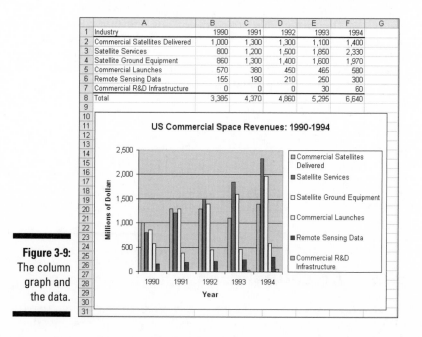

6. **Click Finish to close the Chart Wizard.**

   This puts the graph in the specified location. Figure 3-9 shows how it looks in the original worksheet.

	A	B	C	D	E	F	G
1	Industry	1990	1991	1992	1993	1994	
2	Commercial Satellites Delivered	1,000	1,300	1,300	1,100	1,400	
3	Satellite Services	800	1,200	1,500	1,850	2,330	
4	Satellite Ground Equipment	860	1,300	1,400	1,600	1,970	
5	Commercial Launches	570	380	450	465	580	
6	Remote Sensing Data	155	190	210	250	300	
7	Commercial R&D Infrastructure	0	0	0	30	60	
8	Total	3,385	4,370	4,860	5,295	6,640	

Figure 3-9: The column graph and the data.

To make this graph look like the graph in Figure 3-1, all that's left is to take the gray out of the background and reformat the columns.

Start with the background. Right-click the background to open the menu in Figure 3-10, and select Format Plot Area to open a dialog box from which you can choose the background color.

Figure 3-10:
Right-click
the back-
ground of
the graph to
open this
menu.

| Format Plot Area... |
| Chart Type... |
| Source Data... |
| Chart Options... |
| Location... |
| 3-D View... |
| Chart Window |
| Clear |

Note the next four choices on this menu. Each one opens a dialog box from the Wizard, so you can revisit your original choices. Any changes you make appear on the graph.

To reformat the columns, right-click any column to select all the columns in that data series and open the menu in Figure 3-11. Select the first choice, Format Data Series, to open a multitabbed dialog box. The first tab enables you to change the color of the columns and to insert a pattern. Other tabs allow a variety of edits, like changing the order of the data series and changing the space between columns. You can also add a line that connects the tops of the columns. It's called a *trendline*, and I'll have more to say about it later.

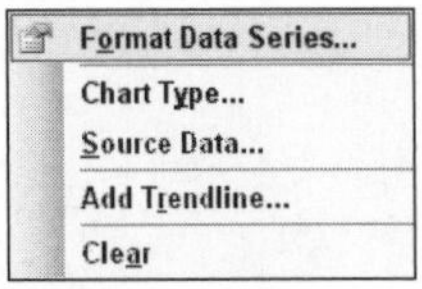

Figure 3-11:
Right-click a
column in
the graph to
open this
menu.

| Format Data Series... |
| Chart Type... |
| Source Data... |
| Add Trendline... |
| Clear |

Stacking the columns

If I had selected Column's second subtype — Stacked Columns — from the Chart Sub-Type box, I would have created a set of columns that presents the same information in a slightly different way. Each column represents the total of all the data series at a point on the x-axis. Each column is divided into segments. Each segment's size is proportional to how much it contributes to the total.

Figure 3-12 shows what I mean and also shows what happens when you tell the Chart Wizard to create a new page for the graph.

Notice that the data series are in reverse order from the way they're set up in the first column graph. The Chart Wizard sets them up in this order for the stacked columns, and in the other order for the clustered columns.

This is a nice way of showing percentage changes over the course of time. If you just want to focus on percentages in one year, another type of graph is more effective. I'll discuss it in a moment, but first I want to tell you. . . .

One more thing

Statisticians often use column graphs to show how frequently something occurs. For example, in a thousand tosses of a pair of dice how many times does a 6 come up? How many tosses result in a 7? The x-axis shows each possible outcome of the dice tosses, and the heights of the columns represent the frequencies. Whenever the heights represent frequencies, your column graph is a *histogram*.

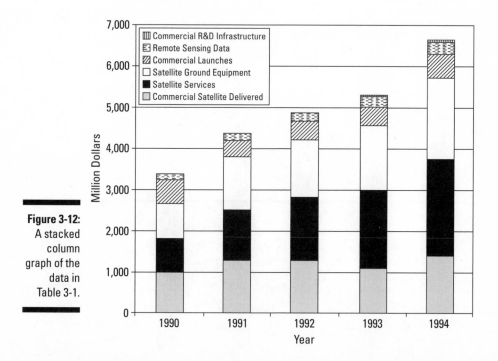

Figure 3-12:
A stacked column graph of the data in Table 3-1.

It's easy enough to use the Chart Wizard to set up a histogram, but Excel makes it easier still. Excel provides a data analysis tool that does everything you need to create a histogram. It's called — believe it or not — Histogram. You provide an array of cells that holds all the data, like the outcomes of many dice tosses, and an array that holds a list of intervals, like the possible outcomes of the tosses (the numbers 2 through 12). Histogram goes through the data array, counts the frequencies within each interval, and then draws the column graph. I describe this tool in greater detail in Chapter 7.

Slicing the Pie

On to the next chart type. To show the percentages that make up one total, a pie graph gets the job done effectively.

Suppose you want to focus on the U.S. commercial space revenues in 1994 — the last column of data in Table 3-1. You'll catch people's attention if you present the data in the form of a pie graph like the one in Figure 3-13.

I filled the slices the same way I filled the columns, with one exception: For Commercial Launches I changed from diagonal lines to a slightly tamer pattern.

US Commercial Space Revenues 1994

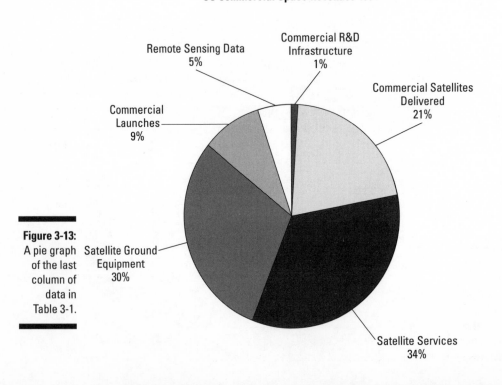

Figure 3-13: A pie graph of the last column of data in Table 3-1.

Here's how to create this graph:

1. **Select the data range, open the Chart Wizard and select Pie from the Chart Type box.**

   The data range is F2:F7 (see Figure 3-3), because the focus is on the data from 1994. Selecting Pie causes a set of subtypes to appear.

2. **Select a subtype.**

   The first subtype, highlighted by default, is the simplest, and that's the one I selected.

3. **Click Next to open the Chart Source Data dialog box.**

   This dialog box opens with your pie chart practically done. The data are organized into columns this time (one column, actually), so the graph is good.

4. **Click the Series tab and type the category information.**

   In the Category Labels box, I entered the cells in Column A that hold the names of the industries (A2 through A7). This is different from the previous example. In that one, I graphed the changes from year to year, so the columns in the data sheet were the categories, and I put them on the x-axis. Here, with the focus on one column, the rows in the data sheet are the categories (and the pie chart has no x-axis). Entering the cell range from Column A into the Category Labels box attaches the industry names to the appropriate slices.

5. **Click Next to open the Chart Options dialog box.**

   The dialog box opens on the Data Labels tab (see Figure 3-14). For a pie graph, it's a good idea to put labels into the graph, along with their percentages (which, after all, is the reason for a pie graph). The figure shows a lot of clutter in the picture, but that's because the space is so limited. Locating the graph onto a separate page solves that problem.

   I used the Titles tab to put the title on the graph and the Legend tab to remove the legend. The legend is unnecessary with the labels on the slices.

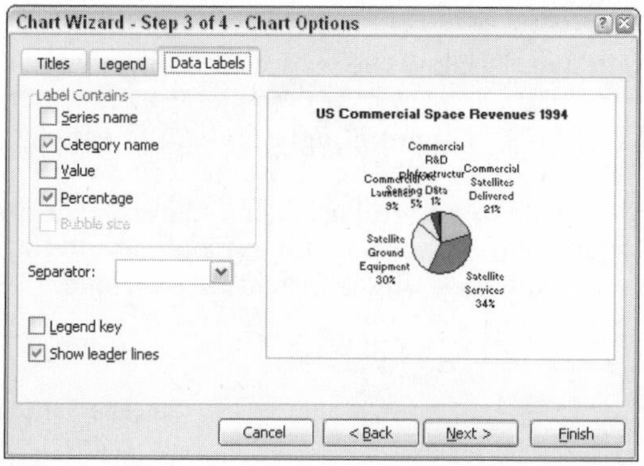

Figure 3-14:
The Chart Options dialog box for the pie chart.

Note that the Show Leader Lines option is selected. A *leader line* connects the category name and the percentage to the slice.

6. Click Next to open the Chart Location dialog box.

I chose to put the pie graph on its own page.

7. Click Finish to put the graph in the selected location.

The result is the pie graph in Figure 3-13, almost. To get it to look exactly that way, you have to put the patterns in the slices and drag and drop the labels around a bit. If you want to format an individual slice, you're not formatting a data series (as in the column graph). Instead, you're formatting a data point. It requires a bit more clicking to select an individual slice and get the right menu.

Pulling the slices apart

One variant of the pie chart is to explode the slices. I'm not particularly fond of this type of graph, but you might be. In some circumstances, it might come in handy.

One of the nice things about Excel's graphics capabilities is that you can "what-if" to your heart's content. So, after you finish creating the pie chart, you can explode it. To do that, right-click on the graph and select Chart Type from the menu that appears. In the Sub-Type box select Exploded Pie. The result is Figure 3-15.

Whenever you set up a pie graph — whether intact or exploded — always keep in mind. . . .

A word from the wise

Social commentator, raconteur, and former baseball player Yogi Berra once went to a restaurant and ordered a whole pizza.

"How many slices should I cut," asked the waitress, "four or eight?"

"Better make it four," said Yogi. "I'm not hungry enough to eat eight."

Yogi's insightful analysis leads to a useful guideline about pie graphs: They're more digestible if they have fewer slices. If you cut a pie graph too fine, you're likely to leave your audience with information overload.

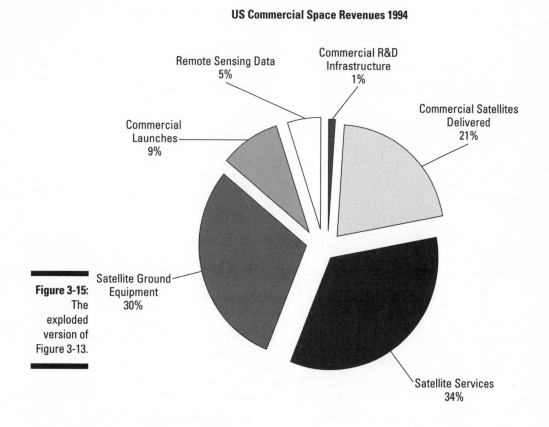

US Commercial Space Revenues 1994

Figure 3-15: The exploded version of Figure 3-13.

Drawing the Line

In the preceding example, I focused on one column of data from Table 3-1. In this one, I focus on one row. The idea is to trace the progress of one space-related industry across the years 1990 through 1994. In this example, I graph the revenues from Satellite Services. The final product is Figure 3-16.

A line graph is a good way to show change over time, when you aren't dealing with too many data series. If you try to graph all six industries on one line graph, it begins to look like spaghetti.

How do you create a graph like Figure 3-16? Follow along:

1. **Select the data range, open the Chart Wizard, and select Line from the Chart Type box.**

   The data range is B3:F3 because the focus is on the data for Satellite Services. Selecting Line causes a set of subtypes to appear.

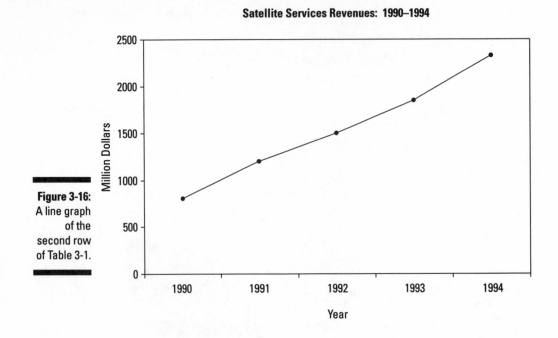

Satellite Services Revenues: 1990–1994

Figure 3-16:
A line graph
of the
second row
of Table 3-1.

2. Select a subtype.

As Figure 3-17 shows, from the Chart Sub-Type box I selected a subtype called Line with Markers Displayed at Each Data Value.

3. Click Next to open the Chart Source Data dialog box.

This dialog box opens with your line graph practically done. The data are organized into rows this time (one row, to be precise).

4. Click the Series tab and type the category information.

In the Category (X) Axis Labels box, I entered the cells in Row 1 that hold the years. Just as a reminder, the way to enter those cells is to click the Category (X) Axis Labels box, then click the first cell in Row 1, press and hold Shift, and then click the last cell in the range. This puts the cell range in the right format for the Wizard, which is:

=Sheet1!$B$1:$F$1

It's better to click buttons and press keys than to have to remember all the formatting details.

5. Click Next to open the Chart Options dialog box.

The dialog box opens on the Data Labels tab. For this graph, I open the Legend tab and deselect the Show Legend option. That gets rid of the legend, which is unnecessary.

I used the Titles tab to put the title on the graph, Year on the x-axis, and Million Dollars on the y-axis. I also opened the Gridlines tab to get rid of the gridlines. To do this, I deselected the Major Gridlines option. Unlike the columns graph, gridlines here seem to interfere.

6. **Click Next to open the Chart Location dialog box.**

   I chose to put the line graph on its own page.

7. **Click Finish to close the Wizard and put the graph in the selected location.**

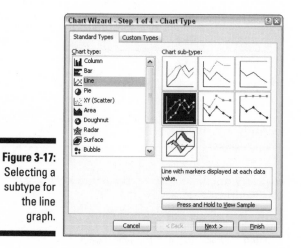

Figure 3-17:
Selecting a
subtype for
the line
graph.

The graph now looks almost like Figure 3-16. To remove the gray background, right-click the background and choose Format Plot Area from the menu that appears. This opens the dialog box you use to change the color from gray to white.

Passing the Bar

Excel's bar chart is a column chart laid on its side. This is the one that reverses the horizontal-vertical convention. Here, the vertical axis holds the independent variable, and it's referred to as the x-axis. The horizontal axis is the y-axis, and it tracks the dependent variable.

When would you use the bar graph? This type of graph is useful when the category names for the independent variable are long, and would look cramped on the horizontal axis in a column graph. It also fits the bill when you want to make a point about reaching a goal, or about the inequities in attaining one. . . .

Table 3-2 shows the data on (what I feel, anyway) is an important social issue. The data, from the U.S. Census Bureau (via the U.S. Statistical Abstract), are for the year 2000. *Percent* means the percentage of children in each income group.

Table 3-2	Children's Use of the Internet at Home (2000)
Family Income	*Percent*
Under $15,000	7.7
$15,000-$19,999	12.9
$20,000-$24,999	15.2
$25,000-$34,999	21.0
$35,000-$49,999	31.8
$50,000-$74,999	39.9
Over $75,000	51.7

The numbers in the table are pretty dramatic. Casting them into a bar chart renders them even more so, as Figure 3-18 shows.

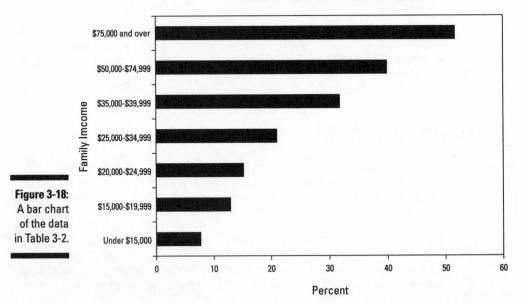

Figure 3-18:
A bar chart
of the data
in Table 3-2.

To produce this graph, start by putting the data in a worksheet. Then:

1. **Select the data range, open the Chart Wizard, and select Bar from the Chart Type box.**

   The data range is B2 through B8.

2. **Select a subtype.**

   I selected the default subtype.

3. **Click Next to open the Chart Source Data dialog box.**

   This dialog box opens with your bar graph almost complete. The data are organized into columns (well, one column).

4. **Click the Series tab and type the category information.**

   In the Category (X) Axis Labels box, I entered the cells in Column 1 that hold the Family Income groups — cells A2 through A7. In this type of graph, the x-axis is the vertical axis, and the y-axis is the horizontal axis. Figure 3-19 shows this dialog box with my entries.

5. **Click Next to open the Chart Options dialog box.**

   I used the Legend tab to deselect the Show Legend option and eliminate the legend.

   I used the Titles tab to put the title on the graph, to put Family Income on the vertical axis (the x-axis, in this chart type), and Percent on the horizontal axis (the y-axis). I also opened the Gridlines tab to get rid of the gridlines. To do this, I deselected the Major Gridlines option.

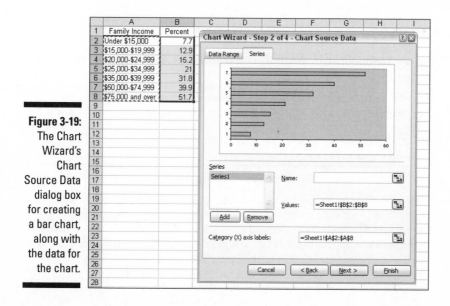

Figure 3-19:
The Chart Wizard's Chart Source Data dialog box for creating a bar chart, along with the data for the chart.

6. **Click Next to open the Chart Location dialog box.**

   I chose to put the bar graph on its own page.

7. **Click Finish to close the Wizard and put the graph in the selected location.**

Some reformatting is necessary to make the finished graph look like Figure 3-18. You have to format the plot area to turn the gray into white, and format the data series to change the color of the bars to black.

The Plot Thickens

You use an important statistical technique called *linear regression* to determine the relationship between one variable, x, and another variable, y. For more information on linear regression, see Chapter 14.

The basis of this technique is a graph that shows individuals measured on both x and y. The graph represents each individual as a point. Because the points seem to scatter around the graph, the graph is called a *scatterplot*.

Suppose you're trying to find out how well a test of aptitude for sales predicts salespeople's productivity. You administer the test to a sample of salespeople and tabulate how much money they make in commissions over a two-month period. Each person's pair of scores (test score and commissions) locates him or her within the scatterplot.

To create a scatterplot, put each pair of scores into a worksheet and follow these steps:

1. **Select the data range, open the Chart Wizard, and select XY Scatter from the Chart Type box.**

   The data range is B2 through C21. Figure 3-20 shows the worksheet with the selected cell array along with the Chart Type box.

2. **Select a subtype.**

   I selected the default subtype, Scatter.

3. **Click Next to open the Chart Source Data dialog box.**

   This dialog box opens with your scatterplot. The Wizard guesses correctly that the data are organized into columns. More impressively, the Series tab shows that the Wizard guesses which column is x and which column is y. Apparently, this is a left-right thing. The column on the left is x. So if you set up your data that way, you save yourself some work. Figure 3-21 shows this dialog box with the Series tab open.

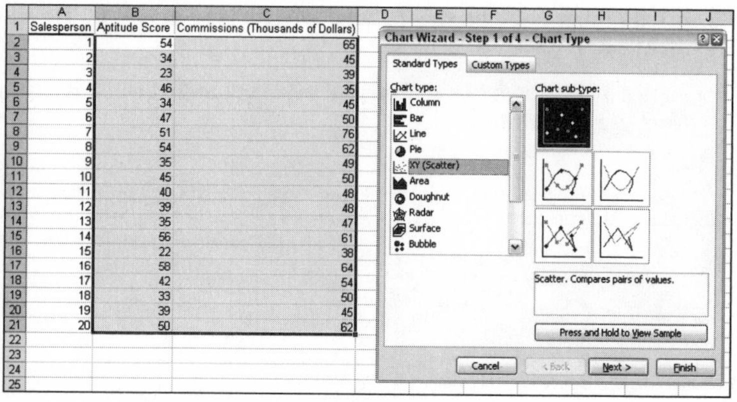

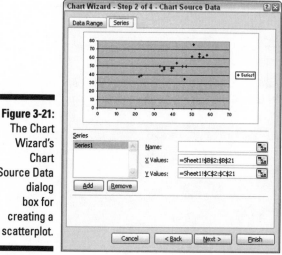

Figure 3-20:
The Chart
Wizard's
Chart Type
dialog
box for
creating a
scatterplot,
along with
the data for
the chart.

Figure 3-21:
The Chart
Wizard's
Chart
Source Data
dialog
box for
creating a
scatterplot.

4. Click Next to open the Chart Options dialog box.

I used the Legend tab to deselect the Show Legend option and eliminate the legend.

I used the Titles tab to put the title on the graph, to put Aptitude Score on the x-axis, and Commission (Thousands of Dollars) on the y-axis. I also opened the Gridlines tab to get rid of the gridlines. To do this, I deselected the Major Gridlines option.

5. Click Next to open the Chart Location dialog box.

I chose to put the scatterplot on the same page as the data.

6. **Click Finish to close the Wizard and put the graph in the selected location.**

Figure 3-22 shows the scatterplot after I changed the background color from gray to white.

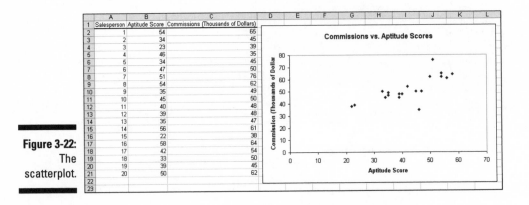

	A	B	C	D	E	F	G	H	I	J	K	L
1	Salesperson	Aptitude Score	Commissions (Thousands of Dollars)									
2	1	54	65									
3	2	34	45									
4	3	23	39									
5	4	46	35									
6	5	34	45									
7	6	47	50									
8	7	51	76									
9	8	54	62									
10	9	35	49									
11	10	45	50									
12	11	40	48									
13	12	39	48									
14	13	35	47									
15	14	56	61									
16	15	22	38									
17	16	58	64									
18	17	42	54									
19	18	33	50									
20	19	39	45									
21	20	50	62									
22												
23												

Figure 3-22:
The
scatterplot.

For the other graphs, that would just about do it, but this one's special. If you right-click any of the points in the scatterplot, you get the menu in Figure 3-23.

Figure 3-23:
Right-
clicking any
point on the
scatterplot
opens this
menu.

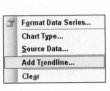

Select Add Trendline to open the two-tabbed Add Trendline dialog box. On the Type tab, select Linear. On the Options tab, select the option Display Equation on Chart and the option Display R-Squared Value on Chart. They're at the bottom of the box (see Figure 3-24).

Click OK to close the Add Trendline dialog box. A couple of additional items are now on the scatterplot, as Figure 3-25 shows. A line passes through the points. Excel refers to it as a *trendline*, but it's really called a *regression line*. A couple of equations are there, too. (For clarity, I dragged them from their original locations.) What do they mean? What are those numbers all about?

You'll just have to read Chapter 14 to find out.

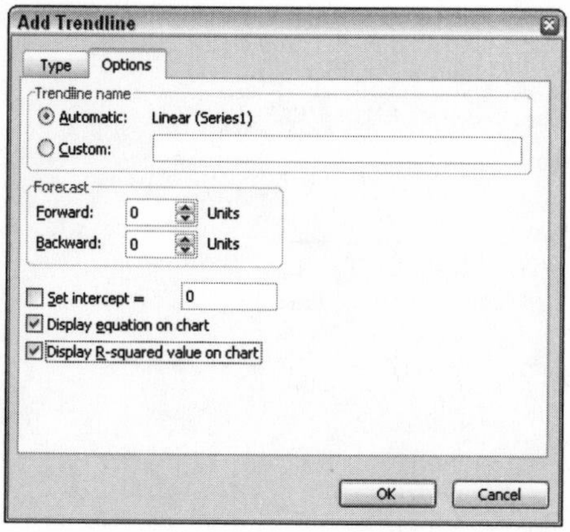

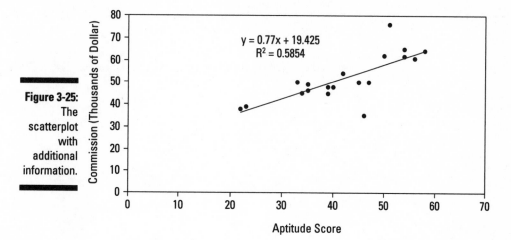

Figure 3-24:
The Options
tab of the
Add
Trendline
dialog box.

Commissions vs. Aptitude Scores

$y = 0.77x + 19.425$
$R^2 = 0.5854$

Figure 3-25:
The
scatterplot
with
additional
information.

Aptitude Score

Chapter 4

Finding Your Center

Statisticians deal with groups of numbers. They often find it helpful to use a single number to summarize a group of numbers. Where would a single summary number come from?

The best bet is to find a number that's somewhere in the middle, and use that number to stand for the whole group. If you look at a group of numbers and try to find one that's somewhere in the middle, you're dealing with that group's *central tendency*. Like good ice cream, central tendency comes in several flavors.

Means: The Lore of Averages

Just about everyone uses averages. The statistical term for an average is *mean*. Sometime in your life, you've undoubtedly calculated one. The mean is a quick way of characterizing your grades, your money, or perhaps your performance in some task or sport over time.

Another reason for calculating means concerns the kind of work that scientists do. Typically, a scientist applies some kind of procedure to a small sample of people or things and measures the results in some way. He or she uses the results from the sample to estimate the effects of the procedure on the population that produced the sample. As it happens, the mean of the sample is the best estimate of the population mean.

Calculating the mean

You probably don't need me to tell you how to calculate a mean, but I'm going to do it anyway. Then, I'll show you the statistical formula. My goal is to help you understand statistical formulas in general, and then set you up for how Excel calculates means.

A mean is just the sum of a bunch of numbers divided by the amount of numbers you added up. Here's an example. Suppose you measure the reading speeds of six children in words per minute, and you find that their speeds are

56, 78, 45, 49, 55, 62

The average reading speed of these six children is

$$\frac{56 + 78 + 45 + 49 + 55 + 62}{6} = 57.5$$

That is, the mean of this sample is 57.5 words per minute.

A first try at a formula might be

$$\text{Mean} = \frac{\text{Sum of Numbers}}{\text{Amount of Numbers You Added Up}}$$

This is unwieldy as formulas go, so statisticians use abbreviations. A commonly used abbreviation for "Number" is x. A typical abbreviation for "Amount of Numbers You Added Up" is N. With these abbreviations, the formula becomes

$$\text{Mean} = \frac{\text{Sum of } X}{N}$$

Another abbreviation, used throughout statistics, stands for *Sum of*. It's the uppercase Greek letter for S. It's pronounced "sigma," and it looks like this: Σ. Here's the formula with the sigma:

$$\text{Mean} = \frac{\Sigma X}{N}$$

What about "mean"? Statisticians abbreviate that, too. M would be a good abbreviation, and some statisticians use it, but most use $\overline{X}$ (pronounced "x bar") to represent the mean. So here's the formula:

$$\overline{X} = \frac{\Sigma X}{N}$$

Is that it? Well, not quite. English letters, like $\overline{X}$, represent characteristics of samples. For characteristics of populations, the abbreviations are Greek letters. For the population mean, the abbreviation is the Greek equivalent of M, which is μ (pronounced like "you," but with "m" in front of it). The formula for the population mean, then, is

$$\mu = \frac{\Sigma X}{N}$$

And that's that.

AVERAGE and AVERAGEA

Excel's AVERAGE worksheet function calculates the mean of a set of numbers. Figure 4-1 shows the data and Function Arguments dialog box for AVERAGE.

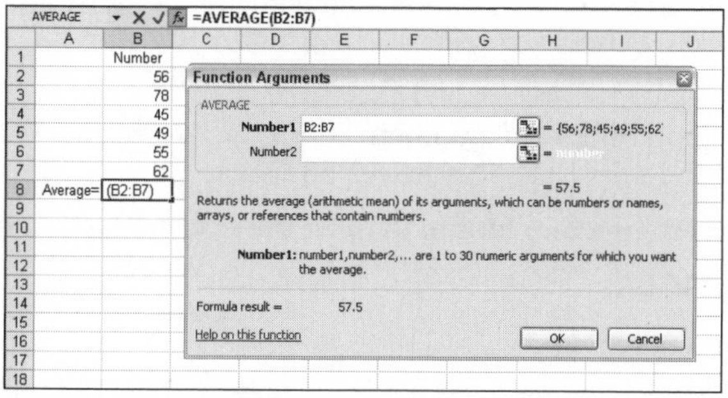

Figure 4-1:
Working
with
AVERAGE.

Here are the steps:

1. **In your worksheet, type your numbers into an array of cells, and select the cell where you want AVERAGE to place the result.**

   For this example, I've entered 56, 78, 45, 49, 55, and 62 into cells B2 through B7, and I've selected B8 for the result.

2. **Click the Insert Function button (the f_x button) to open the Insert Function dialog box.**

3. **Click AVERAGE to open the Function Arguments dialog box.**

   To do this, Select Statistical Functions from the second box (labeled Or Select a Category), and select AVERAGE from the third box (labeled Select a Function).

4. **Identify the data array.**

   If the array of number-containing cells isn't already in the Number1 box, enter it into that box. The mean (57.5 for this example, as shown in Figure 4-1) appears in this dialog box.

5. **Click OK to close the Function Arguments dialog box.**

   This puts the mean into the cell you selected in the worksheet. In this example, that's B8.

AVERAGEA does the same thing as AVERAGE, but with one important difference. When AVERAGE calculates a mean, it ignores cells that contain text and it ignores cells that contain the expressions TRUE or FALSE. AVERAGEA takes text and expressions into consideration when it calculates a mean. As far as

AVERAGEA is concerned, if a cell has text or FALSE, it has a value of 0. If a cell holds the word TRUE, it has a value of 1. AVERAGEA includes these values in the mean.

I'm not sure you'll use this capability during everyday statistical work (I never have), but Excel has worksheet functions like AVERAGEA, VARA, and STDEVA, and I want you to know how they operate. So here are the steps for AVERAGEA:

1. **Type the numbers and select a cell for the result.**

   For this example, enter the numbers 56, 78, 45, 49, 55, and 62 in cells B2 through B7 and select B9. This leaves B8 blank.

2. **Click the Insert Function button to open the Insert Function dialog box.**

3. **Select AVERAGEA.**

4. **Identify the data array.**

   In the Function Arguments dialog box, enter the array (B2:B8 for this example) into the Number1 box. The mean (57.5 for this example) appears in this dialog box. AVERAGEA ignores blank cells, just as AVERAGE does.

5. **Click OK to close the Function Arguments dialog box and put the mean into the selected cell.**

Now for some experimentation with those capabilities for text, TRUE, and FALSE so you can see how each one affects AVERAGEA's result. In B8, type xxx. The mean in B9 changes from 57.5 to 49.28571. Next, type TRUE into B8. The mean in B9 changes to 49.42857. Finally, type FALSE into B8. The mean changes back to 49.28571.

TRIMMEAN

In a retake on a famous quote about statistics, someone said, "There are three kinds of liars: liars, darned liars, and statistical outliers." An *outlier* is an extreme value in a set of scores — so extreme, in fact, that the person who gathered the scores believes that something is amiss.

One example of outliers involves psychology experiments that measure a person's time to make a decision. Measured in thousandths of a second, these reaction times depend on the complexity of the decision — the more complex the decision, the longer the reaction time.

Typically, a person in this kind of experiment goes through many experimental trials — one decision per trial. A trial with an overly fast reaction time (way below the average) might indicate that the person made a quick guess without really considering what he or she was supposed to do. A trial with a very slow reaction time (way above the average) might mean the person wasn't paying attention at first and then buckled down to the task at hand.

Either kind of outlier can get in the way of conclusions based on averaging the data. For this reason, it's often a good idea to eliminate them before you calculate the mean. Statisticians refer to this as *trimming the mean*, and Excel's TRIMMEAN function does this.

Here's how you use TRIMMEAN:

1. **Type your scores into an array and select a cell for the result.**

   For this example, I put these numbers into cells B2 through B11:

   500, 280, 550, 540, 525, 595, 620, 1052, 591, 618

   These scores might result from a psychology experiment that measures reaction time in thousandths of a second (milliseconds).

   I selected B12 for the result.

2. **Click the Insert Function button to open the Insert Function dialog box.**

3. **Select TRIMMEAN to open the Function Arguments dialog box.**

4. **Identify the data array.**

   The data array goes into the Array box. For this example, enter B2:B11.

5. **Identify the percent of scores you want to trim.**

   In the Percent box, type .2. This tells TRIMMEAN to eliminate the extreme 20 percent of the scores before calculating the mean. The extreme 20 percent means the highest 10 percent of scores and the lowest 10 percent of scores. Figure 4-2 shows the dialog box, the array of scores, and the selected cell. The dialog box shows the value of the trimmed mean, 567.375.

6. **Click OK to close the dialog box and put the trimmed mean into B12.**

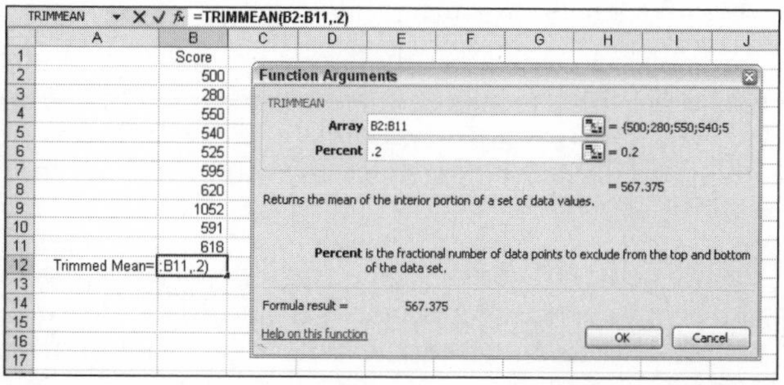

Figure 4-2:
The TRIMMEAN dialog box along with the array of cells and the selected cell.

The label Percent is a little misleading here. You have to express the percent as a decimal. So you type .2 rather than 20 in the Percent box if you want to trim the extreme 20 percent. (Quick question: If you type 0 in the Percent box, what's the answer equivalent to? Answer: AVERAGE(B2:B11).)

What percentage of scores should you trim? That's up to you. It depends on what you're measuring, how extreme your scores can be, and how well you know the area you're studying. When you do trim scores and report a mean, it's important to let people know you've done this and to let them know the percentage you've trimmed.

In the upcoming section on the median, I show you another way to deal with extreme scores.

Other means to an end

This section deals with two types of averages that are different from the one you're familiar with. I tell you about them because you might run into them as you go through Excel's statistical capabilities. (How many different kinds of averages are possible? Ancient Greek mathematicians came up with 11! Modern mathematicians concoct all kinds of means, with exotic names like "super symmetric mean," "power mean," and "logarithmic binary mean.")

Geometric mean

Suppose you have a two-year investment that yields 25 percent the first year and 75 percent the second year. (If you do, I want to know about it!) What's the average annual rate of return?

To answer that question, you might be tempted to find the mean of 25 and 75 (which averages out to 50). But that misses an important point: At the end of the first year, you *multiply* your investment by 1.25 — you don't add 1.25 to it. At the end of the second year, you multiply the first-year result by 1.75.

The regular everyday garden-variety mean won't give you the average rate of return. Instead, you calculate the mean this way:

$$\text{Average Rate of Return} = \sqrt{1.25 \times 1.75} = 1.654$$

The average rate of return is about 65.4 percent, not 50 percent. This kind of average is called the *geometric mean*.

In this example, the geometric mean is the square root of the product of two numbers. For three numbers, the geometric mean is the cube root of the product of the three. For four numbers, it's the fourth root of their product, and so on. In general, the geometric mean of N numbers is the Nth root of their product.

The Excel worksheet function GEOMEAN calculates the geometric mean of a group of numbers. Follow the same steps as you would for AVERAGE, but select GEOMEAN from the Insert Function dialog box.

Harmonic mean

Still another mean is something you run into when you have to solve the kinds of problems that live in algebra textbooks.

Suppose, for example, you're in no particular hurry to get to work in the morning, and you drive from your house to your job at the rate of 40 miles per hour. At the end of the day you'd like to get home quickly, so on the return trip (over exactly the same distance) you drive from your job to your house at 60 miles per hour. What is your average speed for the total time you're on the road?

It's not 50 miles per hour, because you're on the road a different amount of time on each leg of the trip. Without going into this in too much detail, the formula for figuring this one out is:

$$\frac{1}{\text{Average}} = \frac{1}{2}\left[\frac{1}{40} + \frac{1}{60}\right] = \frac{1}{48}$$

The average here is 48. This kind of average is called a *harmonic mean*. I showed it to you for two numbers, but you can calculate it for any amount of numbers. Just put each number in the denominator of a fraction with 1 as the numerator. Mathematicians call this the *reciprocal* of a number: 1/40 is the reciprocal of 40. Add all the reciprocals together and take their average. The result is the reciprocal of the harmonic mean.

In the rare event you ever have to figure one of these out in the real world, Excel saves you from the drudgery of calculation. The worksheet function HARMEAN calculates the harmonic mean of a group of numbers. Follow the same steps as you would for AVERAGE, but in the Insert Function dialog box, select HARMEAN.

Medians: Caught in the Middle

The mean is a useful way to summarize a group of numbers. It's sensitive to extreme values, however: If one number is out of whack relative to the others the mean quickly gets out of whack, too. When that happens, the mean might not be a good representative of the group.

For example, with these numbers as reading speeds (in words per minute) for a group of children

56, 78, 45, 49, 55, 62

the mean is 57.5. Suppose the child who reads at 78 words per minute leaves the group and an exceptionally fast reader replaces him. Her reading speed is 180 words per minute. Now the group's reading speeds are

56, 180, 45, 49, 55, 62

The new average is 74.5. It's misleading because except for the new child, no one else in the group reads nearly that fast. In a case like this, it's a good idea to turn to a different measure of central tendency — the *median*.

Simply put, the median is the middle value in a group of numbers. Arrange the numbers in order, and the median is the value below which half the scores fall and above which half the scores fall.

Finding the median

In our example, the first group of reading speeds (in increasing order) is:

45, 49, 55, 56, 62, 78

The median is right in the middle of 55 and 56 — it's 55.5

What about the group with the new child? That's

45, 49, 55, 56, 62, 180

The median is still 55.5. The extreme value doesn't change the median.

MEDIAN

The worksheet function MEDIAN (you guessed it) calculates the median of a group of numbers. Here are the steps:

1. **Type your data into an array and select a cell for the result.**

   I use 45, 49, 55, 56, 62, and 78 for this example, in cells B2 through B7, with cell B8 selected for the median. I arranged the numbers in increasing order, but you don't have to do that to use MEDIAN.

2. **Click the Insert Function button to open the Insert Function dialog box.**

3. **Select MEDIAN to open the Function Arguments dialog box.**

   The Function Arguments dialog box opens with the data array in the Number1 box. The median appears in that dialog box. (It's 55.5 for this example.) Figure 4-3 shows the dialog box along with the array of cells and the selected cell.

4. **Click OK to close the dialog box and put the median in the selected cell.**

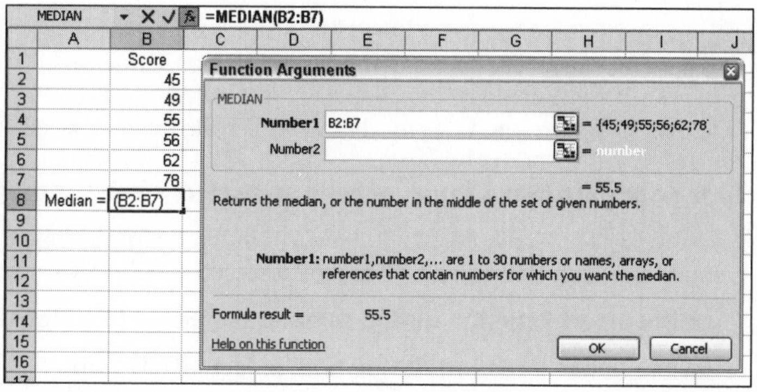

Figure 4-3: The MEDIAN dialog box along with the array of cells and the selected cell.

As an exercise, replace 78 with 180 in A6; you'll see that the median doesn't change.

Statistics À La Mode

One more measure of central tendency is important. This one is the score that occurs most frequently in a group of scores. It's called the *mode*.

Finding the mode

Nothing is complicated about finding the mode. Look at the scores, find the one that occurs most frequently, and you've found the mode. Two scores tie for that honor? In that case, your set of scores has two modes. (The technical name is *bimodal*.)

Can you have more than two modes? Absolutely.

Suppose every score occurs equally often. When that happens, you have no mode.

Sometimes, the mode is the most representative measure of central tendency. Imagine a small company that consists of 30 consultants and two high-ranking officers. Each consultant has an annual salary of $40,000. Each officer has an annual salary of $250,000. The mean salary in this company is $53,125.

Does the mean give you a clear picture of the company's salary structure? If you were looking for a job with that company, would the mean influence your expectations? You're probably better off if you consider the mode, which in this case is $40,000.

MODE

Excel's MODE function finds the mode for you. Follow these steps:

1. **Type your data into a range of cells and select a cell for the result.**

   I use 56, 23, 77, 75, 57, 75, 91, 59, and 75 in this example. The data are in cells B2 through B10, with B11 as the selected cell for the mode.

2. **Click the Insert Function button to open the Insert Function dialog box.**

3. **From the Insert Function dialog box, select MODE and click OK to open the Function Arguments dialog box.** (See Figure 4-4).

   The Function Arguments dialog box opens with an array highlighted in the Number1 box. For this example, the highlighted array is correct, and the mode (75 for this example) appears in the dialog box.

4. **Click OK to close the dialog box and put the mode into the selected cell.**

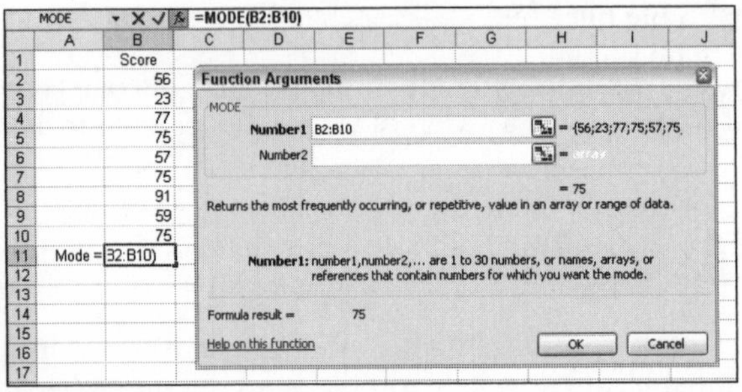

Figure 4-4: The MODE dialog box along with the array of cells and the selected cell.

That's it for measures of central tendency — but central tendency is only part of the story when you're summarizing a set of numbers. In the next chapter, I discuss another important part — variability.

Chapter 5

Deviating from the Average

· ·

· ·

*H*ere are three pieces of wisdom about statisticians:

Piece of Wisdom No. 1: "A statistician is a person who stands in a bucket of ice water, sticks their head in an oven and says 'on average, I feel fine.'" (K. Dunning)

Piece of Wisdom No. 2: "A statistician drowned crossing a stream with an average depth of 6 inches." (Anonymous)

Piece of Wisdom No. 3: "Three statisticians go deer hunting with bows and arrows. They spot a big buck and take aim. One shoots, and his arrow flies off 10 feet to the left. The second shoots and his arrow goes 10 feet to the right. The third statistician jumps up and down yelling, 'We got him! We got him!'" (Bill Butz, quoted by Diana McLellan in *Washingtonian*)

What's the common theme? Calculating the mean is a great way to summarize a group of numbers, but it doesn't supply all the information you typically need. If you just rely on the mean, you might miss something important.

In order to not miss important information, another type of statistic is necessary — a statistic that measures *variation*. It's a kind of average of how much each number in a group differs from the group mean. Several statistics are available for measuring variation. All of them work the same way: The larger the value of the statistic, the more the numbers differ from the mean. The smaller the value, the less they differ.

Measuring Variation

Suppose you measure the heights of a group of children and you find that their heights (in inches) are

48, 48, 48, 48, 48

Then you measure another group and find that their heights are

50, 47, 52, 46, 45

If you calculate the mean of each group, you'll find they're the same — 48 inches. Just looking at the numbers tells you the two groups of heights are different: The heights in the first group are all the same, while the heights in the second vary quite a bit.

Averaging squared deviations: Variance and how to calculate it

One way to show the dissimilarity between the two groups is to examine the deviations in each one. Think of a "deviation" as the difference between a score and the mean of all the scores in a group.

Here's what I'm talking about. Table 5-1 shows the first group of heights and their deviations.

Table 5-1	The First Group of Heights and Their Deviations	
Height	*Height-Mean*	*Deviation*
48	48-48	0
48	48-48	0
48	48-48	0
48	48-48	0
48	48-48	0

One way to proceed is to average the deviations. Clearly, the average of the numbers in the Deviation column is zero.

Table 5-2 shows the second group of heights and their deviations.

Table 5-2	The Second Group of Heights and Their Deviations	
Height	*Height-Mean*	*Deviation*
50	50-48	2
47	47-48	-1
52	52-48	4
46	46-48	-2
45	45-48	-3

What about the average of the deviations in Table 5-2? That's . . . zero!

Hmmm . . . Now what?

Averaging the deviations doesn't help us see a difference between the two groups, because the average of deviations from the mean in any group of numbers is *always* zero. In fact, veteran statisticians will tell you that's a defining property of the mean.

The joker in the deck here is the negative numbers. How do statisticians deal with those?

The trick is to use something you might recall from algebra: A minus times a minus is a plus. Sound familiar?

So, does this mean that you multiply each deviation times itself, and then average the results? Absolutely. Multiplying a deviation times itself is called *squaring a deviation*. The average of the squared deviations is so important that it has a special name: *variance*.

Table 5-3 shows the group of heights from Table 5-2, along with their deviations and squared deviations.

Table 5-3	The Second Group of Heights and Their Squared Deviations		
Height	*Height-Mean*	*Deviation*	*Squared Deviation*
50	50-48	2	4
47	47-48	-1	1
52	52-48	4	16
46	46-48	-2	4
45	45-48	-3	9

The variance — the average of the squared deviations for this group — is $(4 + 1 + 16 + 4 + 9)/5 = 34/5 = 6.8$. This, of course, is very different from the first group, whose variance is zero.

To develop the variance formula for you and show you how it works, I'll use symbols to show all this. x represents the Height heading in the first column of the table and $\overline{X}$ represents the mean. Because a deviation is the result of subtracting the mean from each number,

$$(X - \overline{X})$$

represents a deviation. Multiplying a deviation by itself? That's just

$$(X - \overline{X})^2$$

To calculate variance you square each deviation, add them up, and find the average of the squared deviations. If N represents the amount of squared deviations you have (in our example, five), then the formula for calculating the variance is

$$\text{Variance} = \frac{\sum(X - \overline{X})^2}{N}$$

Σ is the uppercase Greek letter sigma, and it stands for the sum of.

What's the symbol for Variance? As I said in Chapter 1, Greek letters represent population parameters and English letters represent statistics. Imagine that our little group of five numbers is an entire population. Does the Greek alphabet have a letter that corresponds to V in the same way that μ (the symbol for the population mean) corresponds to M?

As a matter of fact, it doesn't. Instead, we use the *lowercase* sigma! It looks like this: σ. Not only that, but because we're talking about squared quantities, the symbol is σ^2.

So the formula for calculating variance is

$$\sigma^2 = \frac{\sum(X - \overline{X})^2}{N}$$

Variance is large if the numbers in a group vary greatly from their mean. Variance is small if the numbers are very similar to their mean.

The variance you just worked through is appropriate if the group of five measurements is a population. Does this mean that variance for a sample is different? It does, and you'll see why in a minute. First, I turn your attention back to Excel.

VARP and VARPA

Excel's two worksheet functions, VARP and VARPA, calculate the population variance.

Start with VARP. Figure 5-1 shows the Function Arguments dialog box for VARP along with data.

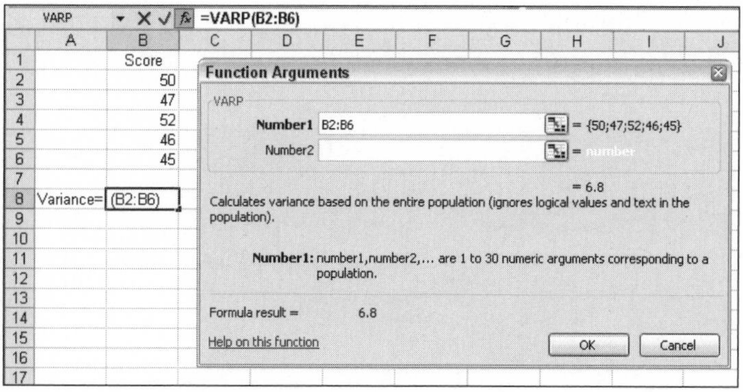

Figure 5-1:
Working
with VARP.

1. **Put your data into a worksheet and select a cell to display the result.**

   Figure 5-1 shows that for this example, I've put the numbers 50, 47, 52, 46, 45 into cells B2 through B6 and selected B8 for the result.

2. **Click the Insert Function button to open the Insert Function dialog box.**

3. **In the Insert Function dialog box select VARP and click OK to open the Function Arguments dialog box for VARP.**

4. **Identify the data array.**

   Type B2:B6 in the Number1 field. The population variance, 6.8, appears in the Function Arguments dialog box.

5. **Click OK to close the dialog box and put the result in the selected cell.**

When VARP calculates the variance in a range of cells, it only sees numbers. If text or logical values are in some of the cells, VARP ignores them.

VARPA, on the other hand, does not. VARPA takes text and logical values into consideration and includes them in its variance calculation. How? If a cell contains text, VARPA sees that cell as containing a value of zero. If a cell contains the logical value FALSE, that's also zero as far as VARPA is concerned. In VARPA's view of the world, the logical value TRUE is 1. Those zeros and ones get added into the mix and affect the mean and the variance.

To see this in action, keep the numbers in cells B2 through B6 and again select cell B8. Follow the steps you followed for VARP, but this time open the VARPA dialog box. In the Value1 field of the VARPA dialog box type B2:B7 (that's B7, *not* B6) and click OK. In cell B8, you see the same result you saw before because VARPA evaluates the blank cell B6 as no entry.

Now type TRUE into cell B7. The result in B8 changes because VARPA evaluates B7 as 1. (See Figure 5-2.)

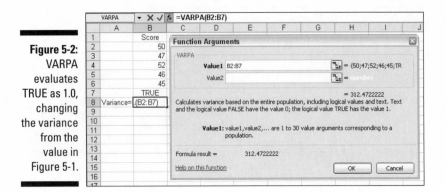

Figure 5-2:
VARPA
evaluates
TRUE as 1.0,
changing
the variance
from the
value in
Figure 5-1.

Next, type FALSE (or any other string of letters except TRUE) into B7. Once again the value in B8 changes. This time, VARPA evaluates B7 as zero.

Sample variance

Earlier, I mentioned that you use this formula to calculate population variance:

$$\sigma^2 = \frac{\sum (X - \overline{X})^2}{N}$$

I also said that sample variance is a little different. Here's the difference. If your set of numbers is a sample drawn from a large population, you're probably interested in using the variance of the sample to estimate the variance of the population.

The formula you used for the variance doesn't quite work as an estimate of the population variance. Although the sample mean works just fine as an estimate of the population mean, this doesn't hold true with variance, for reasons *way* beyond the scope of this book.

How do you calculate a good estimate of the population variance? It's pretty easy. You just use *N-1* in the denominator rather than *N*. (For reasons way beyond our scope.)

Also, because we're working with a characteristic of a sample (rather than of a population), we use the English equivalent of the Greek letter — s rather than σ. This means that the formula for the sample variance is

$$s^2 = \frac{\sum(X - \overline{X})^2}{N - 1}$$

The value of s^2, given the squared deviations in our set of five numbers is

(4 + 1 + 16 + 4 + 9)/4 = 34/4 = 8.5

So, if these numbers,

50, 47, 52, 46, 45,

are an entire population, their variance is 6.4. If they're a sample drawn from a larger population, our best estimate of that population's variance is 8.5

VAR and VARA

The worksheet functions VAR and VARA calculate the sample variance. Repeat the steps from the VARP and VARPA section to open the Function Arguments dialog box for VAR. Figure 5-3 shows the Function Arguments dialog box for VAR with 50, 47, 52, 46, 45 entered into cells B2 through B6.

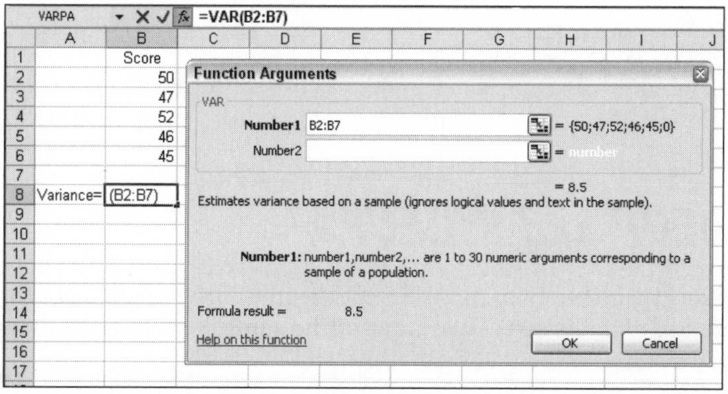

Figure 5-3:
Working
with VAR.

The relationship between VAR and VARA is the same as the relationship between VARP and VARPA: VAR ignores cells that contain logical values (TRUE and FALSE) and text. VARA includes those cells. Once again, TRUE evaluates to 1.0 and FALSE evaluates to 0. Text in a cell causes VARA to see that cell's value as 0.

Back to the Roots: Standard Deviation

After you calculate the variance of a set of numbers, you have a value whose units are different from your original measurements. For example, if your original measurements are in inches, their variance is in square inches. This is because you square the deviations before you average them.

Often, it's more intuitive if you have a variation statistic that's in the same units as the original measurements. It's easy to turn variance into that kind of statistic. All you have to do is take the square root of the variance.

Like the variance, this square root is so important that we give it a special name: *standard deviation.*

Population standard deviation

The standard deviation of a population is the square root of the population variance. The symbol for the population standard deviation is σ (sigma). Its formula is

$$\sigma = \sqrt{\sigma^2} = \sqrt{\frac{\sum(X - \overline{X})^2}{N}}$$

For these measurements (in inches)

50, 47, 52, 46, 45

the population variance is 6.8 square inches, and the population standard deviation is 2.61 inches (rounded off).

STDEVP and STDEVPA

The Excel worksheet functions STDEVP and STDEVPA calculate the population standard deviation. After entering your numbers into your worksheet and selecting a cell,

1. **Type your data into an array and select a cell for the result.**

2. **Click the Insert Function button to open the Insert Function dialog box.**

3. **In the Insert Function dialog box, select STDEVP and click OK to open the Function Arguments dialog box for STDEVP.**

4. **In the Function Arguments dialog box, identify the data array.**

After you enter the data array, the dialog box shows the value of the population standard deviation for the numbers in the data array. Figure 5-4 shows this.

5. **Click OK to close the dialog box and put the result into the selected cell.**

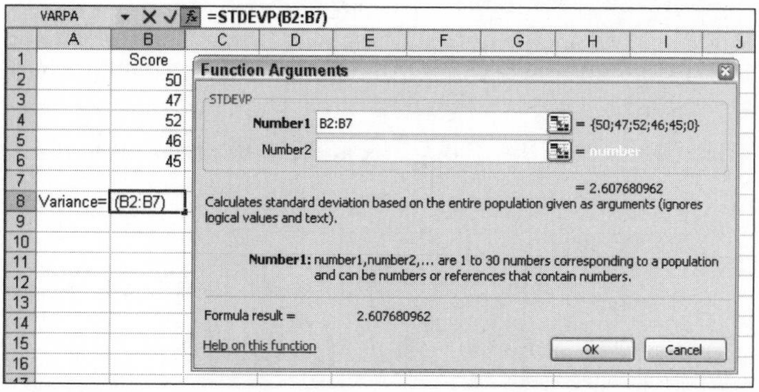

Figure 5-4: The Function Arguments dialog box for STDEVP, along with the data.

Like VARPA, STDEVPA uses any logical values and text values it finds when it calculates the population standard deviation. TRUE evaluates to 1.0 and FALSE evaluates to 0. Text in a cell gives that cell a value of 0.

Sample standard deviation

The standard deviation of a sample — an estimate of the standard deviation of a population — is the square root of the sample variance. Its symbol is *s* and its formula is

$$s = \sqrt{s^2} = \sqrt{\frac{\sum (X - \overline{X})^2}{N-1}}$$

For these measurements (in inches)

50, 47, 52, 46, 45

the population variance is 8.4 square inches, and the population standard deviation is 2.92 inches (rounded off).

STDEV and STDEVA

The Excel worksheet functions STDEV and STDEVA calculate the sample standard deviation. To work with STDEV:

1. **Type your data into an array and select a cell for the result.**

2. **Click the Insert Function button to open the Insert Function dialog box.**

3. **In the Insert Function dialog box, select STDEV and click OK to open the Function Arguments dialog box for STDEV.**

4. **In the Function Arguments dialog box, identify the data array.**

   With the data array entered, the dialog box shows the value of the population standard deviation for the numbers in the data array. Figure 5-5 shows this.

5. **Click OK to close the dialog box and put the result into the selected cell.**

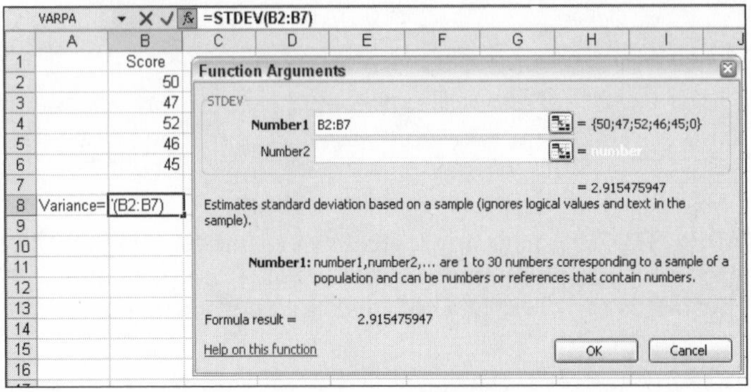

Figure 5-5:
The
Function
Arguments
dialog box
for STDEV.

STDEVA uses text and logical values in its calculations. Cells with text have values of 0, and cells whose values are FALSE also evaluate to 0. Cells that evaluate to TRUE have values of 1.0.

Related Functions

Before we move on, let's take a quick look at a couple of other variation-related worksheet functions. Repeat the steps from the STDEV section to get to the Function Arguments dialog boxes for these functions.

DEVSQ

DEVSQ calculates the sum of the squared deviations from the mean (without dividing by N or by $N-1$). For these numbers

50, 47, 52, 46, 45

that's 34, as Figure 5-6 shows.

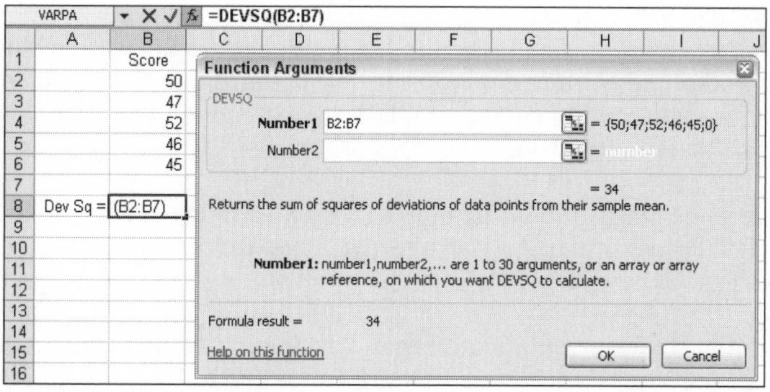

| VARPA | ▼ ✕ ✓ ƒₓ | =DEVSQ(B2:B7) |

Function Arguments dialog box showing DEVSQ function with Number1 B2:B7 = {50;47;52;46;45;0}, Number2 = number, = 34. "Returns the sum of squares of deviations of data points from their sample mean." Number1: number1,number2,... are 1 to 30 arguments, or an array or array reference, on which you want DEVSQ to calculate. Formula result = 34.

Figure 5-6:
The Function Arguments dialog box for DEVSQ.

Average deviation

One more Excel function deals with deviations in a way other than squaring them.

The variance and standard deviation deal with negative deviations by squaring all the deviations before averaging them. How about if we just ignore the minus signs? This is called taking the *absolute value* of each deviation. (That's the way mathematicians say, "How about if we just ignore the minus signs?").

If we do that for the heights

50, 47, 52, 46, 45

we can put the absolute values of the deviations into a table like Table 5-4.

Table 5-4	A Group of Numbers and Their Absolute Deviations	
Height	*Height-Mean*	*|Deviation|*
50	50-48	2
47	47-48	1
52	52-48	4
46	46-48	2
45	45-48	3

In Table 5-4, notice the vertical lines around Deviation in the heading for the third column. Vertical lines around a number symbolize its absolute value. That is, the vertical lines are the mathematical symbol for "How about if we just ignore the minus signs?"

The average of the numbers in the third column is 2.4. This average is called the *average absolute deviation*, and it's a quick and easy way to characterize the spread of measurements around their mean. It's in the same units as the original measurements. So if the heights are in inches, the absolute average deviation is in inches, too.

Like variance and standard deviation, a large average absolute deviation signifies a lot of spread. A small average absolute deviation signifies little spread.

This statistic is less complicated than variance or standard deviation, but is rarely used. Why? For reasons that are (once again) beyond our scope, statisticians can't use it as the foundation for additional statistics you'll meet later. Variance and standard deviation serve that purpose.

AVEDEV

Excel's AVEDEV worksheet function calculates the average absolute deviation of a group of numbers. Figure 5-7 shows the Function Arguments dialog box for AVEDEV, which presents the average absolute deviation for the cells in the indicated range.

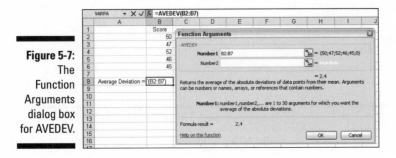

Figure 5-7:
The Function Arguments dialog box for AVEDEV.

Now that you've seen measures of central tendency and variability, the next chapter shows when and how to combine them.

Chapter 6

Meeting Standards and Standings

*I*n my left hand I hold 15 Argentine pesos. In my right, I hold 100 Chilean pesos. Which is worth more? Both currencies are called "*pesos*," right? So shouldn't the 100 be greater than the 15? Not necessarily. "Peso" is just word magic — a coincidence of names. Each one comes out of a different country, and each country has its own economy. To compare the two amounts of money, you have to convert each currency into a standard unit. The most intuitive standard for us is our own currency. How much is each amount worth in dollars and/or cents? As I write this, 15 Argentine pesos are worth about $5. One hundred Chilean pesos are worth about 20 cents.

In this chapter, I show you how to use statistics to create standard units. Standard units show you where a score stands in relation to other scores in a group, and I show you additional ways to determine a score's standing within a group.

Catching Some Zs

As the preceding paragraph shows, a number in isolation doesn't really tell a story. In order to fully understand what a number means, you have to consider the process that produced it. In order to compare one number to another, they both have to be on the same scale.

In some cases, like currency conversion, it's easy to figure out a standard. In others, like temperature conversion or conversion into the metric system, a formula guides you.

When it's not all laid out for you, you can use the mean and the standard deviation to standardize scores that come from different processes. The idea is to take a set of scores and use its mean as a zero-point and its standard deviation as a unit of measure. Then, you compare the deviation of each score from the mean to the standard deviation. You may ask: How big is a particular deviation relative to (something like) an average of all the deviations?

To find out, you divide the score's deviation by the standard deviation. In effect, you transform the score into another kind of score. The transformed score is called a *standard score*, or a *z-score*.

The formula for this is

$$z = \frac{X - \overline{X}}{s}$$

if you're dealing with a sample, and

$$z = \frac{X - \mu}{\sigma}$$

if you're dealing with a population. In either case, *x* represents the score you're transforming into a z-score.

Characteristics of z-scores

A z-score can be positive, negative, or zero. A negative z-score represents a score that's less than the mean and a positive z-score represents a score that's greater than the mean. When the score is equal to the mean, its z-score is zero.

When you calculate the z-score for every score in the set, the mean of the z-scores is 0, and the standard deviation of the z-scores is 1.

Once you do this for several sets of scores, you can legitimately compare a score from one set to a score from another. If the two sets have different means and different standard deviations, comparing without standardizing is like comparing apples with cumquats.

In the examples that follow, I show how to use z-scores to make comparisons.

Bonds vs. The Bambino

Here's an important question that often comes up in the context of serious metaphysical discussions: Who is the greatest home run hitter of all time: Barry Bonds or Babe Ruth? While this is a difficult question to answer, one way to get your hands around it is to look at each player's best season and compare the two. Bonds hit 73 home runs in 2001, and Ruth hit 60 in 1927. On the surface, Bonds appears to be the more productive hitter.

The year 1927 was very different from 2001, however. Baseball (and everything else) experienced huge changes in the years between then and now, and player statistics reflect those changes. It was more difficult to hit a home run in the 1920s than it was in 2001. Still, 73 versus 60? Hmmm . . .

Standard scores can help us decide whose best season was better. To standardize, I took the top 50 home run hitters of 1927 and the top 50 from 2001. I calculated the mean and standard deviation of each group, and then turned Ruth's 60 and Bonds' 73 into z-scores.

The average from 1927 is 12.68 homers with a standard deviation of 10.49. The average from 2001 is 37.02 homers with a standard deviation of 9.64. Although the means differ greatly, the standard deviations are pretty close.

And the z-scores? Ruth's is

$$z = \frac{60 - 12.68}{10.49} = 4.51$$

Bonds' is

$$z = \frac{73 - 37.02}{9.64} = 3.73$$

The clear winner in the z-score best-season home run derby is Babe Ruth. Period.

Just to show you how times have changed, Lou Gehrig hit 47 home runs in 1927 (finishing second to Ruth) for a z-score of 3.27. In 2001, 47 home runs amounted to a z-score of 1.04.

Exam scores

Getting away from sports debates, one practical application of z-scores is the assignment of grades to exam scores. Based on percentage scoring, instructors traditionally evaluate a score of 90 points or higher (out of 100) as an A, 80-89 points as a B, 70-79 points as a C, 60-69 points as a D, and less than 60 points as an F. Then, they average scores from several exams together to assign a course grade.

Is that fair? Just as a peso from Argentina is worth more than a peso from Chile, and a home run was harder to hit in 1927 than in 2001, is a point on one exam worth the same as a point on another? Like peso, isn't that just word magic?

Indeed it is. A point on a difficult exam is, by definition, harder to come by than a point on an easy exam. Because points might not mean the same thing from one exam to another, the fairest thing to do is convert scores from each exam into z-scores before averaging them. That way, you're averaging numbers on a level playing field.

In the courses I teach, I do just that. I often find that a lower score on one exam results in a higher z-score than a higher score from another exam. For example, on an exam where the mean is 65 and the standard deviation is 12, a score of 71 results in a z-score of 0.5. On another exam, with a mean of 69 and a standard deviation of 14, a score of 75 is equivalent to a z-score of .429. (Yes, it's like Ruth's 60 home runs vs. Bonds' 73.) Moral of the story: Numbers in isolation tell you very little. You have to understand the process that produces them.

STANDARDIZE

Excel's STANDARDIZE worksheet function calculates z-scores. Figure 6-1 shows a set of exam scores along with their mean and standard deviation. I used AVERAGE and STDEVP to calculate the statistics. The Function Arguments dialog box for STANDARDIZE is also in the figure.

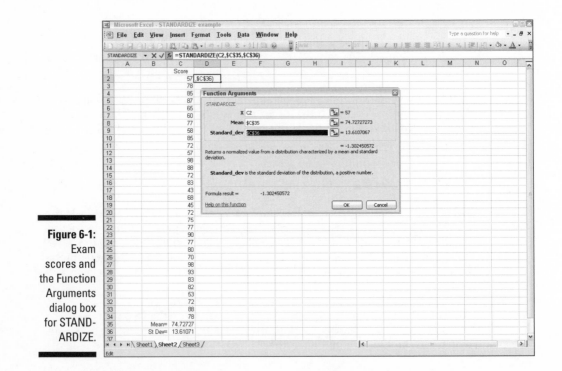

Figure 6-1: Exam scores and the Function Arguments dialog box for STANDARDIZE.

Caching some z's

Because negative z-scores might have connotations that are, well, negative, educators sometimes change the z-score when they evaluate students. In effect, they're hiding the z-score, but the concept is the same — standardization with the standard deviation as the unit of measure.

One popular transformation is called the T-score. (In Chapter 11, I introduce the *t*-test. Similar name, totally different concept.) The T-score eliminates negative scores because a set of T-scores has a mean of 50 and a standard deviation of 10. The idea is to give an exam, grade all the tests, and calculate the mean and standard deviation. Next, turn each score into a z-score. Then, follow this formula:

$$T = (z)(10) + 50$$

People who use the T-score often like to round to the nearest whole number.

SAT scores are another transformation of the z-score. (Some refer to the SAT as a C-score.)

The SAT has a mean of 500 and a standard deviation of 100. After the exams are graded, and their mean and standard deviation calculated, each exam score becomes a z-score in the usual way. This formula converts the z-score into an SAT score:

$$SAT = (z)(100) + 500$$

Rounding to the nearest whole number is part of the procedure here, too.

The IQ score is still another transformed z. Its mean is 100 and (in the Stanford-Binet version) its standard deviation is 16. What's the procedure for computing an IQ score? You guessed it. In a group of IQ scores, calculate the mean and standard deviation, and then calculate the z-score. Then it's

$$IQ = (z)(16) + 100$$

As with the other two, IQ scores are rounded to the nearest whole number.

Here are the steps:

1. **Type the data into an array and select a cell.**

   I selected D2 for this example. Ultimately, I'll line up all the z-scores next to the corresponding exam score.

2. **Click the Insert Function button (labeled *f_x*), to open the Insert Function dialog box.**

3. **From the Insert Function dialog box, select STANDARDIZE to open the Function Arguments dialog box for STANDARDIZE.**

4. **In the Function Arguments dialog box, type the cell number that holds the first exam score into the X box.**

   In this example, that's C2.

5. **In the Mean box, type the cell number that holds the mean.** Make sure it's in absolute reference format.

   This cell is C35 for this example.

   The final step, Step 8, is to drag the cursor and autofill the remaining z-scores.

   To get this done correctly, cell C35 has to be in absolute reference format. In practical terms, that means it has to look like this: $C$35. You can make the change manually, or you can highlight the Mean box and press F4.

6. **In the Standard_dev box type the cell number that holds the standard deviation.** Make sure it's in absolute reference format.

   The appropriate cell in this example is C36. Type it into the Standard_dev box, highlight it, and press F4 to insure that the entry is $C$36.

7. **Click OK to close the Function Arguments dialog box and put the z-score for the first exam score into the selected cell.**

8. **Position the cursor on the selected cell's autofill handle, hold the left mouse button down, and drag the cursor to autofill the remaining z-scores.**

   Figure 6-2 shows the autofilled array of z-scores.

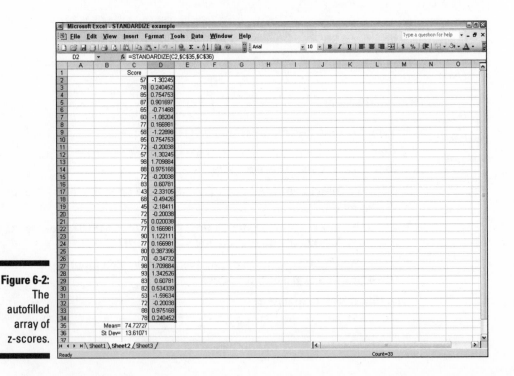

Figure 6-2:
The autofilled array of z-scores.

Where Do You Stand?

Standard scores are designed to show you where a score stands in relation to other scores in the same group. To do this, they use the standard deviation as a unit of measure.

If you don't want to use the standard deviation, you can show a score's relative standing in a simpler way. You can determine the score's rank within the group: The highest score has a rank of 1, the second highest has a rank of 2, and so on.

RANK

With Excel's RANK function you can quickly determine the ranks of all the scores in a group. Figure 6-3 shows the Function Arguments dialog box for RANK along with a group of scores. I've also set up a column for the ranks.

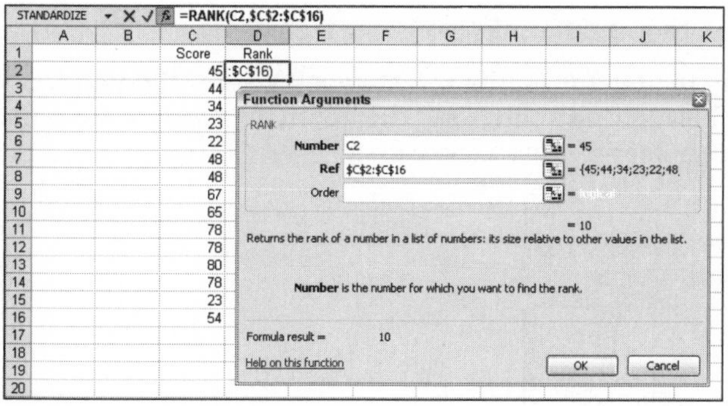

Figure 6-3: Working with RANK.

Here are the steps for using RANK:

1. **Type the data into an array and select a cell.**

   For this example, I've entered the scores into cells C2 through C16, and selected cell D2.

2. **Click the Insert Function button (labeled *f*_x) to open the Insert Function dialog box.**

3. **In the Insert Function dialog box select RANK to open the Argument Functions dialog box for RANK.**

4. **In the Number box, type the cell number that holds the score whose rank you want to insert into the selected cell.**

   For this example, that's C2.

5. **In the Ref box, type the array that contains the scores.** Make sure the array is in absolute reference format.

   The array here is C2 through C16, so type C2:C16 into the Ref box.

   This part is important. After you insert RANK into D2 you're going to drag the cursor through column D and autofill the ranks of the remaining scores. To set up for this, you have to let Excel know you want C2 through C16 to be the array for every score, not just the first one.

   That means the array in the Ref box has to look like this: $C$2:$C$16. You can either add the dollar signs ($) manually or highlight the Ref box and press F4.

6. **In the Order box, indicate the order for sorting the scores.**

   If you want to rank the scores in descending order, leave the Order box alone or type 0 (zero) into that box. To rank the scores in ascending order, type a non-zero value into the Order box.

7. **Click OK to put the rank into the selected cell.**

8. **Position the cursor on the selected cell's autofill handle, hold the left mouse button down and drag the cursor to autofill the ranks of the remaining scores. (See Figure 6-4.)**

	A	B	C	D	E
			Score	Rank	
1			Score	Rank	
2			45	10	
3			44	11	
4			34	12	
5			23	13	
6			22	15	
7			48	8	
8			48	8	
9			67	5	
10			65	6	
11			78	2	
12			78	2	
13			80	1	
14			78	2	
15			23	13	
16			54	7	

D2   *fx* =RANK(C2,$C$2:$C$16)

Figure 6-4: The autofilled ranks.

Take another look at Figure 6-4. In cells D7 and D8 you see a tie. The values in C7 and C8 are equal, so their ranks are the same: Excel assigns 8 as the rank to both. In most statistical applications that deal with ranks, however, ties are assigned the average of the ranks, not the identical ranks. If Excel assigned ranks that way, D7 and D8 would each have the rank 8.5 — the average of 8 and 9. In a help file for this function, Excel provides a correction that assigns averages to ties — but that only works if two scores are tied. For ties involving more than two scores, the correction doesn't work. Bottom line: Be careful when you use RANK.

LARGE and SMALL

You can turn the ranking process inside out by supplying a rank and asking which score has that rank. The worksheet functions LARGE and SMALL handle this from either end. They'll tell you the fifth-largest score, the third-smallest score, or any other rank you're interested in. To get to the Function Arguments dialog boxes for these functions, follow the steps in the RANK section.

Figure 6-5 shows the Function Arguments dialog box for LARGE. In the Array box you enter the array of cells that holds the group of scores. In the K box you enter the position whose value you want to find. To find the seventh-largest score in the array, for example, type 7 into the K box.

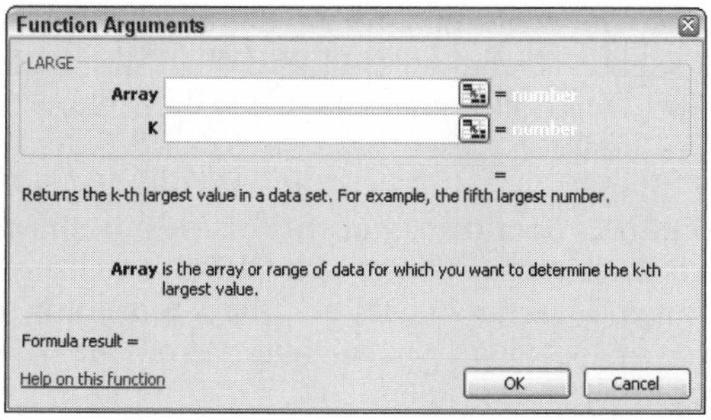

Figure 6-5: The Function Arguments dialog box for LARGE.

SMALL does the same thing, except it finds score positions from the lower end of the group. The Function Arguments dialog box for SMALL also has an Array box and a K box. Entering 7 in the K box returns the seventh-lowest score in the array.

PERCENTILE and PERCENTRANK

Closely related to rank is the *percentile*, which represents a score's standing in the group as the percent of scores below it. If you've taken standardized tests like the SAT, you've encountered percentiles. An SAT score in the 80th percentile is higher than 80 percent of the other SAT scores.

Excel's PERCENTILE function enables you to find the value at any percentile. Figure 6-6 shows the Function Arguments dialog box for PERCENTILE. To get to this Function Arguments dialog box, follow the steps in the RANK section.

The dialog box shows the 75th percentile (the value that's greater than 75 percent of the scores) for the numbers in cells C2 through C16. In this example, the 75th percentile is 72.5.

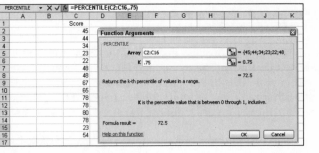

Figure 6-6: The Function Arguments dialog box for PERCEN-TILE.

You have to express the percentile as a decimal in the PERCENTILE Function Arguments dialog box. So, for the 75th percentile, you type .75 into the K box.

In contrast to percentiles, you might be interested in the flip side: Given a value, what percent of scores falls below it? PERCENTRANK handles this. To get to the PERCENTRANK Function Arguments dialog box, follow the steps in the RANK section.

In Figure 6-7, PERCENTRANK Function Arguments dialog box shows the percent rank of 65 for the scores in cells C2 through C16. (It's 0.642, or 64.2 percent.) The Array box holds the array of cells and the X box holds the score (65). The Significance box is optional: You can enter an amount of significant figures for the answer, or you can leave it blank.

Figure 6-7: The Function Arguments dialog box for PERCENT-RANK.

For the X box, you can enter either the value of a score or the label of the cell in which it appears. In this example, C10 in the X box gives you the same result as putting 65 in the X box.

Drawn and quartiled

A few specific percentiles are often used to summarize a group of scores. The median — the 50th percentile (because it's higher than 50 percent of the scores) — is one. Three others are the 25th percentile, the 75th percentile and the 100th percentile (the maximum score).

Because they divide a group of scores into fourths, these particular four percentiles are called *quartiles*. Excel's QUARTILE function calculates them. Selecting QUARTILE from the Insert Function dialog box opens the QUARTILE dialog box, which is shown in the figure.

Function Arguments		☒
QUARTILE		
Array		= *number*
Quart		= *number*
	=	
Returns the quartile of a data set.		
Array is the array or cell range of numeric values for which you want the quartile value.		
Formula result =		
Help on this function	OK	Cancel

The trick is to enter the right kind of numbers into the Quart box — 1 for the 25th percentile, 2 for the 50th, 3 for the 75th, and 4 for the 100th.

Entering 0 into the Quart box gives you the lowest score in the group.

Data analysis tool: Rank and Percentile

As the name of this section indicates, Excel provides a data analysis tool that calculates ranks and percentiles of each score in a group. The Rank and Percentile tool calculates both at the same time, so it saves you some steps over using the separate worksheet functions. (See Chapter 2 to install Excel's data analysis tools.) In Figure 6-8, I've taken the exam scores from the z-score example and opened the Rank and Percentile dialog box.

Here are the steps for using Rank and Percentile:

1. **Type your data into an array.**

   In this example, the data are in cells C2 through C37

2. **In the Tools menu, choose Data Analysis to open the Data Analysis dialog box.**

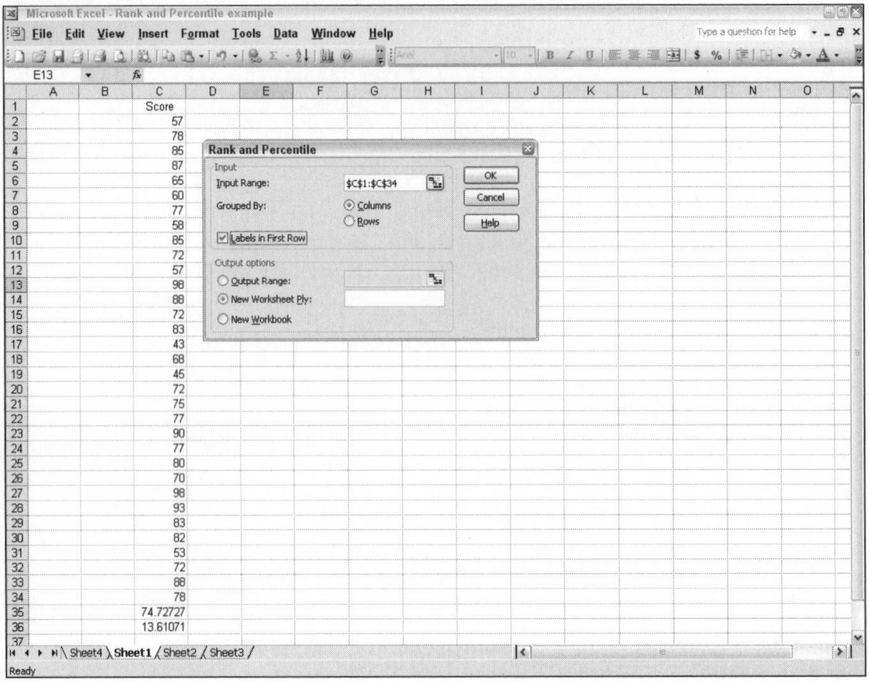

Figure 6-8:
The Rank
and
Percentile
analysis
tool.

3. **In the Data Analysis dialog box select Rank and Percentile.**

4. **Click OK to open the Rank and Percentile dialog box.**

5. **In the dialog box, type the data array into the Input Range box.** Make sure it's in absolute reference format.

   In this example, a label is in the first row, so check the Labels in First Row check box

6. **Click the Columns radio button to indicate the data are organized by columns.**

7. **Click the New Ply radio button to create a new tabbed page in the worksheet and to send the results to the newly created page.**

8. **Click OK to close the dialog box.** Open the newly created page to see the results.

Figure 6-9 shows the new page with the results. The table orders the scores from highest to lowest, as the Score column shows along with the Rank column. The Point column tells you the score's position in the original grouping. For example, the 98 in cell B2 is the 12th score in the original data. The Percent column gives the percentile for each score.

	A	B	C	D
	Point	Score	Rank	Percent
1	Point	Score	Rank	Percent
2	12	98	1	96.80%
3	26	98	1	96.80%
4	27	93	3	93.70%
5	22	90	4	90.60%
6	13	88	5	84.30%
7	32	88	5	84.30%
8	4	87	7	81.20%
9	3	85	8	75.00%
10	9	85	8	75.00%
11	15	83	10	68.70%
12	28	83	10	68.70%
13	29	82	12	65.60%
14	24	80	13	62.50%
15	2	78	14	56.20%
16	33	78	14	56.20%
17	7	77	16	46.80%
18	21	77	16	46.80%
19	23	77	16	46.80%
20	20	75	19	43.70%
21	10	72	20	31.20%
22	14	72	20	31.20%
23	19	72	20	31.20%
24	31	72	20	31.20%
25	25	70	24	28.10%
26	17	68	25	25.00%
27	5	65	26	21.80%
28	6	60	27	18.70%
29	8	58	28	15.60%
30	1	57	29	9.30%
31	11	57	29	9.30%
32	30	53	31	6.20%
33	18	45	32	3.10%
34	16	43	33	0.00%

Figure 6-9:
The output
of the Rank
and
Percentile
analysis
tool.

The concepts and tools from this chapter are good ones to keep in mind. They help you build a context for your data, so you can make effective decisions.

Chapter 7

Summarizing It All

Measures of central tendency and variability are excellent ways of summarizing a set of scores. But they aren't the only ways. Central tendency and variability make up a subset of descriptive statistics. Some descriptive statistics are intuitive — like count, maximum, and minimum. Some are not — like skewness and kurtosis. In this chapter, I discuss descriptive statistics, and I show you Excel's capabilities for calculating them and visualizing them.

Counting Out

The most fundamental descriptive statistic I can imagine is the number of scores in a set of scores. Excel offers four ways to determine that number. Yes, four ways. Count them.

COUNT, COUNTA, COUNTBLANK, and COUNTIF

Given an array of cells, COUNT gives you the amount of those cells that contain numerical data. Figure 7-1 shows that I've entered a group of scores, selected a cell to hold COUNT's result, and opened the Function Arguments dialog box for COUNT. Here are the steps:

1. **Type your data into the worksheet and select a cell for the result.**

   I entered data into columns C, D, and E to show off COUNT's multiargument capability. I selected cell C14 to hold the count.

2. **Click the Insert Function button to open the Insert Function dialog box.**

3. **From the Insert Function dialog box, select COUNT and click OK to open the Function Arguments dialog box for COUNT.**

4. **Identify the data array.**

   - In the Value1 box, enter one of the data columns for this example, like C1:C12.

   - Click in the Value2 box and enter another data column. I entered D1:D6.

   - Click in the Value3 box and enter the last column, which in this example is E1:E2.

5. **Click OK to put the tally in the selected cell.**

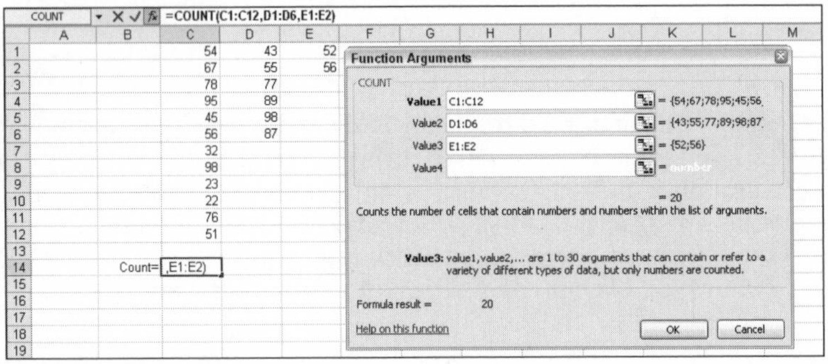

Figure 7-1:
The Function Arguments dialog box for COUNT showing multiple arguments.

COUNTA works like COUNT, except that its tally includes cells that contain text and logical values.

COUNTBLANK counts the number of blank cells in an array. In Figure 7-2, I use the numbers from the preceding example, but I extend the array to include cells D7 through D12 and E3 through E12. The array in the Range box is C1:E12. The Argument Functions dialog box for COUNTBLANK shows the number of blank cells (16 for this example).

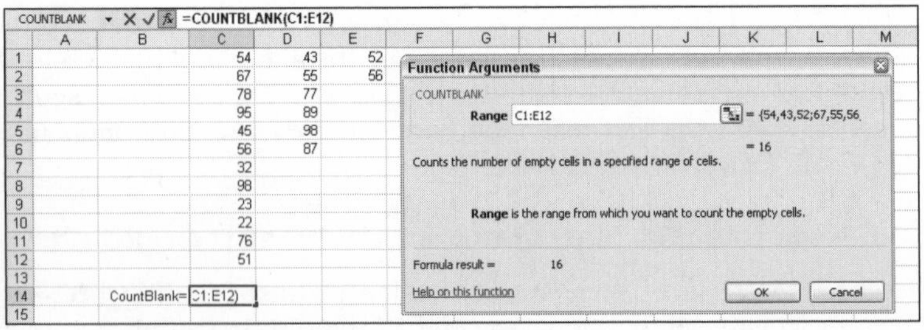

Figure 7-2:
COUNT-BLANK tallies the blank cells in a specified array.

COUNTIF shows the number of cells whose value meets a specified criterion. Figure 7-3 reuses the data once again, showing the Arguments Function dialog box for COUNTIF. Although the range is C1:E12, COUNTIF doesn't include blank cells.

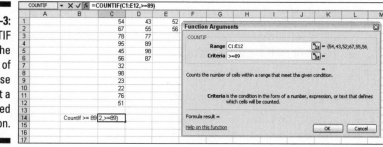

Figure 7-3: COUNTIF tallies the amount of cells whose data meet a specified criterion.

The criterion I used, >= 89, tells COUNTIF to include only the cells whose values are greater than or equal to 89. The count doesn't appear in the dialog box. You have to click OK to put the count into your selected cell (see Figure 7-4).

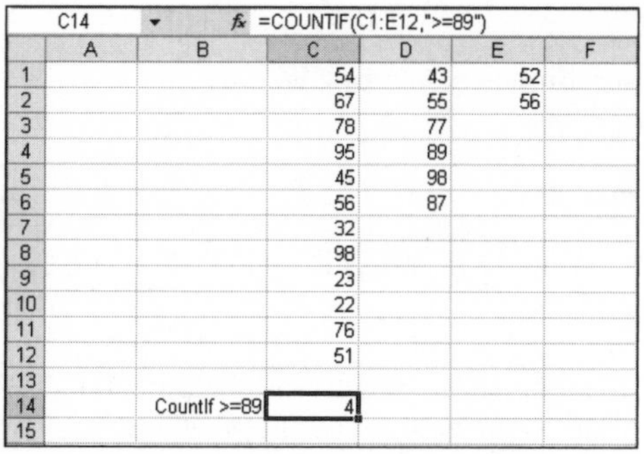

Figure 7-4: Putting COUNTIF's result into a selected cell.

The Long and Short of It

Two more descriptive statistics that probably require no introduction are the maximum and the minimum. These, of course, are the largest value and the smallest value in a group of scores.

MAX, MAXA, MIN, and MINA

Excel has worksheet functions that determine a group's largest and smallest values. I'll show you what MAX is all about. The others work in a similar fashion.

Figure 7-5 reuses the scores from the preceding examples. I selected a cell to hold their maximum value, and opened the Function Arguments dialog box for MAX. Here are the steps:

1. **Type your data into the worksheet, and select a cell to hold the maximum.**

   I entered data into columns C, D, and E to show off MAX's multiargument capability. For this example, I selected cell C14.

2. **Select a cell to hold the maximum.**

   I selected C14 for this example.

3. **Click the Insert Function button to open the Insert Function dialog box.**

4. **From the Insert Function dialog box, select MAX and click OK to open the Function Arguments dialog box for MAX.**

5. **Identify the data array.**

   • In the Number1 box enter one of the data columns, like C1:C12.

   • Click in the Number2 box. This creates and opens the Number3 box. In the Number2 box enter another array, D1:D6.

   • Click in the Number3 box and enter the last array, which in this example is E1:E2.

6. **Click OK to put the maximum score in C14.**

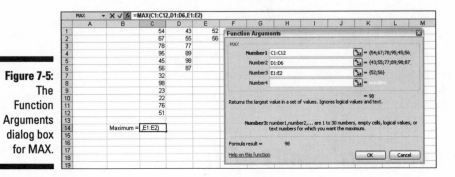

Figure 7-5:
The
Function
Arguments
dialog box
for MAX.

MAX ignores any text or logical values it encounters along the way. MAXA takes text and logical values into account when it finds the maximum. If MAXA encounters the logical value TRUE, it converts that value to 1. MAXA converts FALSE, or any text other than TRUE, to 0.

MIN and MINA work the same way as MAX and MAXA, except that they find the minimum instead of the maximum. Take care when you use MINA, because the conversions of logical values and text to 0 and 1 influence the result. With the numbers in the preceding example, the minimum is 22. If you enter FALSE or other text into a cell in any of the arrays, MINA gives 0 as the minimum. If you enter TRUE, MINA gives 1 as the minimum.

Getting Esoteric

In this section, I discuss some little-used statistics that are related to the mean and the variance. For most people, the mean and the variance are enough to describe a set of data. These other statistics, *skewness* and *kurtosis*, go just a bit further. You might use them someday if you have a huge set of data and you want to provide some in-depth description.

Think of the mean as *locating* a group of scores by showing you where their center is. This is the starting point for the other statistics. With respect to the mean

- ✔ The variance tells you how the scores are *spread out*.
- ✔ Skewness indicates how *symmetrically* the scores are distributed.
- ✔ Kurtosis shows you whether or not your scores are distributed with a *peak* in the neighborhood of the mean.

Skewness and kurtosis are related to the mean and variance in fairly involved mathematical ways. The variance involves the sum of squared deviations of scores around the mean. Skewness depends on cubing the deviations around the mean before you add them all up. Kurtosis takes it all to a higher power — the fourth power, to be exact. I get more specific in the subsections that follow.

Skewness: positive and negative

Where do zero, positive, and negative skew come from? They come from this formula:

$$skewness = \frac{\sum (X - \overline{X})^3}{(N-1)s^3}$$

In the formula, $\overline{X}$ is the mean of the scores, N is the number of scores, and s is the standard deviation.

I include this formula for completeness. If you're ever concerned with skewness, you probably won't use this formula anyway because Excel's SKEW function does the work for you.

SKEW

Figures 7-6, 7-7, and 7-8 show three histograms. (See Chapter 3 for more about histograms.) The first is symmetric, the other two are not. The symmetry and the asymmetry are reflected in the skewness statistic.

For the symmetric histogram, the skewness is zero. For the second histogram, the one that tails off to the right, the value of the skewness statistic is positive. It's also said to be skewed to the right. For the third histogram, which tails off to the left, the value of the skewness statistic is negative. It's also said to be skewed to the left.

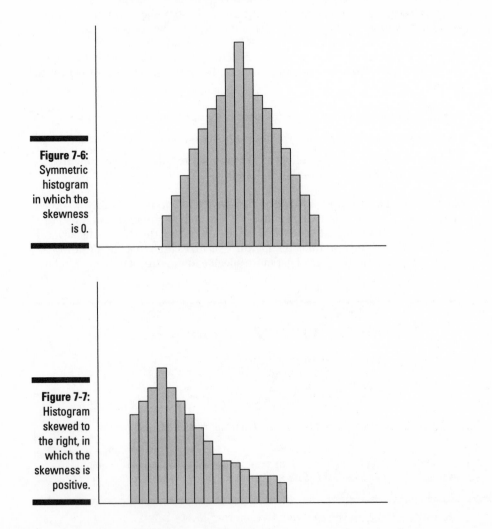

Figure 7-6:
Symmetric histogram in which the skewness is 0.

Figure 7-7:
Histogram skewed to the right, in which the skewness is positive.

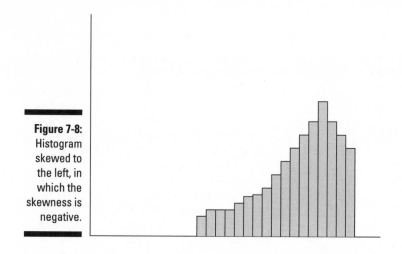

Figure 7-8:
Histogram
skewed to
the left, in
which the
skewness is
negative.

To use SKEW:

1. **Type your numbers into a worksheet and select a cell for the result.**

   For this example, I've entered scores into the first 10 rows of columns C, D, E, and F. (See Figure 7-9.) I selected cell G12 for the result.

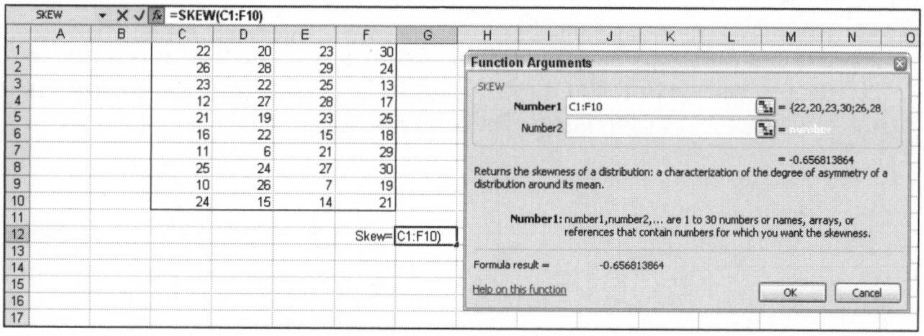

Figure 7-9:
Using the
SKEW
function.

2. **Click the Insert Function button to open the Insert Function dialog box, and choose SKEW.**

3. **Click OK to open the Function Arguments dialog box for SKEW.**

4. **Identify the data array.**

   In the Number1 box, enter the array of cells that holds the data. For this example, the array is C1:F10. With the data array entered, the Function Arguments dialog box shows the skewness, which for this example is negative, -0.656813864.

5. **Click OK to put the skewness value into the selected cell.**

Kurtosis: positive and negative

Negative? Wait a second. How can that be? I mentioned earlier that kurtosis involves the sum of fourth powers of deviations from the mean. Because 4 is an even number, even the fourth power of a negative deviation is positive. If you're adding all positive numbers, how can kurtosis ever be negative?

Here's how. The formula for kurtosis is

$$kurtosis = \frac{\sum (X - \overline{X})^4}{(N-1)s^4} - 3$$

where $\overline{X}$ is the mean of the scores, N is the number of scores, and s is the standard deviation.

Uh . . . why 3? The 3 comes into the picture because that's the kurtosis of something special called the *standard normal distribution*. (I discuss the normal distribution at length in Chapter 8.) Technically, statisticians refer to this formula as *kurtosis excess* — meaning that it shows the kurtosis in a set of scores that's in excess of the standard normal distribution's kurtosis. If you're about to ask the question, "Why is the kurtosis of the standard normal distribution equal to 3?" don't ask.

This is another formula you'll probably never use because Excel's KURT function takes care of business.

KURT

Figures 7-10 and 7-11 show two histograms. The first has a peak at its center, the second is flat. The first is said to be *leptokurtic*. Its kurtosis is positive. The second is *platykurtic*. Its kurtosis is negative.

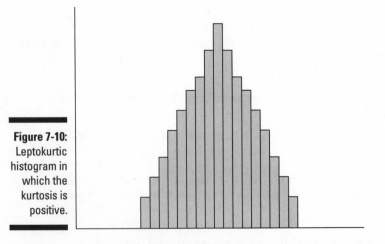

Figure 7-10:
Leptokurtic histogram in which the kurtosis is positive.

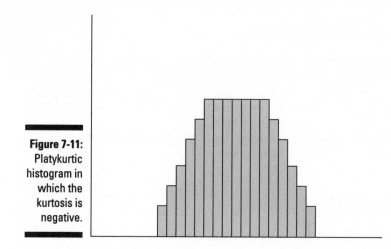

Figure 7-11:
Platykurtic
histogram in
which the
kurtosis is
negative.

Figure 7-12 shows the scores from the preceding example, a selected cell, and
the Function Arguments dialog box for KURT.

To use KURT:

1. **Type your numbers into a worksheet and select a cell for the result.**

   For this example, I've entered scores into the first 10 rows of columns C,
   D, E, and F. I selected cell G12 for the result.

2. **Click the Insert Function button to open the Insert Function dialog
   box, and choose KURT.**

3. **Click OK to open the Function Arguments dialog box for KURT.**

4. **Identify the data array.**

   In the Number1 box, enter the array of cells that holds the data. For this
   example, the array is C1:F10. With the data array entered, the Function
   Arguments dialog box shows the kurtosis, which for this example is negative, -0.290808675.

5. **Click OK to put the kurtosis into the selected cell.**

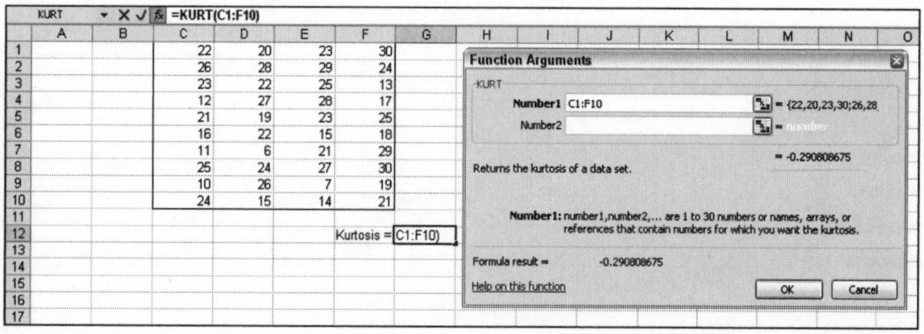

Figure 7-12:
Using KURT
to calculate
kurtosis.

Tuning in the Frequency

While the calculations for skewness and kurtosis are all well and good, it's helpful to see how the scores are distributed. To do this, you create a *frequency distribution,* a table that divides the possible scores into intervals, and shows the number (the frequency) of scores that fall into each interval.

Excel gives you two ways to create a frequency distribution. One is a worksheet function, the other is a data analysis tool.

FREQUENCY

I showed you the FREQUENCY worksheet function in Chapter 2 when I introduced array functions. Here, I give you another look. This function creates a frequency distribution. When you set up a frequency distribution, you typically organize it into intervals, and show the number of scores that occurs in each interval. An interval has a lower bound and an upper bound, so for each interval the frequency distribution shows the number of scores greater than the lower bound and less than the upper bound.

In the upcoming example, I reuse the data from the skewness and kurtosis discussions so you can see what the distribution of those scores looks like.

Figure 7-13 shows the data once again, along with a selected array, labeled Frequency. I've also added the label Intervals to a column, and in that column I put the interval boundaries. Each number in that column is the upper bound of an interval. (What about the lower bounds? Excel interprets the upper bound of an interval as the lower bound of the next interval.) The figure also shows the Argument Functions dialog box for FREQUENCY.

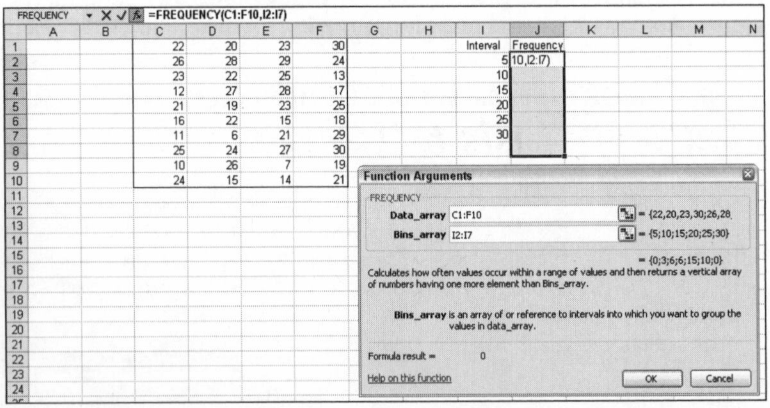

Figure 7-13: Finding the frequencies in an array of cells.

1. **Type the scores into an array of cells.**

2. **Type the intervals into an array.**

3. **Select an array for the frequencies.**

   I've put Frequency as the label at the top of column I, so J2 through J7 will hold the resulting frequencies.

4. **Click the Insert Function button to open the Insert Function dialog box.**

5. **Select FREQUENCY and click OK to open the Function Arguments dialog box.**

6. **Identify the data array.**

   In the Data_array box enter the cells that hold the scores. In this example, that's C1:F10.

7. **Identify the intervals array.**

   FREQUENCY refers to intervals as bins and holds the intervals in the Bins_array box. For this example, I2:I7 goes into the Bins_array box. After you identify both arrays, the Insert Function dialog box shows the frequencies inside a pair of curly brackets.

8. **Press Ctrl+Shift+Enter to close the Function Arguments dialog box.**

   Use this keystroke combination because FREQUENCY is an array function.

When you close the Function Arguments dialog box, the frequencies go into the appropriate cells, as seen in Figure 7-14.

Figure 7-14:
FRE-
QUENCY's
frequencies.

	A	B	C	D	E	F	G	H	I	J
									Interval	Frequency
1			22	20	23	30			5	0
2			26	28	29	24			10	3
3			23	22	25	13			15	6
4			12	27	28	17			20	6
5			21	19	23	25			25	15
6			16	22	15	18			30	10
7			11	6	21	29				0
8			25	24	27	30				
9			10	26	7	19				
10			24	15	14	21				
11										

J2   *fx* {=FREQUENCY(C1:F10,I2:I7)}

Data analysis tool: Histogram

Here's another way to create a frequency distribution — with the Histogram data analysis tool. (See Chapter 2 to install Excel's data analysis tools.) To show you that the two methods are equivalent, I use the data from the FREQUENCY example. Figure 7-15 shows the data along with the Histogram dialog box.

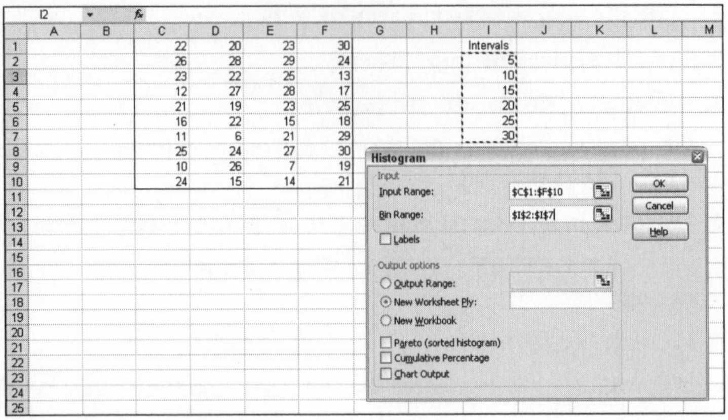

Figure 7-15:
The
Histogram
analysis
tool.

1. **Type the scores into an array, and type intervals into another array.**

2. **Click on the Tools menu, and select Data Analysis to open the Data Analysis dialog box.**

3. **From the Data Analysis dialog box, select Histogram to open the Histogram dialog box.**

4. **In the Input Range box, type the data array.**

   The data array is C1 through F10 in this example. The easiest way to enter this array is to click C1, press and hold Shift, and then click F10. Excel puts the absolute reference format ($C$1:$F$10) into the Input Range box.

5. **In the Bin Range box, type the array that holds the intervals.**

   In this example, that's I2 through I7. Click on I2, press and hold Shift, and then click I8. The absolute reference format ($I$2:$I$7) appears in the Bin Range box.

6. **Click the New Worksheet Ply radio button to create a new tabbed page and to put the results on the new page.**

7. **Click the Chart Output check box to create a histogram and visualize the results.**

8. **Click OK to close the dialog box.**

Figure 7-16 shows Histogram's output. The table matches up with what FRE-QUENCY produces. The size of the histogram is somewhat smaller when it first appears. I used the mouse to stretch the histogram and give it the appearance you see in the figure. The histogram shows that the distribution does tail off to the left (consistent with the negative skewness statistic) and seems to not have a distinctive peak (consistent with the negative kurtosis statistic).

By the way, the other check box options on the Histogram dialog box are Pareto and Cumulative Percentage. The Pareto chart sorts the intervals in order from highest frequency to lowest before creating the graph. Cumulative percentage shows the percentage of scores in an interval combined with the percentages in all the preceding intervals. Checking this box also puts a cumulative percentage line in the histogram.

Figure 7-16:
The
Histogram
tool's output
(after I
stretched
the chart).

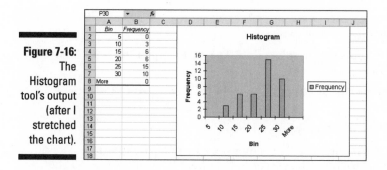

Can You Give Me a Description?

If you're dealing with individual descriptive statistics, the worksheet functions I've discussed get the job done nicely. If you want an overall report that presents just about all the descriptive statistical information in one place, use the data analysis tool I describe in the next section.

Data analysis tool: Descriptive Statistics

In Chapter 2, I showed you the Descriptive Statistics tool to introduce Excel's data analysis tools. Here's a slightly more complex example. Figure 7-17 shows three columns of scores and the Descriptive Statistics dialog box. I've labeled the columns First, Second, and Third so you can see how this tool incorporates labels.

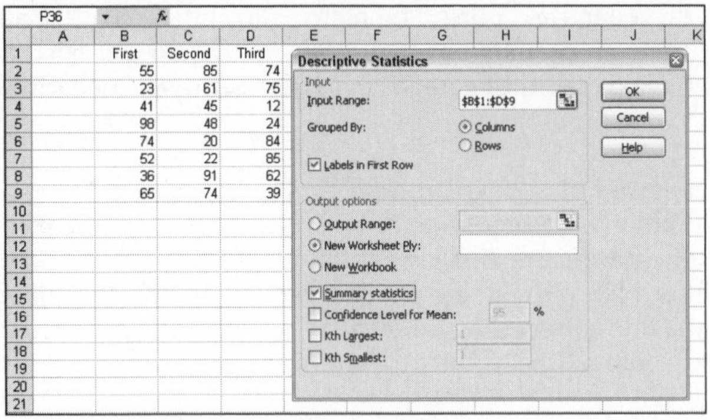

Figure 7-17:
The
Descriptive
Statistics
tool at work.

1. **Type the data into an array.**

2. **From the Tools menu, select Data Analysis to open the Data Analysis dialog box.**

3. **Choose Descriptive Statistics to open the Descriptive Statistics dialog box.**

4. **Identify the data array.**

   • In the Input Range box, enter the columns. The easiest way to do this is to move the cursor to the upper-left cell (B1), press Shift, and click the lower-right cell (D9). That puts $B$1:$D$9 into Input Range.

   • Click the Columns radio button to indicate the data are organized by columns.

   • Check the Labels in First Row check box because the Input Range includes the column headings.

5. **Click the New Worksheet Ply radio button to create a new tabbed sheet within the current worksheet, and to send the results to the newly created sheet.**

6. **Click the Summary Statistics check box, and leave the others unchecked.**

7. **Click OK to close the dialog box.**

   The new tabbed sheet (ply) opens, displaying statistics that summarize the data.

As Figure 7-18 shows, the statistics summarize each column separately. When this page first opens, the columns that show the statistic-names are too narrow, so the figure shows what the page looks like after I widened the columns.

The Descriptive Statistics tool gives values for these statistics: mean, standard error, median, mode, standard deviation, sample variance, kurtosis, skewness, range, minimum, maximum, sum, and count. Except for standard error and range, I've discussed all of them.

Figure 7-18:
The
Descriptive
Statistics
tool's
output.

	A	B	C	D	E	F
1	*First*		*Second*		*Third*	
2						
3	Mean	55.5	Mean	55.75	Mean	56.875
4	Standard Error	8.34309	Standard Error	9.49765	Standard Error	9.990062
5	Median	53.5	Median	54.5	Median	68
6	Mode	#N/A	Mode	#N/A	Mode	#N/A
7	Standard Deviation	23.59782	Standard Deviation	26.86341	Standard Deviation	28.25616
8	Sample Variance	556.8571	Sample Variance	721.6429	Sample Variance	798.4107
9	Kurtosis	0.288278	Kurtosis	-1.38727	Kurtosis	-1.30336
10	Skewness	0.567053	Skewness	-0.10605	Skewness	-0.66104
11	Range	75	Range	71	Range	73
12	Minimum	23	Minimum	20	Minimum	12
13	Maximum	98	Maximum	91	Maximum	85
14	Sum	444	Sum	446	Sum	455
15	Count	8	Count	8	Count	8
16						

Range is just the difference between the maximum and the minimum. Standard error is more involved, and I defer the explanation until Chapter 9. For now, I'll just say that standard error is the standard deviation divided by the square root of the sample size and leave it at that.

By the way, one of the check boxes left unchecked in the example's Step 6 provides something called the Confidence Limit for Mean, which I also defer until Chapter 9. The remaining two check boxes, Kth Largest and Kth Smallest, work like the functions LARGE and SMALL.

Instant Statistics

Need descriptive statistics in a hurry? No problem. Select the array you're interested in and . . . voilà; the statistics you're looking for appear instantly in the status bar at the bottom of the worksheet.

Figure 7-19 shows what I'm talking about. With an array of numbers selected, the sum shows up in the status bar. Right-click on the sum and you see the other available choices: Average, Count, Count Nums, Max, and Min. The difference between Count and Count Nums is that Count includes everything in an array, and Count Nums only includes numerical data in its tally.

With this chapter and the previous two, you have an extensive set of tools for describing a distribution of data. In the next chapter, I discuss a family of distributions that's near and dear to the hearts of statisticians.

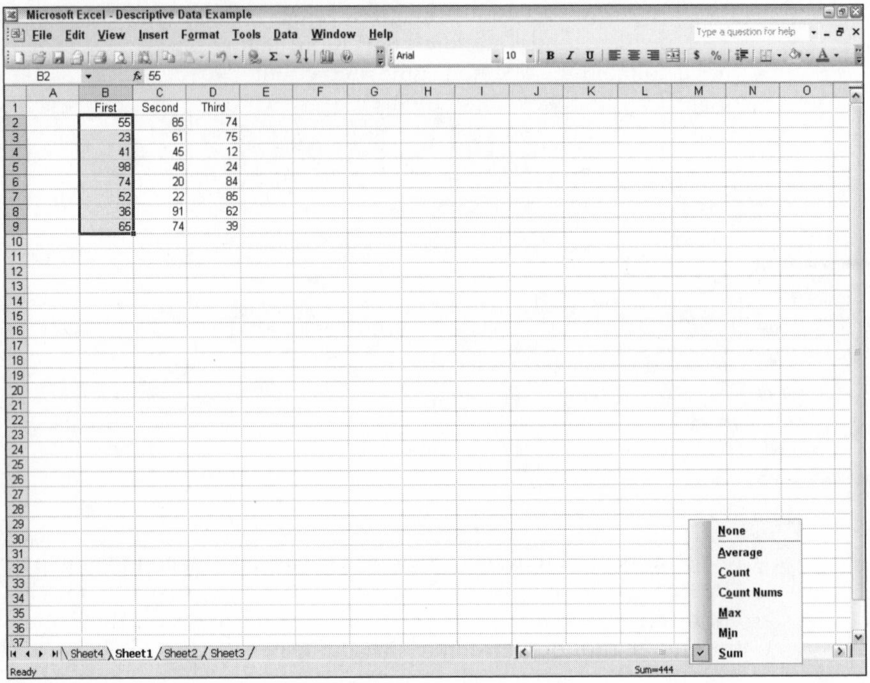

Figure 7-19: Statistics on the status bar.

Chapter 8

What's Normal?

A main job of statisticians is to estimate population characteristics. The job becomes easier if they can make some assumptions about the populations they study.

One particular assumption works over and over again: A specific attribute, trait, or ability is distributed throughout a population so that most people have an average or near-average amount of the attribute, and progressively fewer people have increasingly extreme amounts of the attribute. In this chapter, I discuss this assumption and what it means for statistics. I also describe Excel functions related to this assumption.

Hitting the Curve

When you measure something in the physical world like length or weight, you deal with objects you can see and touch. Statisticians, social scientists, market researchers, and businesspeople, on the other hand, often have to measure things they can't see or put their hands around. Traits like intelligence, musical ability, or willingness to buy a new product fall into this category.

These kinds of traits are usually distributed throughout the population so that most people are around the average — with progressively fewer people represented toward the extremes. Because this happens so often, it's become an assumption about how most traits are distributed.

It's possible to capture the most-people-are-about-average assumption in a graphic way. Figure 8-1 shows the familiar *bell curve* that characterizes how a variety of attributes are distributed. The area under the curve represents the population. The horizontal axis represents measurements of the ability under consideration. A vertical line drawn down the center of the curve would correspond to the average of the measurements.

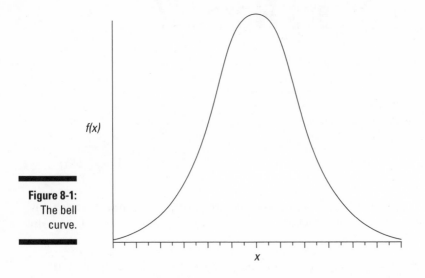

Figure 8-1:
The bell
curve.

So if we assume it's possible to measure a trait like intelligence and if we assume this curve represents how intelligence is distributed in the population, we can say this: The bell curve shows that most people have about average intelligence, very few have very little intelligence, and very few are geniuses. That seems to fit nicely with our intuitions about intelligence, doesn't it?

Digging deeper

On the horizontal axis of Figure 8-1 you see x, and on the vertical axis $f(x)$. What do these symbols mean? The horizontal axis, as I just mentioned, represents measurements, so think of each measurement as an x.

The explanation of $f(x)$ is a little more involved. A mathematical relationship between x and $f(x)$ creates the bell curve and enables us to visualize it. The relationship is rather complex, and I won't burden you with it. Just understand that $f(x)$ represents the height of the curve for a specified value of x. You supply a value for x (and for a couple of other things), and that complex relationship I mentioned returns a value of $f(x)$.

Now for some specifics. The bell curve is formally called the *normal distribution*. The term $f(x)$ is called *probability density*, so the normal distribution is an example of a *probability density function*. Rather than give you a technical definition of probability density, I ask you to think of probability density as something that turns the area under the curve into probability. Probability of . . . what? I discuss that in the next section.

Parameters of a normal distribution

People often speak of *the* normal distribution. That's a misnomer. It's really a family of distributions. The members of the family differ from one another in terms of two parameters — yes, *parameters* because I'm talking about populations. Those two parameters are the mean (μ) and the standard deviation (σ). The mean tells you where the center of the distribution is, and the standard deviation tells you how spread out the distribution is around the mean. The mean is in the middle of the distribution. Every member of the normal distribution family is symmetric — the left side of the distribution is a mirror image of the right.

The characteristics of the normal distribution are well known to statisticians. More importantly, you can apply those characteristics to your work.

How? This brings me back to probability. You can find some useful probabilities if you can do four things:

- If you can lay out a line that represents the scale of the attribute you're measuring
- If you can indicate on the line where the mean of the measurements is
- If you know the standard deviation
- If you know (or if you can assume) the attribute is normally distributed throughout the population

I'll work with IQ scores to show you what I mean. Scores on the Stanford-Binet IQ test follow a normal distribution. The mean of the distribution of these scores is 100 and the standard deviation is 16. Figure 8-2 shows this distribution.

As the figure shows, I've drawn a line for the IQ scale. Each point on the line represents an IQ score. With 100 (the mean) as the reference point, I've marked off every 16 points (the standard deviation). I've drawn a dotted line from the mean up to *f(100)* (the height of the normal distribution where x = 100) and a dotted line from each standard deviation point.

The figure also shows the proportion of area bounded by the curve and the horizontal axis, and by successive pairs of standard deviations. It also shows the proportion beyond three standard deviations on either side (52 and 148).

Note that the curve never touches the horizontal. It gets closer and closer, but it never touches. (Mathematicians say the curve is *asymptotic* to the horizontal.)

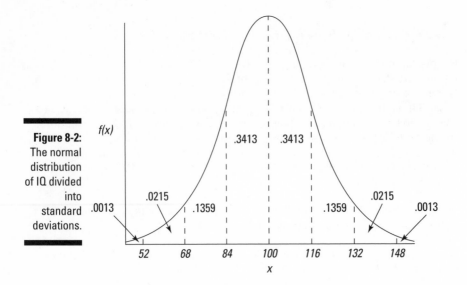

Figure 8-2:
The normal distribution of IQ divided into standard deviations.

So between the mean and one standard deviation — between 100 and 116 — lie .3413 (or 34.13 percent) of the scores in the population. Another way to say this: The probability that an IQ score is between 100 and 116 is .3413. At the extremes — in the tails of the distribution — lie .0013 (.13 percent) of the scores on either side.

The proportions of area under a normal distribution shown in Figure 8-2 hold for every member of the normal distribution family, not just for Stanford-Binet IQ scores. For example, in a sidebar in Chapter 6, I mention SAT scores, which have a mean of 500 and a standard deviation of 100. They're normally distributed, too. That means 34.13 percent of SAT scores are between 500 and 600, 34.13 percent between 400 and 500, and . . . well, you can use Figure 8-2 as a guide for other proportions.

NORMDIST

Figure 8-2 only shows areas partitioned by scores at the standard deviations. What about the proportion of IQ scores between 100 and 125? Or between 75 and 91? Or greater than 118? If you've ever taken a course in statistics, you might remember homework problems that involve finding proportions of areas under the normal distribution. You might also remember relying on tables of the normal distribution to solve them.

Excel's NORMDIST worksheet function enables you to find normal distribution areas without relying on tables. NORMDIST finds a *cumulative area*. You supply a score, a mean, and a standard deviation for a normal distribution, and NORMDIST returns the proportion of area to the left of the score (also called *cumulative proportion* or *cumulative probability*). For example, Figure 8-2 shows that in the IQ distribution .8413 of the area is to the left of 116.

How did I get that proportion? All the proportions to the left of 100 add up to .5000. (All the proportions to the right of 100 add up to .5000, too.) Add that .5000 to the .3413 between 100 and 116 and you have .8413.

Restating this in another way, the probability of an IQ score less than or equal to 116 is .8413.

In Figure 8-3, I use NORMDIST to find this proportion. Here are the steps:

1. **Select a cell for NORMDIST's answer.**

   For this example, I selected C2.

2. **Click the Insert Function button (labeled *f*) to open the Insert Function dialog box.**

3. **In the Insert Function dialog box, select NORMDIST to open the Function Arguments dialog box for NORMDIST.**

4. **In the X box, type the score for which you want to find the cumulative area.**

   In this example, that's 116.

5. **In the Mean box, type the mean of the distribution, and in the Standard_dev box, type the standard deviation.**

   Here, the mean is 100 and the standard deviation is 16.

6. **In the Cumulative box, type True**. This tells NORMDIST that you want to find the cumulative area. The dialog box shows the result.

Figure 8-3 shows the cumulative area is .84134476. If you enter FALSE in the Cumulative box, NORMDIST returns the height of the normal distribution at 116.

To find the proportion of IQ scores greater than 116, subtract the result from 1.0. (Just for the record, that's .15865524.)

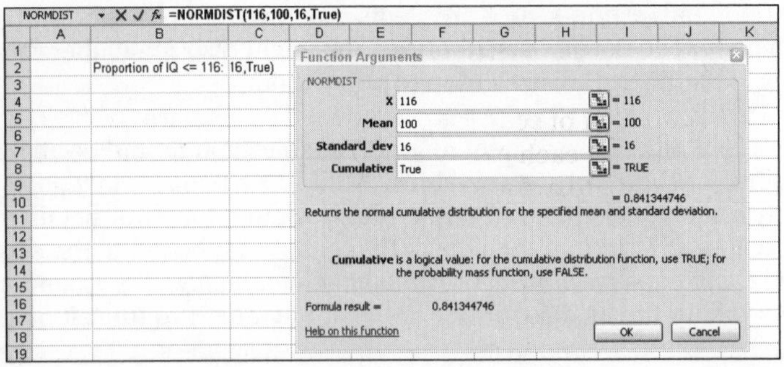

Figure 8-3:
Working
with
NORMDIST.

How about the proportion of IQ scores between 116 and 125? Apply NORMDIST for each score and subtract the results. For this particular example, the formula is

=NORMDIST(125,100,16,TRUE)-NORMDIST(116,100,16,TRUE)

The answer, by the way, is .09957.

NORMINV

NORMINV is the flip side of NORMDIST. You supply a cumulative probability, a mean, and a standard deviation, and NORMINV returns the score that cuts off the cumulative probability. For example, if you supply .5000 along with a mean and a standard deviation, NORMINV returns the mean.

This function is useful if you have to calculate the score for a specific percentile in a normal distribution. Repeat the steps from the NORMDIST section to open the Function Arguments dialog box for NORMINV, shown in Figure 8-4. Enter .75 as the cumulative probability, 500 as the mean, and 100 as the standard deviation. Because the SAT follows a normal distribution with 500 as its mean and 100 as its standard deviation, the result corresponds to the score at the 75th percentile for the SAT. (For more on percentiles, see Chapter 6.)

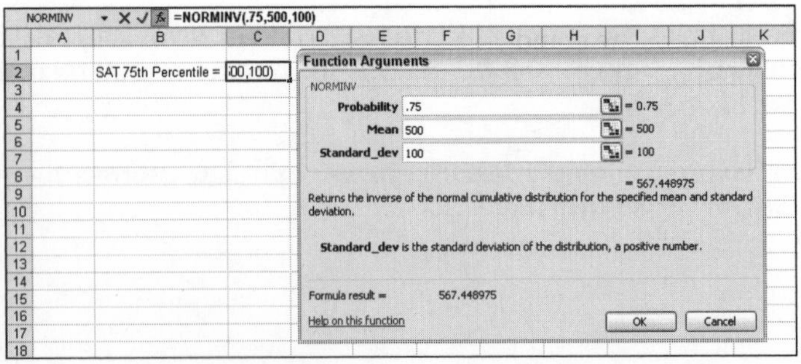

Figure 8-4:
Working
with
NORMINV.

A Distinguished Member of the Family

To standardize a set of scores so that you can compare them to other sets of scores, you convert each one to a *z-score* (see Chapter 6). The formula for converting a score to a z-score (also known as a standard score) is:

$$z = \frac{x - \mu}{\sigma}$$

The idea is to use the standard deviation as a unit of measure. For example, the Stanford-Binet version of the IQ test has a mean of 100 and a standard deviation of 16. The Wechsler version has a mean of 100 and a standard deviation of 15. How does a Stanford-Binet score of, say, 110, stack up against a Wechsler score of 110?

An easy way to answer this question is to put the two versions on a level playing field by standardizing both scores. For the Stanford-Binet

$$z = \frac{110 - 100}{16} = .625$$

For the Wechsler

$$z = \frac{110 - 100}{15} = .667$$

So 110 on the Wechsler is a slightly higher score than 110 on the Stanford-Binet.

Now, if you convert all the scores in a normal distribution (such as either version of the IQ), you have a normal distribution of z-scores. Any set of z-scores (normally distributed or not) has a mean of 0 and a standard deviation of 1. If a normal distribution has those parameters it's a *standard normal distribution* — a normal distribution of standard scores.

This is the member of the normal distribution family that most people have heard of. It's the one they remember most from statistics courses, and it's the one that most people are thinking about when they say *the* normal distribution. It's also what people think of when they hear *z-scores*. This distribution leads many to the mistaken idea that converting to z-scores somehow transforms a set of scores into a normal distribution.

Figure 8-5 shows the standard normal distribution. It looks like Figure 8-2, except that I've substituted zero for the mean and standard deviation units in the appropriate places.

In the next two sections, I describe Excel's functions for working with the standard normal distribution.

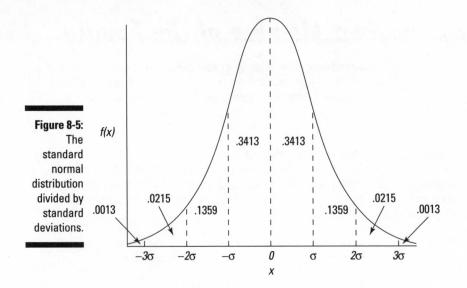

Figure 8-5:
The
standard
normal
distribution
divided by
standard
deviations.

NORMSDIST

NORMSDIST is like its counterpart NORMDIST, except that it's designed for a normal distribution whose mean is 0 and whose standard deviation is 1. Repeat the steps from the NORMDIST section to open the Function Arguments dialog box for NORMSDIST. You supply a z-score and it returns the area to the left of the z-score — the probability that a z-score is less than or equal to the one you supplied.

Figure 8-6 shows the Function Arguments dialog box with 1 as the z-score. The dialog box presents .841344746, the probability that a z-score is less than or equal to 1.00 in a standard normal distribution. Clicking OK puts that result into a selected cell.

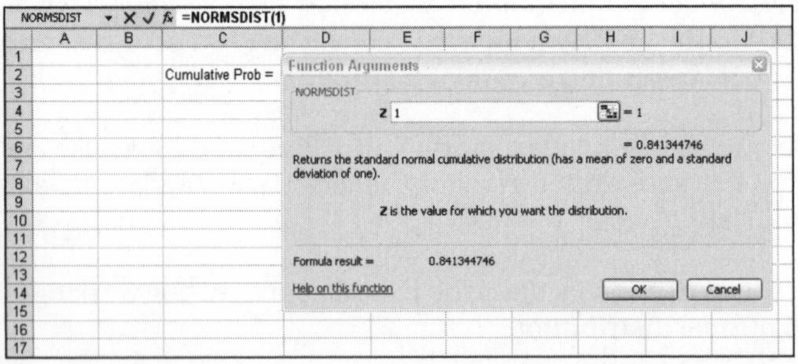

Figure 8-6:
Working
with
NORMS-
DIST.

OK, just because you asked . . .

The relationship between *x* and *f(x)* for the normal distribution is, as I mentioned, a pretty complex one. Here's the equation:

$$f(x) = \frac{1}{\sigma\sqrt{2\pi}} e^{\left[\frac{-(x-\mu)^2}{2\sigma^2}\right]}$$

If you supply values for μ (the mean), σ (the standard deviation), and *x* (a score), the equation gives you back a value for *f(x)*, the height of the normal distribution at *x*. π and e are important constants in mathematics. π is approximately 3.1416 (the ratio of a circle's circumference to its diameter). e is approximately 2.71828. It's related to something called *natural logarithms* and to a variety of other mathematical concepts. (See Chapter 20.)

In a standard normal distribution, μ = 0 and σ = 1, so the equation becomes

$$f(z) = \frac{1}{\sqrt{2\pi}} e^{\left[\frac{-z^2}{2}\right]}$$

I changed the *x* to *z* because you deal with z-scores in this member of the normal distribution family.

In Excel, you can set up a range of cells that contain standard scores, enter the formula that follows to capture the preceding equation, and autofill another range of cells with the formula results. Then, you can use the Chart Wizard (see Chapter 3) to trace the standard normal distribution. In the accompanying figure, I've done just that.

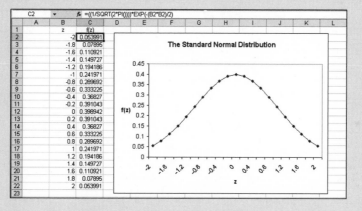

The Formula Bar shows the Excel formula that corresponds to the normal distribution equation:

=((1/SQRT(2*PI()))*EXP(-(B2*B2)/2)

PI() is an Excel function that gives the value of π. The function EXP() raises e to the power indicated by what's in the parentheses that follow it.

I showed you all this because I wanted you to see the equation of the normal distribution. The

NORMDIST worksheet function offers a much easier way to autofill the f(z) values. Enter this formula into C2

=NORMDIST(B2,0,1,FALSE)

autofill column C, and you have the same values as in the figure.

NORMSINV

NORMSINV is the flip side of NORMSDIST. Repeat the steps from the NORM-DIST section to open the Function Arguments dialog box for NORMSINV. You supply a cumulative probability and NORMSINV returns the z-score that cuts off the cumulative probability. For example, if you supply .5000, NORMSINV returns 0, the mean of the standard normal distribution.

Figure 8-7 shows the Function Arguments dialog box for NORMSINV, with .75 as the cumulative probability. The dialog box shows the answer, .67448975, the z-score at the 75th percentile of the standard normal distribution.

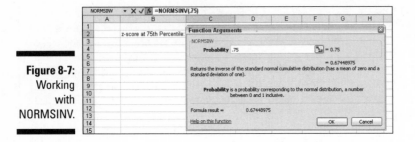

Figure 8-7:
Working
with
NORMSINV.

Part III
Drawing Conclusions from Data

The 5th Wave — By Rich Tennant

"Unless there's a corrupt cell in our spreadsheet analysis concerning the importance of trunk space, this should be a big seller next year."

In this part . . .

*P*art III deals with using statistical methods to make
inferences about data. Here's where you find out how
to use the data from samples to draw conclusions about
populations — the essence of statistical analysis. I begin
with the extremely important concept of sampling distributions. I move on to estimation and confidence limits, and
then to statistical tests geared at one sample, two samples,
and more. Part III ends with discussions of regression and
correlation — the statistics of relationships.

The statistical methods in this part are computationally
intensive. Fortunately, Excel has specialized features for
doing the calculations. The seven chapters in this part
describe Excel functions and tools for inferential statistics.

Chapter 9

The Confidence Game: Estimation

*P*opulations and samples are pretty straightforward ideas. A population is a huge collection of individuals from which you draw a sample. Assess the members of the sample on some trait or attribute, calculate statistics that summarize that sample, and you're in business.

In addition to summarizing the scores in the sample, you can use the statistics to create estimates of the population parameters. This is no small accomplishment. On the basis of a small percentage of individuals from the population, you can draw a picture of the population.

A question emerges, however: How much confidence can you have in the estimates you create? In order to answer this, you must have a context in which to place your estimates. How probable are they? How likely is the true value of a parameter to be within a particular lower bound and upper bound?

In this chapter, I introduce the context for estimates, show how that plays into confidence in those estimates, and describe an Excel function that enables you to calculate your confidence level.

What Is a Sampling Distribution?

Imagine that you have a population, and you draw a sample from this population. You measure the individuals of the sample on a particular attribute and calculate the sample mean. Return the sample members to the population. Draw another sample, assess the new sample's members, and then calculate *their* mean. Repeat this process again and again, always using the same number of individuals as you had in the original sample. If you could do this an infinite amount of times (with the same-sized sample each time), you'd have an infinite amount of sample means. Those sample means form a distribution of their own. This distribution is called *the sampling distribution of the mean.*

For a sample mean, this is the context I mentioned at the beginning of this chapter. Like any other number, a statistic makes no sense by itself. You have to know where it comes from in order to understand it. Of course, a statistic *comes from* a calculation performed on sample data. In another sense, a statistic is part of a sampling distribution.

In general, *a sampling distribution is the distribution of all possible values of a statistic for a given sample size.*

I italicized that definition because it's extremely important. After many years of teaching statistics, I can tell you that this concept usually sets the boundary line between people who understand statistics and people who don't.

So . . . if you understand what a sampling distribution is, you'll understand what the field of statistics is all about. If you don't, you won't. It's almost that simple.

If you don't know what a sampling distribution is, statistics will be a cookbook type of subject for you: Whenever you have to apply statistics, you'll plug numbers into formulas and hope for the best. On the other hand, if you're comfortable with the idea of a sampling distribution, you'll grasp the big picture of inferential statistics.

To help clarify the idea of a sampling distribution, take a look at Figure 9-1. It summarizes the steps in creating a sampling distribution of the mean.

A sampling distribution — like any other group of scores — has a mean and a standard deviation. The symbol for the mean of the sampling distribution of the mean (yes, I know that's a mouthful) is $\mu_{\bar{x}}$.

The standard deviation of a sampling distribution is a pretty hot item. It has a special name — *standard error*. For the sampling distribution of the mean, the standard deviation is called *the standard error of the mean*. Its symbol is $\sigma_{\bar{x}}$.

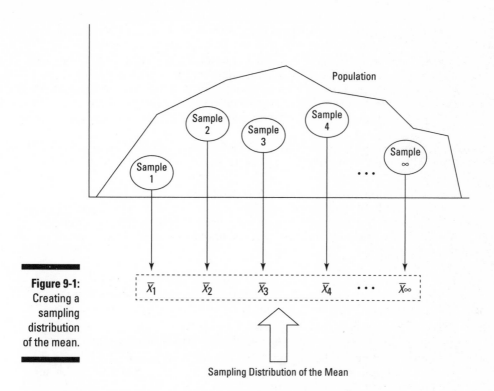

Figure 9-1:
Creating a
sampling
distribution
of the mean.

Sampling Distribution of the Mean

An *EXTREMELY* Important Idea:
The Central Limit Theorem

The situation I asked you to imagine is one that never happens in the real
world. You never take an infinite amount of samples and calculate their
means, and you never create a sampling distribution of the mean. Typically,
you draw one sample and calculate its statistics.

So, if you have only one sample, how can you ever know anything about a
sampling distribution — a theoretical distribution that encompasses an infinite number of samples? Is this all just a wild goose chase?

No, it's not. You can figure out a lot about a sampling distribution because of
a great gift from mathematicians to the field of statistics. This gift is called
the *Central Limit Theorem*.

According to the Central Limit Theorem

✔ The sampling distribution of the mean is approximately a normal distribution if the sample size is large enough.

Large enough means about 30 or more.

✔ The mean of the sampling distribution of the mean is the same as the population mean.

In equation form that's

$$\mu_{\bar{x}} = \mu$$

✔ The standard deviation of the sampling distribution of the mean (also known as the standard error of the mean) is equal to the population standard deviation divided by the square root of the sample size.

The equation here is

$$\sigma_{\bar{x}} = \frac{\sigma}{\sqrt{N}}$$

Notice that the Central Limit Theorem says nothing about the population. All it says is that if the sample size is large enough, the sampling distribution of the mean is a normal distribution, with the indicated parameters. The population that supplies the samples doesn't have to be a normal distribution for the Central Limit Theorem to hold.

What if the population is a normal distribution? In that case, the sampling distribution of the mean is a normal distribution regardless of the sample size.

Figure 9-2 shows a general picture of the sampling distribution of the mean partitioned into standard error units.

Simulating the Central Limit Theorem

It almost doesn't sound right. How can a population that's not normally distributed result in a normally distributed sampling distribution?

To give you an idea of how the Central Limit Theorem works, I created a simulation. This simulation creates something like a sampling distribution of the mean for a very small sample, based on a population that's not normally distributed. As you'll see, even though the population is not a normal distribution, and even though the sample is small, the sampling distribution of the mean looks quite a bit like a normal distribution.

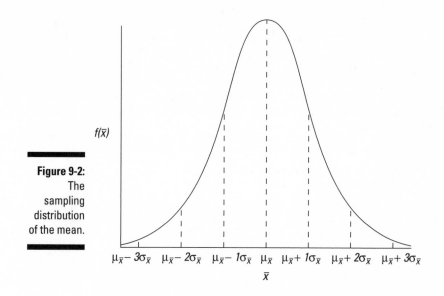

$f(\bar{x})$

$\mu_{\bar{x}}- 3\sigma_{\bar{x}}$ $\mu_{\bar{x}}- 2\sigma_{\bar{x}}$ $\mu_{\bar{x}}- 1\sigma_{\bar{x}}$ $\mu_{\bar{x}}$ $\mu_{\bar{x}}+ 1\sigma_{\bar{x}}$ $\mu_{\bar{x}}+ 2\sigma_{\bar{x}}$ $\mu_{\bar{x}}+ 3\sigma_{\bar{x}}$

$\bar{X}$

Imagine a huge population that consists of just three scores — 1, 2, and 3 —
and each one is equally likely to appear in a sample. (That kind of population
is definitely *not* a normal distribution.) Imagine, also, that you can randomly
select a sample of three scores from this population. Table 9-1 shows all the
possible samples and their means.

**Table 9-1   All Possible Samples of Three Scores (and their Means)
from a Population Consisting of the Scores 1, 2, and 3**

Sample	Mean	Sample	Mean	Sample	Mean
1,1,1	1.00	2,1,1	1.33	3,1,1	1.67
1,1,2	1.33	2,1,2	1.67	3,1,2	2.00
1,1,3	1.67	2,1,3	2.00	3,1,3	2.33
1,2,1	1.33	2,2,1	1.67	3,2,1	2.00
1,2,2	1.67	2,2,2	2.00	3,2,2	2.33
1,2,3	2.00	2,2,3	2.33	3,2,3	2.67
1,3,1	1.67	2,3,1	2.00	3,3,1	2.33
1,3,2	2.00	2,3,2	2.33	3,3,2	2.67
1,3,3	2.33	2,3,3	2.67	3,3,3	3.00

If you look closely at the table, you can almost see what's about to happen in the simulation. The sample mean that appears most frequently is 2.00. The sample means that appear least frequently are 1.00 and 3.00. Hmmm . . .

In the simulation, I randomly select a score from the population, and then randomly select two more. That group of three scores is a sample. Then, I calculate the mean of that sample. I repeat this process for a total of 60 samples, resulting in 60 sample means. Finally, I graph the distribution of the sample means.

What does the simulated sampling distribution of the mean look like? Figure 9-3 shows a worksheet that answers that question.

Figure 9-3: Simulating the sampling distribution of the mean (N=3) from a population consisting of the scores 1, 2, and 3. The simulation consists of 60 samples.

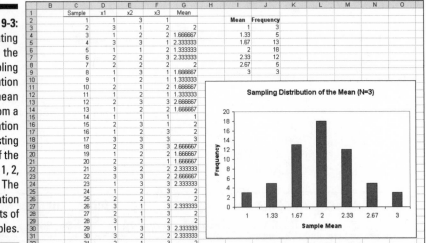

In the worksheet, each row is a sample. The columns labeled x1, x2, and x3 show the three scores for each sample. Column G shows the average for the sample in each row. Column I shows all the possible values for the sample mean, and column J shows how often each mean appears in the 60 samples. Columns I and J, and the graph, show that the distribution has its maximum frequency when the sample mean is 2.00. The frequencies generally tail off as the sample means get farther and farther away from 2.00.

The point of all this is that the population looks nothing like a normal distribution, and the sample size is very small. Even under those constraints, the sampling distribution of the mean based on 60 samples begins to look very much like a normal distribution.

What about the parameters the Central Limit Theorem predicts for the sampling distribution? Start with the population. The population mean is 2.00 and the population standard deviation is .67. (This kind of population requires some slightly fancy mathematics for figuring out the parameters. The math is a little beyond where we are, so I'll leave it at that.)

On to the sampling distribution. The mean of the 60 means is 1.97, and their standard deviation (an estimate of the standard error of the mean) is .51. Those numbers closely approximate the Central Limit Theorem-predicted parameters for the sampling distribution of the mean, 2.00 (equal to the population mean), and .47 (the standard deviation, .67, divided by the square root of 3, the sample size).

In case you're interested in doing the simulation summarized in Figure 9-3, here are the steps:

1. **Select a cell for your first randomly selected number.**

   I selected cell D2.

2. **Use the worksheet function RANDBETWEEN to select 1, 2, or 3.**

   This simulates drawing a number from a population consisting of the numbers 1, 2, and 3 where you have an equal chance of selecting each number. You can either use the Insert Function button and the Insert Function dialog box, or just type

   =RANDBETWEEN(1,3)

   in D2 and press Enter. The first argument is the smallest number RAND-BETWEEN returns, and the second argument is the largest number.

3. **Select the cell to the right of the original cell and type another random number between 1 and 3.** Do this again for a third random number in the cell to the right of the second one.

   The easiest way to do this is to autofill the two cells to the right of the original cell. In my worksheet those two cells are E2 and F2.

4. **Consider these three cells to be a sample and calculate their mean in the cell to the right of the third cell.**

   Again, you can use the Insert Function button and the Insert Function dialog box or just type

   =AVERAGE(D2:F2)

   in cell G2 and press Enter.

5. **Repeat this process for as many samples as you want to include in the simulation.** Have each row correspond to a sample.

I used 60 samples. The quick and easy way to get this done is to select the first row of three randomly selected numbers and their mean, and then autofill the remaining rows. The set of sample means in column G is the simulated sampling distribution of the mean. Use AVERAGE and STDEVP to find its mean and standard deviation.

To see what this simulated sampling distribution looks like use the array function FREQUENCY on the sample means in column G. Follow these steps:

1. **Type the possible values of the sample mean into an array.**

   I used column I for this. I expressed the possible values of the sample mean in fraction form (3/3, 4/3, 5/3, 6/3, 7/3, 8/3, and 9/3) as I entered them into the cells I3 through I9. Excel converts them to decimal form.

2. **Select an array for the frequencies of the possible values of the sample mean.**

   I used column J to hold the frequencies, selecting cells J3 through J9.

3. **Click the Insert Function button to open the Insert Function dialog box.**

4. **Select FREQUENCY and click OK to open the Function Arguments dialog box for FREQUENCY.**

5. **Identify the data array.**

   In the Data_array box, enter the cells that hold the sample means. In this example, that's G2:G61.

6. **Identify the array that holds the possible values of the sample mean.**

   FREQUENCY holds this array in the Bins_array box. For my worksheet, I3:I9 goes into the Bins_array box. After you identify both arrays, the Insert Function dialog box shows the frequencies inside a pair of curly brackets. (See Figure 9-4.)

7. **Press Ctrl+Shift+Enter to close the Function Arguments dialog box and show the frequencies.**

   Use this keystroke combination because FREQUENCY is an array function. (For more on FREQUENCY, see Chapter 7.)

Finally, use the Excel Chart Wizard (see Chapter 3) to produce the graph of the frequencies. Your graph will probably look somewhat different from mine.

By the way, Excel repeats the random selection process whenever you do something that causes Excel to recalculate the worksheet. The effect is that the numbers change as you work through this. For example, each time you save the file those numbers change and the graph changes, too.

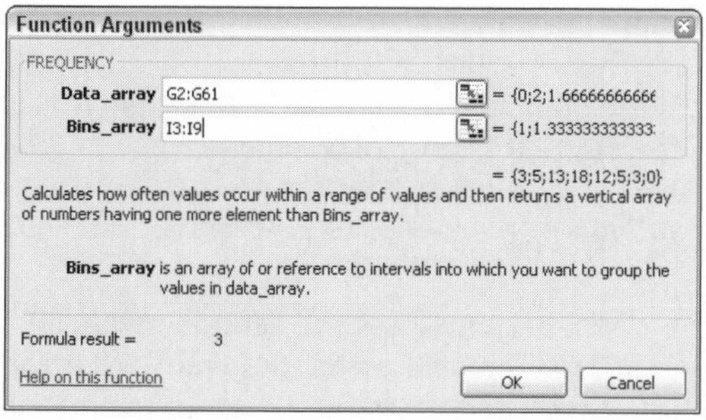

Figure 9-4:
The
Function
Arguments
dialog
box for
FREQUENCY
in the
Simulated
Sampling
Distribution
worksheet.

Function Arguments		⊠

FREQUENCY

Data_array G2:G61 ▦ = {0;2;1.6666666666€

Bins_array I3:I9 ▦ = {1;1.333333333333:

= {3;5;13;18;12;5;3;0}

Calculates how often values occur within a range of values and then returns a vertical array of numbers having one more element than Bins_array.

Bins_array is an array of or reference to intervals into which you want to group the values in data_array.

Formula result =   3

Help on this function   [OK]   [Cancel]

The Limits of Confidence

I told you about sampling distributions because they help you answer the question I posed at the beginning of this chapter: How much confidence can you have in the estimates you create?

The idea is to calculate a statistic and then use that statistic to establish upper and lower bounds for the population parameter with, say, 95 percent confidence. You can only do this if you know the sampling distribution of the statistic and the standard error. In the next section, I show you how to do this for the mean.

Finding confidence limits for a mean

The FarBlonJet Corporation, a manufacturer of navigation systems, has developed a new battery to power its portable model. To help market the system, FarBlonJet wants to know how long, on average, each battery lasts before it burns out.

The company would like to estimate that average with 95 percent confidence. It tests a sample of 100 batteries and finds that the sample mean is 60 hours with a standard deviation of 20 hours. The Central Limit Theorem, remember, says that with a large enough sample (30 or more), the sampling distribution of the mean approximates a normal distribution. The standard error of the mean (the standard deviation of the sampling distribution of the mean) is

$$\sigma_{\bar{x}} = \frac{\sigma}{\sqrt{N}}$$

The sample size, N, is 100. What about σ? That's unknown, so you have to estimate it. If you know σ, that would mean you know μ, and establishing confidence limits would be unnecessary.

The best estimate of σ is the standard deviation. In this case that's 20. This leads to an estimate of the standard error of the mean

$$s_{\bar{x}} = \frac{s}{\sqrt{N}} = \frac{20}{\sqrt{100}} = \frac{20}{10} = 2$$

The best estimate of the population mean is the sample mean, 60. Armed with this information — estimated mean, estimated standard error of the mean, normal distribution — you can envision the sampling distribution of the mean, which I've done in Figure 9-5. Consistent with Figure 9-2, each standard deviation is a standard error of the mean.

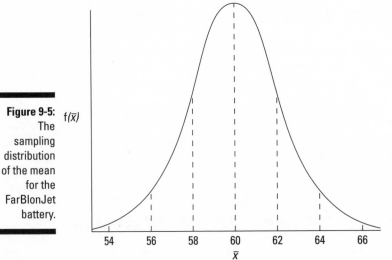

Figure 9-5: The sampling distribution of the mean for the FarBlonJet battery.

$f(\bar{x})$

54 56 58 60 62 64 66

$\bar{x}$

Now that you have the sampling distribution, you can establish the 95 percent confidence limits for the mean. This means that, starting at the center of the distribution, how far out to the sides do you have to extend until you have 95 percent of the area under the curve? (For more on area under the normal distribution and what it means, see Chapter 8.)

One way to answer this question is to work with the standard normal distribution and find the z-score that cuts off 47.5 percent on the right side and 47.5 percent on the left side (yes, Chapter 8 again). The one on the right is a positive z-score; the one on the left is a negative z-score. Then, multiply each z-score by the standard error. Add each result to the sample mean to get the upper confidence limit and the lower confidence limit.

It turns out that the z-score is 1.96 for the boundary on the right side of the standard normal distribution and -1.96 for the boundary on the left. You can calculate those values (difficult), get them from a table of the normal distribution that you typically find in a statistics textbook (easier), or use the Excel worksheet function I describe in the next section to do all the calculations (much easier). The point is that the upper bound in the sampling distribution is 63.92 (60 + 1.96$s_{\bar{x}}$), and the lower bound is 56.08 (60 - 1.96$s_{\bar{x}}$). Figure 9-6 shows these bounds on the sampling distribution.

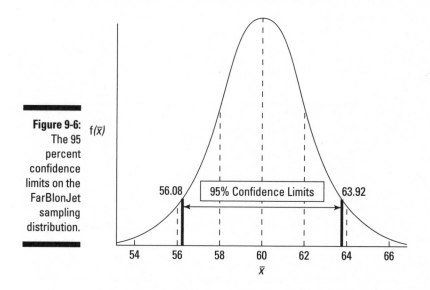

Figure 9-6: The 95 percent confidence limits on the FarBlonJet sampling distribution.

This means you can say with 95 percent confidence that the FarBlonJet battery lasts, on the average, between 56.08 hours and 63.92 hours. Want a narrower range? You can either reduce your confidence level (to, say, 90 percent) or test a larger sample of batteries.

CONFIDENCE

The CONFIDENCE worksheet function does the lion's share of the work in constructing confidence intervals. You supply the confidence level, the standard deviation, and the sample size. CONFIDENCE returns the result of multiplying the appropriate z-score by the standard error of the mean. To determine the upper bound of the confidence limit, you add that result to the sample mean. To determine the lower bound, you subtract that result from the sample mean.

To show you how it works, I'll go through the FarBlonJet batteries example again. Here are the steps:

1. **Select a cell.**

2. **Click the Insert Function button to open the Insert Function dialog box.**

3. **In the Insert Function dialog box, select CONFIDENCE and click OK to open the Function Arguments dialog box for CONFIDENCE. (See Figure 9-7.)**

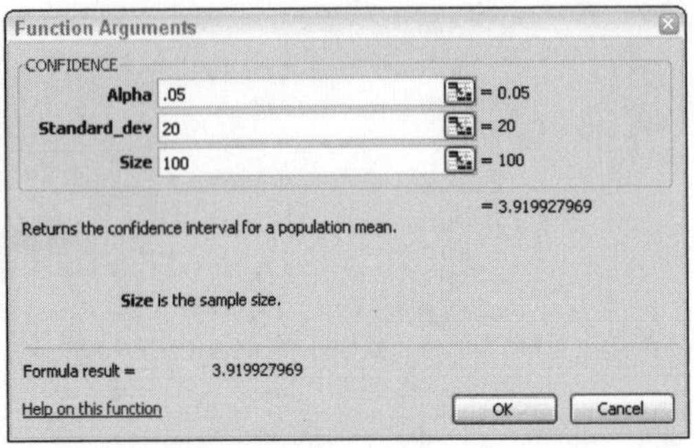

Figure 9-7: The Function Arguments dialog box for CONFI-DENCE.

4. **In the Alpha box, type the result of subtracting your confidence level from 1.00.**

Yes, that's a little confusing. Instead of typing .95 for the 95 percent confidence limit, you type .05. Think of it as the percentage of area *beyond* the confidence limits rather than the area *within* the confidence limits. And why is it labeled Alpha? I'll get into that in Chapter 10.

5. **In the Standard_dev box, type the standard deviation of your sample.**

For this example, the standard deviation is 20.

6. **In the Size box, type the number of subjects in your sample.**

The example specifies 100 batteries tested. After you type that number, the answer (3.919928) appears in the dialog box.

7. **Click OK to put the answer into your selected cell.**

8. **Add the answer to the sample mean (60) to determine the upper confidence limit (63.92) and subtract the answer from the mean to determine the lower confidence limit (56.08).**

Fit to a t

The Central Limit Theorem specifies (approximately) a normal distribution for large samples. Many times, however, you don't have the luxury of large sample sizes, and the normal distribution isn't appropriate. What do you do?

For small samples, the sampling distribution of the mean is a member of a family of distributions called the *t-distribution*. The parameter that distinguishes members of this family from one another is called *degrees of freedom*.

TIP

Think of degrees of freedom as the denominator of your variance estimate. For example, if your sample consists of 25 individuals, the sample variance that estimates population variance is

$$s^2 = \frac{\sum(x-\bar{x})^2}{N-1} = \frac{\sum(x-\bar{x})^2}{25-1} = \frac{\sum(x-\bar{x})^2}{24}$$

The number in the denominator is 24, and that's the value of the degrees of freedom parameter. In general, degrees of freedom (df) = N - 1 (N is the sample size) when you use the *t*-distribution as I'm about to do in this section.

Figure 9-8 shows two members of the *t*-distribution family (df =3 and df = 10), along with the normal distribution for comparison. As the figure shows, the greater the df, the more closely *t* approximates a normal distribution.

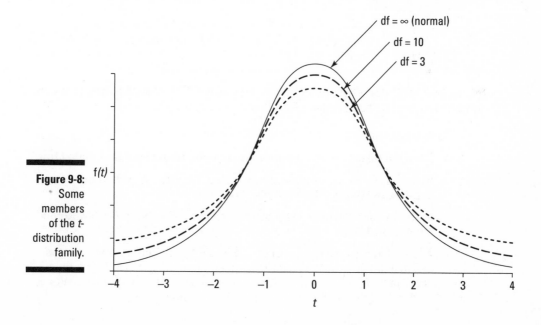

Figure 9-8: Some members of the *t*-distribution family.

So . . . to determine the 95 percent confidence level if you have a small sample, work with the member of the *t*-distribution family that has the appropriate df. Find the value that cuts off 47.5 percent of the area on the right side of the distribution and 47.5 percent of the area on the left side of the distribution. The one on the right is a positive value; the one on the left is negative. Then, multiply each value by the standard error. Add each result to the mean to get the upper confidence limit and the lower confidence limit.

In the FarBlonJet batteries example, suppose the sample consists of 25 batteries, with a mean of 60 and a standard deviation of 20. The estimate for the standard error of the mean is

$$s_{\bar{x}} = \frac{s}{\sqrt{N}} = \frac{20}{\sqrt{25}} = \frac{20}{5} = 4$$

The df = *N* - 1 = 24. The value that cuts off 47.5 percent of the area on the right of this distribution is 2.064 and on the left -2.064. As I said earlier, you can calculate these values (difficult), look them up in a table that's in statistics textbooks (easier), or use the Excel function I describe in the next section (much easier).

The point is that the upper confidence limit is 68.256 (60 + 2.064$s_{\bar{x}}$) and the lower confidence limit is 51.744 (60 - 2.064$s_{\bar{x}}$). With a sample of 25 batteries, you can say with 95 percent confidence that the average life of a FarBlonJet battery is between 51.744 hours and 68.256 hours. Notice that with a smaller sample, the range is wider for the same level of confidence that I used in the previous example.

TINV

Excel's TINV worksheet function finds the value in the *t*-distribution that cuts off the desired area. Working with it is short and sweet:

1. **Select a cell.**

2. **Click the Insert Function button to open the Insert Function dialog box.**

3. **In the Insert Function dialog box, select TINV to open the Function Arguments dialog box for TINV.**

4. **In the Probability box, type the result of subtracting your confidence level from 1.00.**

   As I said in the description of the CONFIDENCE function, that's confusing. Instead of typing .95 for the 95 percent confidence limit, you type .05. Think of it as the percentage of area *beyond* the confidence limits rather than the area *within* the confidence limits.

5. **In the Deg_freedom box type the degrees of freedom.**

   For this example, df = 24.

6. **The answer appears in the dialog box. Click OK to close the dialog box and put the answer in the selected cell.** (See Figure 9-9.)

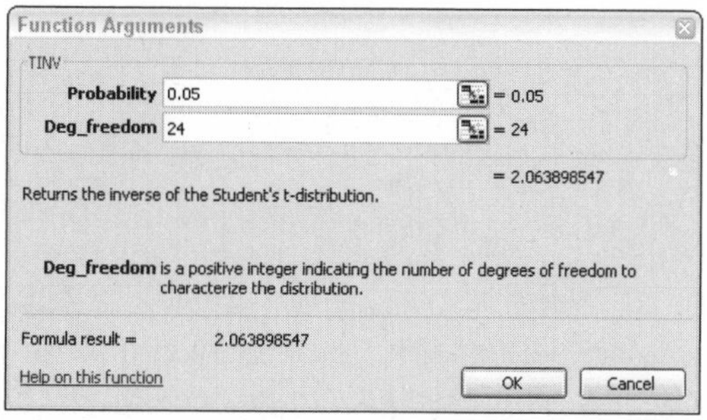

Figure 9-9:
The
Function
Arguments
dialog box
for TINV.

Function Arguments		⊠
TINV		
Probability	0.05	= 0.05
Deg_freedom	24	= 24

= 2.063898547

Returns the inverse of the Student's t-distribution.

Deg_freedom is a positive integer indicating the number of degrees of freedom to characterize the distribution.

Formula result =   2.063898547

Help on this function   OK   Cancel

You still have to multiply TINV's answer by the standard error of the mean and do the arithmetic to find the upper and lower limits.

I advise against using the CONFIDENCE worksheet function if your sample size is less than 30 and if you can't assume your population is a normal distribution. Why? CONFIDENCE always assumes a normally distributed sampling distribution, and that's not always appropriate. So if your confidence level is 95 percent, for example, CONFIDENCE multiplies the standard error by 1.96 regardless of the sample size. The result is that the confidence interval is too narrow for a small sample size.

Chapter 10

One-Sample Hypothesis Testing

. . . .

In This Chapter

▶ Introducing hypothesis tests

▶ Testing hypotheses about means

▶ Testing hypotheses about variances

. . . .

*W*hatever your occupation, you often have to assess whether something out of the ordinary has happened. Sometimes you start with a
sample from a population about whose parameters you know a great deal.
You have to decide whether that sample is like the rest of the population or if
it's different.

Measure that sample and calculate its statistics. Finally, compare those statistics with the population parameters. Are they the same? Are they different?
Does the sample represent something that's off the beaten path? Proper use
of statistics helps you decide.

Sometimes you don't know the parameters of the population you're dealing
with. Then what? In this chapter, I discuss statistical techniques and worksheet functions for dealing with both cases.

Hypotheses, Tests, and Errors

A *hypothesis* is a guess about the way the world works. It's a tentative explanation of some process, whether that process is natural or artificial. Before
studying and measuring the individuals in a sample, a researcher formulates
hypotheses that predict what the data should look like.

Generally, one hypothesis predicts that the data won't show anything new or
interesting. Dubbed the *null hypothesis* (abbreviated H_0), this hypothesis holds
that if the data deviate from the norm in any way, that deviation is due strictly
to chance. Another hypothesis, the *alternative hypothesis* (abbreviated H_1),
explains things differently. According to the alternative hypothesis, the data
show something important.

After gathering the data, it's up to the researcher to make a decision. The way the logic works, the decision centers around the null hypothesis. The researcher must decide to either reject the null hypothesis or to not reject the null hypothesis. *Hypothesis testing* is the process of formulating hypotheses, gathering data, and deciding whether to reject or not reject the null hypothesis.

Nothing in the logic involves *accepting* either hypothesis. Nor does the logic entail any decisions about the alternative hypothesis. It's all about rejecting or not rejecting H_0.

Regardless of the reject-don't reject decision, an error is possible. One type of error occurs when you believe that the data show something important and you reject H_0, and in reality the data are due just to chance. This is called a Type I error. At the outset of a study, you set the criteria for rejecting H_0. In so doing, you set the probability of a Type I error. This probability is called *alpha* (α).

The other type of error occurs when you don't reject H_0 and the data are really due to something out of the ordinary. For one reason or another, you happened to miss it. This is called a Type II error. Its probability is called *beta* (β). Table 10-1 summarizes the possible decisions and errors.

Table 10-1		Decisions and Errors in Hypothesis Testing	
		"True State" of the World	
		H_o is True	*H_1 is True*
	Reject H_0	Type I Error	Correct Decision
Decision			
	Do Not Reject H0	Correct Decision	Type II Error

Note that you never know the true state of the world. All you can ever do is measure the individuals in a sample, calculate the statistics, and make a decision about H_0.

Hypothesis tests and sampling distributions

In Chapter 9, I discussed sampling distributions. A sampling distribution, remember, is the set of all possible values of a statistic for a given sample size.

Also in Chapter 9, I discussed the Central Limit Theorem. This theorem tells you that the sampling distribution of the mean approximates a normal distribution if the sample size is large (for practical purposes, at least 30). This holds whether or not the population is normally distributed. If the population is a normal distribution, the sampling distribution is normal for any sample size. Two other points from the Central Limit Theorem:

- ✔ The mean of the sampling distribution of the mean is equal to the population mean.

  The equation for this is

$$\mu_{\bar{x}} = \mu$$

- ✔ The standard error of the mean (also known as the standard deviation of the sampling distribution) is equal to the population standard deviation divided by the square root of the sample size.

  This equation is

$$\sigma_{\bar{x}} = \frac{\sigma}{\sqrt{N}}$$

The sampling distribution of the mean figures prominently into the type of hypothesis testing I discuss in this chapter. Theoretically, when you test a null hypothesis vs. an alternative hypothesis, each hypothesis corresponds to a separate sampling distribution.

Figure 10-1 shows what I mean. The figure shows two normal distributions. I placed them arbitrarily. Each normal distribution represents a sampling distribution of the mean. The one on the left represents the distribution of possible sample means if the null hypothesis is truly how the world works. The one on the right represents the distribution of possible sample means if the alternative hypothesis is truly how the world works.

Of course, when you do a hypothesis test, you never know which distribution produces the results. You work with a sample mean — a point on the horizontal axis. It's your job to decide which distribution the sample mean is part of. You set up a *critical value* — a decision criterion. If the sample mean is on one side of the critical value, you reject H_0. If not, you don't.

In this vein, the figure also shows α and β. These, as I mentioned earlier, are the probabilities of decision errors. The area that corresponds to α is in the H_0 distribution. I shaded it. It represents the probability that a sample mean comes from the H_0 distribution, but it's so extreme that you reject H_0.

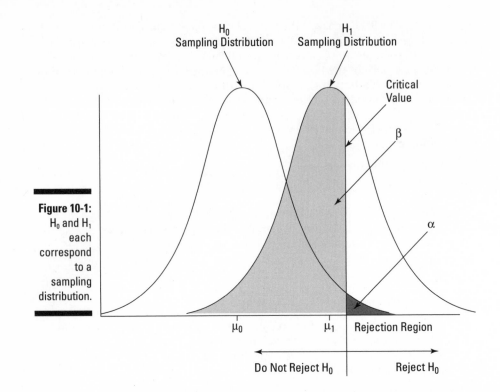

Figure 10-1:
H_0 and H_1
each
correspond
to a
sampling
distribution.

Where you set the critical value determines α. In most hypothesis testing, you set α at .05. This means you're willing to tolerate a Type I error (incorrectly rejecting H_0) 5 percent of the time. Graphically, the critical value cuts off 5 percent of the area of the sampling distribution. By the way, if you're talking about the 5 percent of the area that's in the right tail of the distribution (as in Figure 10-1), you're talking about the upper 5 percent. If it's the 5 percent in the left tail you're interested in, that's the lower 5 percent.

The area that corresponds to β is in the H_1 distribution. I filled it with diagonal stripes. This area represents the probability that a sample mean comes from the H_1 distribution, but it's close enough to the center of the H_0 distribution that you don't reject H_0. You don't get to set β. The size of this area depends on the separation between the means of the two distributions, and that's up to the world we live in — not you.

These sampling distributions are appropriate when your work corresponds to the conditions of the Central Limit Theorem: If you know the population you're working with is a normal distribution, or if you have a large sample.

Catching Some Zs Again

Here's an example of a hypothesis test that involves a sample from a normally distributed population. Because the population is normally distributed, any sample size results in a normally distributed sampling distribution. Because it's a normal distribution, you use *z-scores* in the hypothesis test:

$$z = \frac{\bar{x} - \mu}{\frac{\sigma}{\sqrt{N}}}$$

One more "because": Because you use the z-score in the hypothesis test, it's called the *test statistic*.

Suppose you think that people living in a particular ZIP code have higher-than-average IQs. You take a sample of 16 people from that ZIP code, give them IQ tests, tabulate the results, and calculate the statistics. For the population of IQ scores, it's well known $\mu = 100$ and $\sigma = 16$ (for the Stanford-Binet version).

The hypotheses are:

H_0: $\mu_{zipcode} \leq 100$

H_1: $\mu_{zipcode} > 100$

Assume $\alpha = .05$. That's the shaded area in the tail of the H_0 distribution in Figure 10-1.

Why the $\leq$ in H_0? You use that symbol because you'll only reject H_0 if the sample mean is larger than the hypothesized value. Anything else is evidence in favor of not rejecting H_0.

Suppose the sample mean is 107.75. Can you reject H_0?

The test involves turning 107.75 into a standard score in the sampling distribution of the mean:

$$z = \frac{\bar{x} - \mu_0}{\frac{\sigma}{\sqrt{N}}} = \frac{107.75 - 100}{\frac{16}{\sqrt{16}}} = \frac{7.75}{\frac{16}{4}} = \frac{7.75}{4} = 1.94$$

Is the value of the test statistic large enough to enable you to reject H_0 with $\alpha = .05$? It is. The critical value — the value of z that cuts off 5% of the area in a standard normal distribution — is 1.64. (After years of working with the standard normal distribution, I happen to know this. Read Chapter 8, learn about Excel's NORMSINV function, and you can have information like that at your fingertips, too.) The calculated value, 1.94, exceeds 1.64, so it's in the rejection region. The decision is to reject H_0.

This means that if H_0 is true, the probability of getting a test statistic value that's at least this large is less than .05. That's strong evidence in favor of rejecting H_0. In statistical parlance, any time you reject H_0 the result is said to be "statistically significant."

This type of hypothesis testing is called *one-tailed* because the rejection region is in one tail of the sampling distribution.

A hypothesis test can be one-tailed in the other direction. Suppose you had reason to believe that people in that zip code had lower than average IQ. In that case, the hypotheses are:

H_0: $\mu_{zipcode} \geq 100$

H_1: $\mu_{zipcode} < 100$

For this hypothesis test, the critical value of the test statistic is -1.64 if α =.05.

A hypothesis test can be *two-tailed*, meaning that the rejection region is in both tails of the H_0 sampling distribution. That happens when the hypotheses look like this:

H_0: $\mu_{zipcode} = 100$

H_1: $\mu_{zipcode} \neq 100$

In this case, the alternate hypothesis just specifies that the mean is different from the null hypothesis value, without saying whether it's greater or whether it's less. Figure 10-2 shows what the two-tailed rejection region looks like for α = .05. The 5 percent is divided evenly between the left tail (also called the lower tail) and the right tail (the upper tail).

For a standard normal distribution, incidentally, the z-score that cuts off 2.5 percent in the right tail is 1.96. The z-score that cuts off 2.5 percent in the left tail is -1.96. (Again, I happen to know these values after years of working with the standard normal distribution.) The z-score in the preceding example, 1.94, does not exceed 1.96. The decision in the two-tailed case is to *not* reject H_0.

This brings up an important point. A one-tailed hypothesis test can reject H_0, while a two-tailed test on the same data might not. A two-tailed test indicates that you're looking for a difference between the sample mean and the null hypothesis mean, but you don't know in which direction. A one-tailed test shows that you have a pretty good idea of how the difference should come out. For practical purposes, this means you should try to have enough knowledge to be able to specify a one-tailed test.

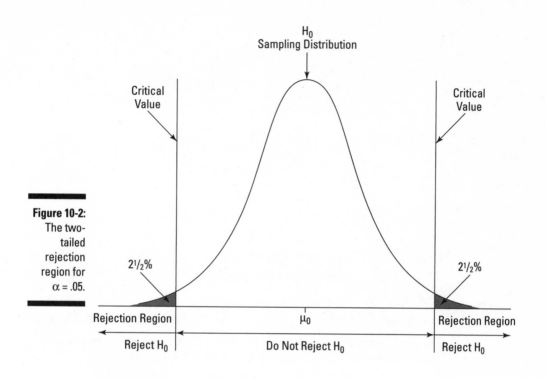

Figure 10-2:
The two-
tailed
rejection
region for
$\alpha = .05$.

ZTEST

Excel's ZTEST worksheet function does the calculations for hypothesis tests involving z-scores in a standard normal distribution. You provide sample data, a null hypothesis value, and a population standard deviation. ZTEST returns the probability in one tail of the H_0 sampling distribution.

This is a bit different from the way things work when you apply the formulas I just showed you. The formula calculates a z-score. Then it's up to you to see where that score stands in a standard normal distribution with respect to probability. ZTEST eliminates the middleman (the need to calculate the z-score) and goes right to the probability.

Figure 10-3 shows the data and the Function Arguments dialog box for ZTEST. The data are IQ scores for 16 people in the ZIP code example in the preceding section. That example, remember, tests the hypothesis that people in a particular ZIP code have a higher-than-average IQ.

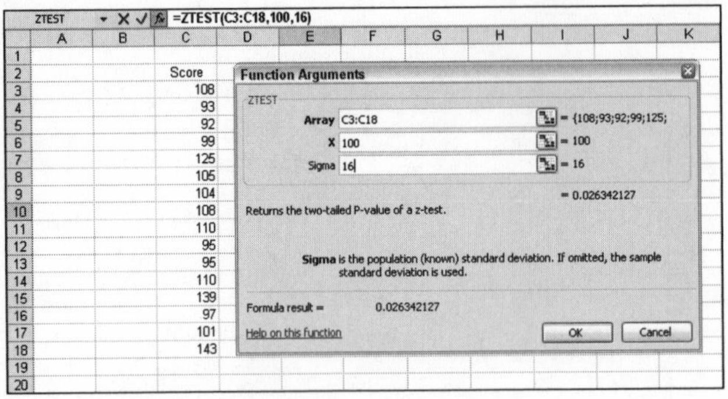

Figure 10-3:
Data and
the Function
Arguments
dialog box
for ZTEST.

Here are the steps:

1. **Type your data into an array of cells.**

   The data in this example are in cells C3 through C18.

2. **Select a cell.**

3. **Click the Insert Function button to open the Insert Function dialog box.**

4. **In the Insert Function dialog box, select ZTEST and click OK to open the Function Arguments dialog box for ZTEST.** (See Figure 10-3.)

5. **In the Array box, type the array of cells that hold the data.**

   For this example, that's C3:C18.

6. **In the X box, type the H_0 mean.**

   For this example, the mean is 100 — the mean of IQ scores in the population.

7. **In the Sigma box, type the population standard deviation.**

   The population standard deviation for IQ is 16. After you type that number, the answer (0.026342) appears in the dialog box.

8. **Click OK to put the answer into your selected cell.**

With $\alpha = .05$, and a one-tailed test (H_1: $\mu > 100$), the decision is to reject H_0, because the answer (0.026) is less than .05. Note that with a two-tailed test (H_1: $\mu \neq 100$), the decision is to not reject H_0. That's because 2×0.026 is greater than .05 — just barely greater (.052), but if you draw the line at .05, you cannot reject H_0.

t for One

In the preceding example, I worked with IQ scores. The population of IQ scores is a normal distribution with a well-known mean and standard deviation. This enabled me to work with the Central Limit Theorem and describe the sampling distribution of the mean as a normal distribution. I was then able to use z as the test statistic.

In the real world, however, you typically don't have the luxury of working with such well-defined populations. You usually have small samples, and you're typically measuring something that isn't as well known as IQ. The bottom line is that you often don't know the population parameters, nor do you know whether or not the population is normally distributed.

When that's the case, you use the sample data to estimate the population standard deviation, and you treat the sampling distribution of the mean as a member of a family of distributions called the *t*-distribution. You use *t* as a test statistic. In Chapter 9, I introduced this distribution, and mentioned that you distinguish members of this family by a parameter called *degrees of freedom* (df).

The formula for the test statistic is

$$t = \frac{\bar{x} - \mu}{\frac{s}{\sqrt{N}}}$$

Think of df as the denominator of the estimate of the population variance. For the hypothesis tests in this section, that's N-1, where N is the number of scores in the sample. The higher the df, the more closely the *t*-distribution resembles the normal distribution.

Here's an example. FarKlempt Robotics, Inc. markets microrobots. It claims that its product averages four defects per unit. A consumer group believes that this average is actually higher. The consumer group takes a sample of nine FarKlempt microrobots and finds an average of seven defects, with a standard deviation of 3.16. The hypothesis test is:

$H_0: \mu \leq 4$

$H_1: \mu > 4$

$\alpha = .05$

The formula is:

$$t = \frac{\bar{x} - \mu}{\frac{s}{\sqrt{N}}} = \frac{7 - 4}{\frac{3.16}{\sqrt{9}}} = \frac{3}{\frac{3.16}{3}} = 2.85$$

Can you reject H_0? The Excel function in the next section tells you.

TDIST

You use the worksheet function TDIST to decide whether or not your calculated t value is in the region of rejection. You supply a value for *t*, a value for df, and whether the test is one-tailed or two-tailed. TDIST returns the probability of obtaining a *t* value at least as high as yours if H_0 is true. If that probability is less than your α, you reject H_0.

The steps are:

1. **Select a cell.**

2. **Click the Insert Function button to open the Insert Function dialog box.**

3. **In the Insert Function dialog box, select TDIST and click OK to open the Function Arguments dialog box for TDIST.** (See Figure 10-4.)

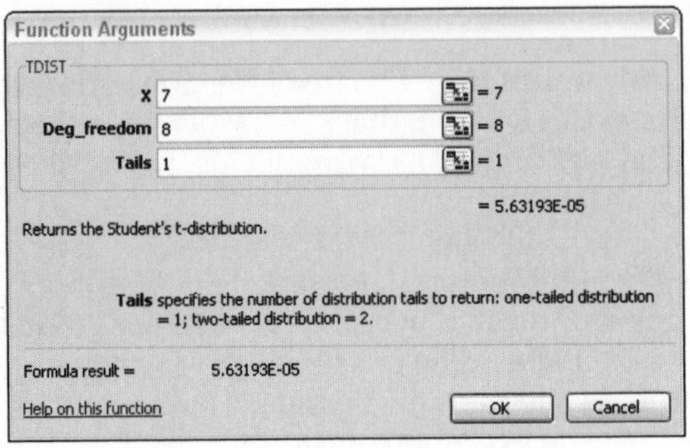

Figure 10-4: The Function Arguments dialog box for TDIST.

4. **In the X box, type the sample mean.**

   For this example, the sample mean is 7.

5. **In the Deg_freedom box, type the degrees of freedom.**

   The degrees of freedom for this example is 8 (9 scores - 1).

6. **In the Tails box, type 1 (for a one-tailed test) or 2 (for a two-tailed test).**

   In this example, it's a one-tailed test. After you type 1, the dialog box shows the probability in the tail of the t-distribution beyond your t value.

7. **Click OK to close the dialog box and put the answer in the selected cell.**

The value in the dialog box in Figure 10-4 is less than .05, so the decision is to reject H_0.

Testing a Variance

So far, I've told you about one-sample hypothesis testing for means. You can also test hypotheses about variances.

This sometimes comes up in the context of manufacturing. For example, suppose FarKlempt Robotics, Inc. produces a part that has to be a certain length with a very small variability. You can take a sample of parts, measure them, find the sample variability, and perform a hypothesis test against the desired variability.

The family of distributions for the test is called *chi-square*. Its symbol is χ^2. I won't go into all the mathematics. I'll just tell you that, again, df is the parameter that distinguishes one member of the family from another. Figure 10-5 shows two members of the chi-square family.

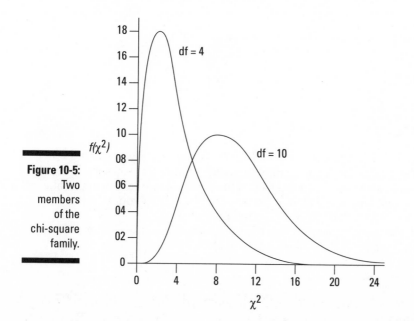

Figure 10-5:
Two members of the chi-square family.

The formula for this test statistic is

$$\chi^2 = \frac{(N-1)\,s^2}{\sigma^2}$$

N is the number of scores in the sample, s^2 is the sample variance, and σ^2 is the population variance under H_0.

With this test, you have to assume that what you're measuring has a normal distribution.

Suppose the process for the FarKlempt part has to have at most a standard deviation of 1.5 inches for its length. (Notice I said *standard deviation*. This allows me to speak in terms of inches. If I said *variance* the units would be square inches.) After measuring a sample of 26 parts, you find a standard deviation of 1.8 inches.

The hypotheses are:

H_0: $\sigma^2 \leq 2.25$ (remember to square the at-most standard deviation of 1.5 inches)

H_1: $\sigma^2 > 2.25$

$\alpha = .05$

Working with the formula,

$$\chi^2 = \frac{(N-1)\,s^2}{\sigma^2} = \frac{(26-1)(1.8)^2}{(1.5)^2} = \frac{(25)(3.24)}{2.25} = 36$$

Can you reject H_0? Read on.

CHIDIST

After calculating a value for your chi-square test statistic, you use the CHIDIST worksheet function to make a judgment about it. You supply the chi-square value and the df, and it tells you the probability of obtaining a value at least that high if H_0 is true. If that probability is less than your α, reject H_0.

To show you how it works, I apply the information from the example in the preceding section. Follow these steps:

1. **Select a cell.**

2. **Click the Insert Function button to open the Insert Function dialog box.**

3. **In the Insert Function dialog box, select CHIDIST and click OK to open the Function Arguments dialog box for CHIDIST.** (See Figure 10-6.)

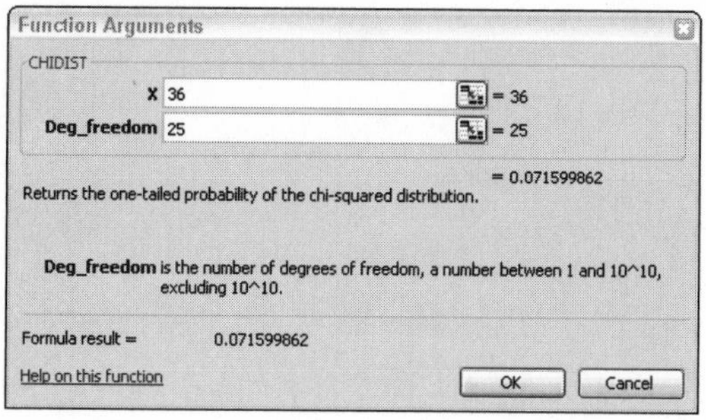

Figure 10-6:
The
Function
Arguments
dialog box
for CHIDIST.

4. **In the X box, type the calculated chi-square value.**

   For this example, that value is 36.

5. **In the Deg_freedom box, type the degrees of freedom.**

   The degrees of freedom for this example is 25 (26 - 1). After you type the df, the dialog box shows the one-tailed probability of obtaining at least this value of chi-square if H_0 is true.

6. **Click OK to close the dialog box and put the answer in the selected cell.**

The value in the dialog box in Figure 10-6 is greater than .05, so the decision is to not reject H_0. (Can you conclude the process is within acceptable limits of variability? See the sidebar "A point to ponder.")

CHIINV

CHIINV is the flip side of CHIDIST. You supply a probability and df, and CHIINV tells you the corresponding value of chi-square. If you want to know the value you have to exceed in order to reject H_0 in the preceding example, follow these steps:

1. **Select a cell.**

2. **Click the Insert Function button to open the Insert Function dialog box.**

3. **In the Insert Function dialog box, select CHIINV and click OK to open the Function Arguments dialog box for CHIINV.** (See Figure 10-7.)

4. **In the Probability box, type the probability you're interested in.**

   For the hypothesis test in this example, that's .05.

5. **In the Deg_freedom box, type the degrees of freedom.**

   The value for degrees of freedom in this example is 25 (26 - 1). After you type the df, the dialog box shows the value (37.65248) that cuts off the upper 5 percent of the area in this chi-square distribution.

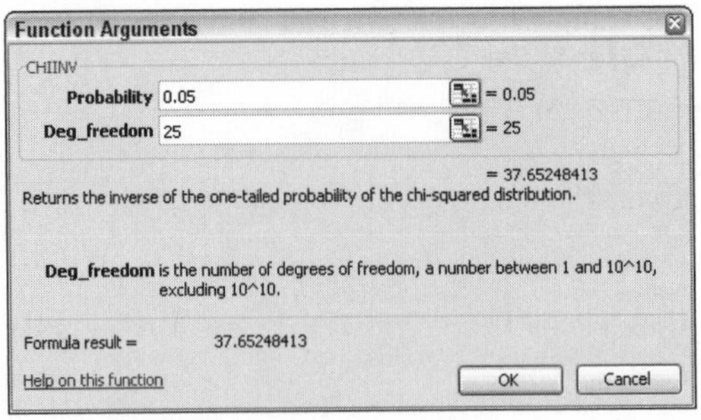

Figure 10-7:
The
Function
Arguments
dialog box
for CHIINV.

6. **Click OK to close the dialog box and put the answer in the selected cell.**

As the dialog box in Figure 10-7 shows, the calculated value (36) didn't miss the cutoff value by much. A miss is still a miss (to paraphrase "As Time Goes By"), and you cannot reject H_0.

A point to ponder

Retrace the preceding example. FarKlempt Robotics wants to show that its manufacturing process is within acceptable limits of variability. The null hypothesis, in effect, says the process is acceptable. The data do not present evidence for rejecting H_0. The value of the test statistic just misses the critical value. Does that mean the manufacturing process is within acceptable limits?

Statistics are an aid to common sense, not a substitute. If the data are just barely within acceptability, that should set off alarms.

Usually, you try to reject H_0. This is a rare case when not rejecting H_0 is more desirable, because nonrejection implies something positive — the manufacturing process is working properly. Can you still use hypothesis testing techniques in this situation?

Yes, you can — with a notable change. Instead of a small value of α, like .05, you choose a large value, like .20. This stacks the deck *against* not rejecting H_0 — small values of the test statistic can lead to rejection. If α is .20 in this example, the critical value is 30.6752. (Use CHINV to verify that.) Because the obtained value, 36, is higher than this critical value the decision with this α is to reject H_0.

Using a high α is not often done. When the desired outcome is to *not* reject H_0, I strongly advise it.

Chapter 11

Two-Sample Hypothesis Testing

• •

In This Chapter

▶ Testing differences between means of two samples

▶ Testing means of paired samples

▶ Testing hypotheses about variances

• •

*I*n business, in education, and in scientific research the need often arises to compare one sample with another. Sometimes the samples are independent; sometimes they're matched in some way. Each sample comes from a different population. The objective is to decide whether or not the populations they come from are different from one another.

Usually, this involves tests of hypotheses about population means. You can also test hypotheses about population variances. In this chapter, I show you how to carry out these tests. I also discuss useful worksheet functions and data analysis tools that help you get the job done.

Hypotheses Built for Two

As in the one-sample case (Chapter 10), hypothesis testing with two samples starts with a null hypothesis (H_0) and an alternative hypothesis (H_1). The null hypothesis specifies that any differences you see between the two samples are due strictly to chance. The alternative hypothesis says, in effect, that any differences you see are real and not due to chance.

It's possible to have a one-tailed test, in which the alternative hypothesis specifies the direction of the difference between the two means, or a two-tailed test in which the alternative hypothesis does not specify the direction of the difference.

For a one-tailed test, the hypotheses look like this:

H_0: $\mu_1 - \mu_2 = 0$

H_1: $\mu_1 - \mu_2 > 0$

or like this:

H_0: $\mu_1 - \mu_2 = 0$

H_1: $\mu_1 - \mu_2 < 0$

For a two-tailed test, the hypotheses are:

H_0: $\mu_1 - \mu_2 = 0$

H_1: $\mu_1 - \mu_2 \neq 0$

The zero in these hypotheses is the typical case. It's possible, however, to test for any value — just substitute that value for zero.

To carry out the test, you first set α, the probability of a Type I error that you're willing to live with (see Chapter 9). Then, you calculate the mean and standard deviation of each sample, subtract one mean from the other, and use a formula to convert the result into a test statistic. Compare the test statistic to a sampling distribution of test statistics. If it's in the rejection region that α specifies (see Chapter 10), reject H_0. If not, don't reject H_0.

Sampling Distributions Revisited

In Chapter 9, I introduce the idea of a sampling distribution — a distribution of all possible values of a statistic for a particular sample size. In that chapter, I describe the sampling distribution of the mean. In Chapter 10, I show its connection with one-sample hypothesis testing.

For this type of hypothesis testing, another sampling distribution is necessary. This one is *the sampling distribution of the difference between means*.

The sampling distribution of the difference between means is the distribution of all possible values of differences between pairs of sample means with the sample sizes held constant from pair to pair. (Yes, that's a mouthful.) *Held constant from pair to pair* means that the first sample in the pair always has the same size, and the second sample in the pair always has the same size. The two sample sizes are not necessarily equal.

Within each pair, each sample comes from a different population. All the samples are independent of one another, so that picking individuals for one sample has no effect on picking individuals for another.

Figure 11-1 shows the steps in creating this sampling distribution. This is something you never do in practice. It's all theoretical. As the figure shows, the idea is to take a sample out of one population and a sample out of another, calculate their means, and subtract one mean from the other. Return the samples to the populations, and repeat over and over and over. The result of the process is a set of differences between means. This set of differences is the sampling distribution.

Applying the Central Limit Theorem

Like any other set of numbers, this sampling distribution has a mean and a standard deviation. As is the case with the sampling distribution of the mean (Chapters 9 and 10), the Central Limit Theorem applies here.

According to the Central Limit Theorem, if the samples are large, the sampling distribution of the difference between means is approximately a normal distribution. If the populations are normally distributed, the sampling distribution is a normal distribution even if the samples are small.

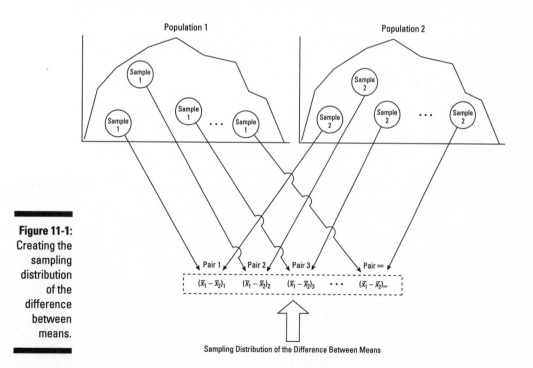

Figure 11-1:
Creating the sampling distribution of the difference between means.

The Central Limit Theorem also has something to say about the mean and standard deviation of this sampling distribution. Suppose the parameters for the first population are μ_1 and σ_1, and the parameters for the second population are μ_2 and σ_2. The mean of the sampling distribution is

$$\mu_{\bar{x}_1 - \bar{x}_2} = \mu_1 - \mu_2$$

The standard deviation of the sampling distribution is

$$\sigma_{\bar{x}_1 - \bar{x}_2} = \sqrt{\frac{\sigma_1^2}{N_1} + \frac{\sigma_2^2}{N_2}}$$

N_1 is the number of individuals in the sample from the first population; N_2 is the number of individuals in the sample from the second.

This standard deviation is called *the standard error of the difference between means*.

Figure 11-2 shows the sampling distribution along with its parameters, as specified by the Central Limit Theorem.

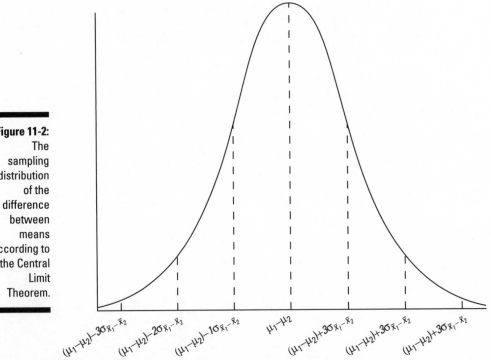

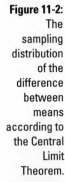

Figure 11-2:
The sampling distribution of the difference between means according to the Central Limit Theorem.

Zs once more

Because the Central Limit Theorem says that the sampling distribution is approximately normal for large samples (or for small samples from normally distributed populations), you use the z-score as your test statistic. Another way to say "use the z-score as your test statistic" is "perform a z-test." Here's the formula:

$$z = \frac{\left(\overline{x}_1 - \overline{x}_2\right) - \left(\mu_1 - \mu_2\right)}{\sigma_{\overline{x}_1 - \overline{x}_2}}$$

The term $(\mu_1 - \mu_2)$ represents the difference between the means in H_0.

This formula converts the difference between sample means into a standard score. Compare the standard score against a standard normal distribution — a normal distribution with $\mu = 0$ and $\sigma = 1$. If the score is in the rejection region defined by α, reject H_0. If it's not, don't reject H_0.

You use this formula when you know the value of σ_1^2 and σ_2^2.

Here's an example. Imagine a new training technique designed to increase IQ. Take a sample of 25 people and train them under the new technique. Take another sample of 25 people and give them no special training. Suppose we find that the sample mean for the new technique sample is 107, and for the no-training sample it's 101.2. The hypothesis test is:

H_0: $\mu_1 - \mu_2 = 0$

H_1: $\mu_1 - \mu_2 > 0$

I'll set α at .05

The IQ is known to have a standard deviation of 16, and I assume that standard deviation would be the same in the population of people trained on the new technique. Of course, that population doesn't exist. The assumption is that if it did, it should have the same value for the standard deviation as the regular population of IQ scores. Does the mean of that (theoretical) population have the same value as the regular population? H_0 says it does. H_1 says it's larger.

The test statistic is

$$z = \frac{\left(\overline{x}_1 - \overline{x}_2\right) - \left(\mu_1 - \mu_2\right)}{\sigma_{\overline{x}_1 - \overline{x}_2}} = \frac{\left(\overline{x}_1 - \overline{x}_2\right) - \left(\mu_1 - \mu_2\right)}{\sqrt{\dfrac{\sigma_1^2}{N_1} + \dfrac{\sigma_2^2}{N_2}}} = \frac{(107 - 101.2)}{\sqrt{\dfrac{16^2}{25} + \dfrac{16^2}{25}}} = \frac{5.8}{4.53} = 1.28$$

With $\alpha = .05$, the critical value of z — the value that cuts off the upper 5 percent of the area under the standard normal distribution — is 1.64. (You can use the worksheet function ZTEST from Chapter 10 to verify this.) The calculated value of the test statistic is less than the critical value, so the decision is to not reject H_0. Figure 11-3 summarizes this.

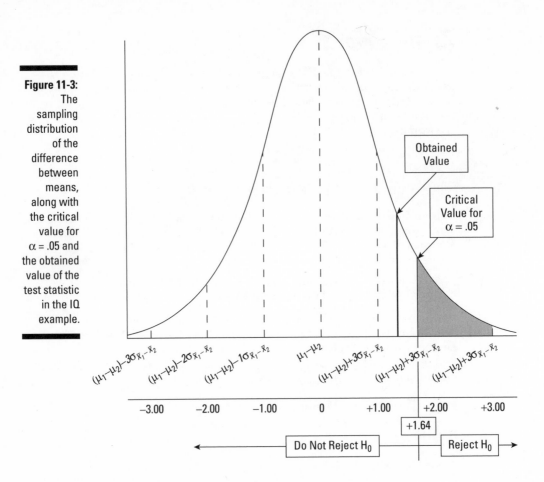

Data analysis tool: z-Test: Two Sample for Means

Excel provides a data analysis tool that makes it easy to do tests like the one in the IQ example. It's called z-Test: Two Sample for Means. Figure 11-4 shows the dialog box for this tool along with sample data that correspond to the IQ example.

To use this tool, follow these steps:

1. **Type the data for each sample into a separate data array.**

   For this example, the data in the New Technique sample are in column E and the data for the No Training sample are in column G.

2. **From the Tools menu, select Data Analysis to open the Data Analysis dialog box.**

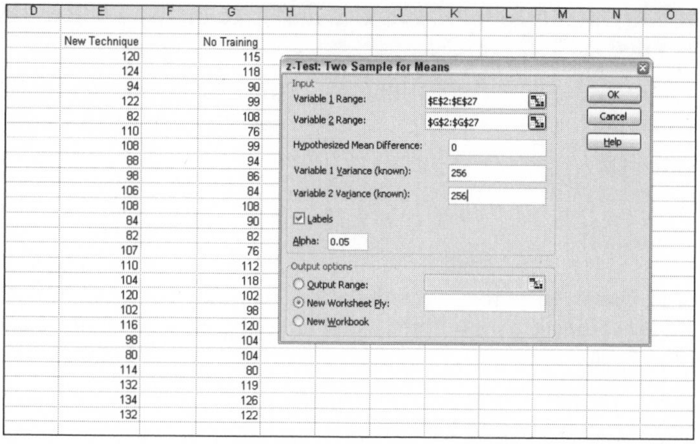

Figure 11-4:
The z-Test
data
analysis tool
and data
from two
samples.

3. **In the Data Analysis dialog box, scroll down the Analysis Tools list and select z-Test: Two Sample for Means.** Click OK to open the z-Test: Two Sample for Means dialog box (see Figure 11-4).

4. **In the Variable 1 Range box, type the cell range that holds the data for one of the samples.**

   For the example, the New Technique data are in $E$2:$E$27. (Note the dollar signs [$] for absolute referencing.)

5. **In the Variable 2 Range box, type the cell range that holds the data for the other sample.**

   The No Training data are in $G$2:$G$27.

6. **In the Hypothesized Mean Difference box, type the difference between μ_1 and μ_2 that H_0 specifies.**

   In this example, that difference is 0.

7. **In the Variable 1 Variance (known) box, type the variance of the first sample.**

   The standard deviation of the population of IQ scores is 16, so this variance is 256.

8. **In the Variable 2 Variance (known) box, type the variance of the second sample.**

   In this example, this variance is also 256.

9. **If the cell ranges include column headings, select the Labels option.**

   I included the headings in the ranges, so I checked the box.

10. **The Alpha box has 0.05 as a default.** I used the default value, consistent with the value of α in the example.

11. **In the Output Options, select an option to indicate where you want the results.**

I selected New Worksheet Ply to put the results on a new page in the worksheet.

12. **Click OK.**

Because I selected New Worksheet Ply, a newly created page opens with the results.

Figure 11-5 shows the tool's results after I expanded the columns. Rows 4, 5, and 7 hold values you input into the dialog box. Row 6 counts the number of scores in each sample.

Figure 11-5:
Results of
the z-Test
data
analysis
tool.

	A	B	C	D
1	z-Test: Two Sample for Means			
2				
3		New Technique	No Training	
4	Mean	107	101.2	
5	Known Variance	256	256	
6	Observations	25	25	
7	Hypothesized Mean Difference	0		
8	z	1.281631041		
9	P(Z<=z) one-tail	0.099986053		
10	z Critical one-tail	1.644853627		
11	P(Z<=z) two-tail	0.199972106		
12	z Critical two-tail	1.959963985		
13				

The value of the test statistic is in cell B8. The critical value for a one-tailed test is in B10, and the critical value for a two-tailed test is in B12.

Cell B9 displays the proportion of area that the test statistic cuts off in one tail of the standard normal distribution. Cell B11 doubles that value — it's the proportion of area cut off by the positive value of the test statistic (in the tail on the right side of the distribution) plus the proportion cut off by the negative value of the test statistic (in the tail on the left side of the distribution).

t for Two

The example in the preceding section involves a situation you rarely encounter — known population variances. If you know a population's variance, you're likely to know the population mean. If you know the mean, you probably don't have to perform hypothesis tests about it.

Not knowing the variances takes the Central Limit Theorem out of play. This means you can't use the normal distribution as an approximation of the sampling distribution of the difference between means. Instead, you use the *t*-distribution, a family of distributions I introduce in Chapter 9 and apply to one-sample hypothesis testing in Chapter 10. The members of this family of distributions differ from one another in terms of a parameter called degrees of freedom (df). Think of df as the denominator of the variance estimate you use when you calculate a value of *t* as a test statistic. Another way to say "calculate a value of *t* as a test statistic" is "perform a *t*-test."

Unknown population variances lead to two possibilities for hypothesis testing. One possibility is that although the variances are unknown, you have reason to assume they're equal. The other possibility is that you cannot assume they're equal. In the subsections that follow, I discuss these possibilities.

Like peas in a pod: Equal variances

When you don't know a population variance, you use the sample variance to estimate it. If you have two samples, you average (sort of) the two sample variances to arrive at the estimate.

Putting sample variances together to estimate a population variance is called *pooling*. With two sample variances, here's how you do it:

$$s_p^2 = \frac{(N_1 - 1)\, s_1^2 + (N_2 - 1)\, s_2^2}{(N_1 - 1) + (N_2 - 1)}$$

In this formula s_p^2 stands for the pooled estimate. Notice that the denominator of this estimate is $(N_1 - 1) + (N_2 - 1)$. Is this the df? Absolutely!

The formula for calculating *t* is

$$t = \frac{\left(\overline{x}_1 - \overline{x}_2\right) - \left(\mu_1 - \mu_2\right)}{s_p \sqrt{\dfrac{1}{N_1} + \dfrac{1}{N_2}}}$$

On to an example. FarKlempt Robotics is trying to choose between two machines to produce a component for its new microrobot. Because speed is of the essence, each machine produces ten copies of the component, and FarKlempt times each production run. The hypotheses are:

H_0: $\mu_1 - \mu_2 = 0$

H_1: $\mu_1 - \mu_2 \neq 0$

FarKlempt Robotics sets α at .05. This is a two-tailed test, because the company doesn't know in advance which machine might be faster.

Table 11-1 presents the data for the production times in minutes.

Table 11-1	Sample Statistics From the FarKlempt Machine Study	
	Machine 1	**Machine 2**
Mean Production Time	23.00	20.00
Standard Deviation	2.71	2.79
Sample Size	10	10

The pooled estimate of σ^2 is

$$s_p^{\,2} = \frac{(n_1 - 1)\,s_1^2 + (n_2 - 1)\,s_2^2}{(n_1 - 1) + (n_2 - 1)} = \frac{(10 - 1)(2.71)^2 + (10 - 1)(2.79)^2}{(10 - 1) + (10 - 1)} =$$

$$\frac{(9)(2.71)^2 + (9)(2.79)^2}{9 + 9} = \frac{66 + 70}{18} = 7.56$$

The estimate of σ is 2.75, the square root of 7.56.

The test statistic is

$$t = \frac{\left(\bar{x}_1 - \bar{x}_2\right) - \left(\mu_1 - \mu_2\right)}{s_p \sqrt{\dfrac{1}{n_1} + \dfrac{1}{n_2}}} = \frac{(23 - 20)}{2.75\sqrt{\dfrac{1}{10} + \dfrac{1}{10}}} = \frac{3}{1.23} = 2.44$$

For this test statistic, df = 18, the denominator of the variance estimate. In a
t-distribution with 18 df, the critical value is 2.10 for the right-side (upper)
tail and -2.10 for the left-side (lower) tail. If you don't believe me, apply TINV
(Chapter 9). The calculated value of the test statistic is greater than 2.10, so
the decision is to reject H_0. The data provide evidence that Machine 2 is significantly faster than Machine 1. (You can use the word *significant* whenever
you reject H_0.)

Like p's and q's: Unequal variances

The case of unequal variances presents a challenge. As it happens, when variances are not equal the t distribution with $(N_1 - 1) + (N_2 - 1)$ degrees of freedom
is not as close an approximation to the sampling distribution as statisticians
would like.

Statisticians meet this challenge by reducing the degrees of freedom. To
accomplish the reduction, they use a fairly involved formula that depends
on the sample standard deviations and the sample sizes.

Because the variances aren't equal, a pooled estimate is not appropriate. So
you calculate the t-test in a different way:

$$t = \frac{\left(\overline{x}_1 - \overline{x}_2\right) - \left(\mu_1 - \mu_2\right)}{\sqrt{\dfrac{s_1^2}{n_1} + \dfrac{s_2^2}{n_2}}}$$

You evaluate the test statistic against a member of the t-distribution family that has the reduced degrees of freedom.

TTEST

The worksheet function TTEST eliminates the muss, fuss, and bother of working through the formulas for the *t*-test.

Figure 11-6 shows the data for the FarKlempt machines example I showed you earlier. The figure also shows the Function Arguments dialog box for TTEST.

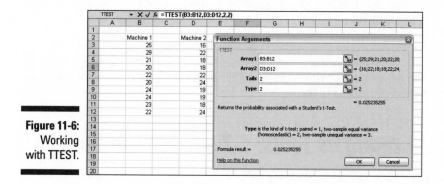

Figure 11-6:
Working
with TTEST.

Follow these steps:

1. **Type the data for each sample into a separate data array.**

   For this example, the data for the Machine 1 sample are in column B and the data for the Machine 2 sample are in column D.

2. **Select a cell.**

3. **Click the Insert Function button to open the Insert Function dialog box.**

4. **In the Insert Function dialog box, select TTEST and click OK to open the Function Arguments dialog box for TTEST.**

5. **In the Array1 box, type the sequence of cells that holds the data for one of the samples.**

   In this example, the Machine 1 data are in B3:B12.

6. **In the Array2 box, type the sequence of cells that holds the data for the other sample.**

   The Machine 2 data are in D3:D12.

7. **In the Tails box, type 1 (for a one-tailed test) or 2 (for a two-tailed test).**

   In this example, it's a two-tailed test.

8. **In the Type box, type a number for the type of *t*-test.**

   The choices are 1 for a paired test (which you'll learn about in the section "A Matched Set: Hypothesis Testing for Paired Samples"), 2 for two samples assuming equal variances, and 3 for two samples assuming unequal variances. I typed 2. When you type an entry for this box, the dialog box shows the probability associated with the *t* value for the data. It does not show the value of *t*.

9. **Click OK to put the answer in the selected cell.**

   The value in the dialog box in Figure 11-6 is less than .05, so the decision is to reject H_0.

By the way, if you type 3 into the Type box (indicating unequal variances) you see a very slight adjustment in the probability from the equal variance test. The adjustment is small because the sample variances are almost equal and the sample sizes are the same.

Data analysis tools: t-Test: Two Sample

Excel provides data analysis tools that carry out *t*-tests. One tool works for the equal variance cases, another for the unequal variances case. As you'll see, when you use these tools you end up with more information than TTEST gives you.

Here's an example that applies the equal variances t-Test tool to the data from the FarKlempt machines example. Figure 11-7 shows the data along with the dialog box for t-Test: Two-Sample Assuming Equal Variances.

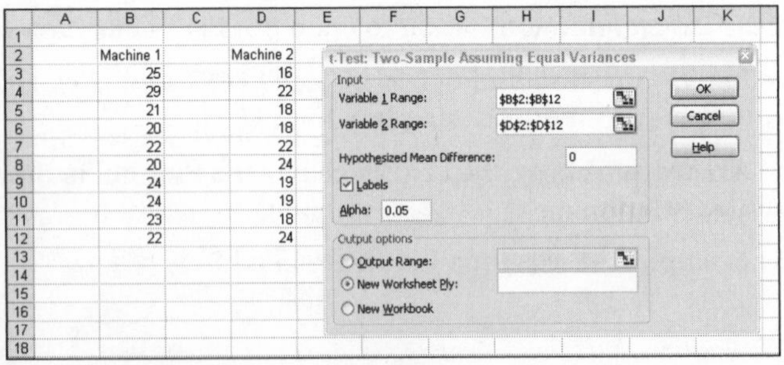

Figure 11-7:
The equal variances t-Test data analysis tool and data from two samples.

To use this tool, follow these steps:

1. **Type the data for each sample into a separate data array.**

   For this example, the data in the Machine 1 sample are in column B and the data for the Machine 2 sample are in column D.

2. **From the Tools menu, select Data Analysis to open the Data Analysis dialog box.**

3. **In the Data Analysis dialog box, scroll down the Analysis Tools list and select t-Test: Two Sample Assuming Equal Variances. Click OK to open this tool's dialog box.**

   This is the dialog box in Figure 11-7.

4. **In the Variable 1 Range box, type the cell range that holds the data for one of the samples.**

   For the example, the Machine 1 data are in $B$3:$B$12. (Note the dollar signs [$] for absolute referencing.)

5. **In the Variable 2 Range box, type the cell range that holds the data for the other sample.**

   The Machine 2 data are in $D$3:$D$12.

6. **In the Hypothesized Mean Difference box, type the difference between μ_1 and μ_2 that H_0 specifies.**

   In this example, that difference is 0.

7. **If the cell ranges include column headings, select the Labels option.**

   I included the headings in the ranges, so I checked the option.

8. **The Alpha box has 0.05 as a default.** I used the default value, which is consistent with how I usually set α.

9. **In the Output Options, select an option to indicate where you want the results.**

   I selected New Worksheet Ply to put the results on a new page in the worksheet.

10. **Click OK.**

    Because I selected New Worksheet Ply, a newly created page opens with the results.

Figure 11-8 shows the tool's results, after I expanded the columns. Rows 4 through 7 hold sample statistics. Cell B8 shows the H_0-specified difference between the population means, and B9 shows the degrees of freedom.

The remaining rows provide *t*-related information. The calculated value of the test statistic is in B10. Cell B11 gives the proportion of area the positive value of the test statistic cuts off in the upper tail of the *t*-distribution with the indicated df. Cell B12 gives the critical value for a one-tailed test: That's the value that cuts off the proportion of the area in the upper tail equal to α.

Cell B13 doubles the proportion in B11. This cell holds the proportion of area from B11 added to the proportion of area that the negative value of the test statistic cuts off in the lower tail. Cell B14 shows the critical value for a two-tailed test: That's the positive value that cuts off α/2 in the upper tail. The corresponding negative value (not shown) cuts off α/2 in the lower tail.

Figure 11-8:
Results of the Equal Variances t-Test data analysis tool.

	A	B	C	D
1	t-Test: Two-Sample Assuming Equal Variances			
2				
3		Machine 1	Machine 2	
4	Mean	23	20	
5	Variance	7.333333333	7.777777778	
6	Observations	10	10	
7	Pooled Variance	7.555555556		
8	Hypothesized Mean Difference	0		
9	df	18		
10	t Stat	2.44046765		
11	P(T<=t) one-tail	0.012617628		
12	t Critical one-tail	1.734063592		
13	P(T<=t) two-tail	0.025235255		
14	t Critical two-tail	2.100922037		
15				

The samples in the example I used have the same number of scores and approximately equal variances, so applying the unequal variances version of the t-Test tool to that data set won't show much of a difference from the equal variances case.

Instead, I created another example, summarized in Table 11-2. The samples in this example have different sizes and widely differing variances.

Table 11-2	Sample Statistics for the Unequal Variances t-Test Example	
	Sample 1	*Sample 2*
Mean	52.50	41.33
Standard Deviation	499.71	41.87
Sample Size	8	6

To show you the difference between the equal variances tool and the unequal variances tool, I ran both on the data and put the results side by side. Figure 11-9 shows the results from both tools. To run the Unequal

Variances tool, you go through the same steps as for the Equal Variances version with one exception: In the Data Analysis Tools dialog box, you select t-Test: Two Sample Assuming Unequal Variances.

Figure 11-9:
Results of
the Equal
Variances
t-Test data
analysis tool
and the
Unequal
Variances
t-Test data
analysis tool
for the data
summa-
rized in
Table 11-2.

	A	B	C	D	E	F	G
1	t-Test: Two-Sample Assuming Equal Variances				t-Test: Two-Sample Assuming Unequal Variances		
2							
3		Sample 1	Sample 2			Sample 1	Sample 2
4	Mean	52.5	41.33333333		Mean	52.5	41.33333333
5	Variance	499.7142857	41.86666667		Variance	499.7142857	41.86666667
6	Observations	8	6		Observations	8	6
7	Pooled Variance	308.9444444			Hypothesized Mean Difference	0	
8	Hypothesized Mean Difference	0			df	9	
9	df	12			t Stat	1.340022972	
10	t Stat	1.176359213			P(T<=t) one-tail	0.106542486	
11	P(T<=t) one-tail	0.13113026			t Critical one-tail	1.833112923	
12	t Critical one-tail	1.782287548			P(T<=t) two-tail	0.213084972	
13	P(T<=t) two-tail	0.26226052			t Critical two-tail	2.262157158	
14	t Critical two-tail	2.178812827					
15							

Figure 11-9 shows one obvious difference between the two tools: The Unequal Variances Tool shows no pooled estimate of σ^2, because the *t*-test for that case doesn't use one. Another difference is in the df. As I pointed out, in the unequal variances case you reduce the df based on the sample variances and the sample sizes. For the equal variances case, the df in this example is 12; for the unequal variances case it's 9.

The effects of these differences show up in the remaining statistics. The *t* values, critical values, and probabilities are different.

A Matched Set: Hypothesis Testing for Paired Samples

In the hypothesis tests I've described so far, the samples are independent of one another. Choosing an individual for one sample has no bearing on the choice of an individual for the other.

Sometimes, the samples are matched. The most obvious case is when the same individual provides a score under each of two conditions — as in a before-after study. For example, suppose ten people participate in a weight-loss program. They weigh in before they start the program and again after one month on the program. The important data is the set of before-after differences. Table 11-3 shows the data:

Table 11-3	Data for the Weight-Loss Example		
Person	**Weight Before Program**	**Weight After One Month**	**Difference**
1	198	194	4
2	201	203	-2
3	210	200	10
4	185	183	2
5	204	200	4
6	156	153	3
7	167	166	1
8	197	197	0
9	220	215	5
10	186	184	2
Mean			2.9
Standard Deviation			3.25

The idea is to think of these differences as a sample of scores and treat them as you would in a one-sample t-test (Chapter 10).

You carry out a test on these hypotheses:

H_0: $\mu_d \leq 0$

H_1: $\mu_d > 0$

The d in the subscripts stands for Difference. Set $\alpha = .05$.

The formula for this kind of t-test is:

$$t = \frac{\bar{d} - \mu_d}{s_{\bar{d}}}$$

In this formula, $\bar{d}$ is the mean of the differences. To find $s_{\bar{d}}$, you calculate the standard deviation of the differences and divide by the square root of the number of pairs:

$$s_{\bar{d}} = \frac{s}{\sqrt{N}}$$

The df is N-1.

From Table 11-3,

$$t = \frac{\bar{d} - \mu_d}{s_{\bar{d}}} = \frac{2.9}{\frac{3.25}{\sqrt{10}}} = 2.82$$

With df = 9 (Number of pairs - 1), the critical value for α = .05 is 2.26. (Use TINV, Chapter 9, to verify.) The calculated value exceeds this value, so the decision is to reject H_0.

TTEST for matched samples

Earlier, I described the worksheet function TTEST and showed you how to use it with independent samples. This time, I use it for the matched samples weight-loss example. Figure 11-10 shows the Function Arguments dialog box for TTEST along with data from the weight-loss example.

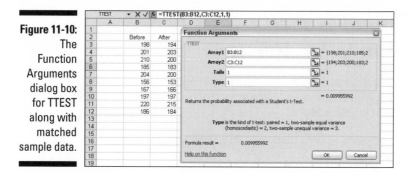

Figure 11-10: The Function Arguments dialog box for TTEST along with matched sample data.

Here are the steps to follow:

1. **Type the data for each sample into a separate data array.**

   For this example, the data for the Before sample are in column B and the data for the After sample are in column D.

2. **Select a cell.**

3. **Click the Insert Function button to open the Insert Function dialog box.**

4. **In the Insert Function dialog box, select TTEST and click OK to open the Function Arguments dialog box for TTEST.**

5. **In the Array1 box, type the sequence of cells that holds the data for one of the samples.**

   In this example, the Before data are in B3:B12.

6. **In the Array2 box, type the sequence of cells that holds the data for the other sample.**

   The After data are in D3:D12.

7. **In the Tails box, type 1 (for a one-tailed test) or 2 (for a two-tailed test).**

   In this example, it's a one-tailed test.

8. **In the Type box, type a number for the type of *t*-test.**

   The choices are 1 for a paired test, 2 for two samples assuming equal variances, and 3 for two samples assuming unequal variances. I typed 1. When you type an entry for this box, the dialog box shows the probability associated with the t value for the data. It does not show the value of t.

9. **Click OK to put the answer in the selected cell.**

   The value in the dialog box in Figure 11-10 is less than .05, so the decision is to reject H_0.

Data analysis tool: t-Test: Paired Two Sample for Means

Excel provides a data analysis tool that takes care of just about everything for matched samples. It's called t-Test: Paired Two Sample for Means. In this section, I use it on the weight-loss data.

Figure 11-11 shows the data along with the dialog box for t-Test: Paired Two Sample for Means.

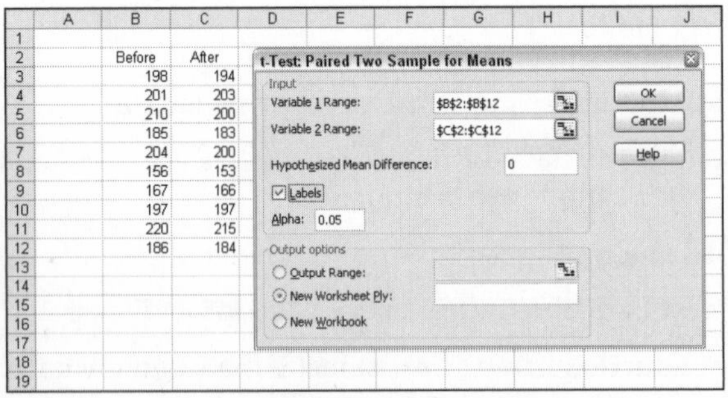

Figure 11-11:
The Paired
Two Sample
t-Test data
analysis tool
and data
from
matched
samples.

Here are the steps to follow:

1. **Type the data for each sample into a separate data array.**

   For this example, the data in the Before sample are in column B and the data for the After sample are in column D.

2. **From the Tools menu, select Data Analysis to open the Data Analysis dialog box.**

3. **In the Data Analysis dialog box, scroll down the Analysis Tools list and select t-Test: Paired Two Sample for Means. Click OK to open this tool's dialog box.**

   This is the dialog box in Figure 11-11.

4. **In the Variable 1 Range box, type the cell range that holds the data for one of the samples.**

   For the example, the Before data are in $B$2:$B$12. (Note the dollar signs [$] for absolute referencing.)

5. **In the Variable 2 Range box, type the cell range that holds the data for the other sample.**

   The After data are in $D$2:$D$12.

6. **In the Hypothesized Mean Difference box, type the difference between μ_1 and μ_2 that H_0 specifies.**

   In this example, that difference is 0.

7. **If the cell ranges include column headings, click the Labels option to select it.**

   I included the headings in the ranges, so I checked the box.

8. **The Alpha box has 0.05 as a default.** I used the default value.

9. **In the Output Options, select an option to indicate where you want the results.**

   I selected New Worksheet Ply to put the results on a new page in the worksheet.

10. **Click OK.**

    Because I selected New Worksheet Ply, a newly created page opens with the results.

Figure 11-12 shows the tool's results after I expanded the columns. Rows 4 through 7 hold sample statistics. The only item that's new is the number in cell B7, the Pearson Correlation Coefficient. This is a number between -1 and +1 that indicates the strength of the relationship between the data in the first sample and the data in the second. If this number is close to 1 (as in the example), high scores in one sample are associated with high scores in the other, and low scores in one are associated with low scores in the other. If the number is close to -1, high scores in the first sample are associated with low scores in the second, and low scores in the first are associated with high scores in the second. If the number is close to zero, scores in the first sample are unrelated to scores in the second. Because the two samples consist of scores on the same people, you expect a high value. (I describe this topic in much greater detail in Chapter 15.)

Cell B8 shows the H_0-specified difference between the population means, and B9 shows the degrees of freedom.

The remaining rows provide t-related information. The calculated value of the test statistic is in B10. Cell B11 gives the proportion of area the positive value of the test statistic cuts off in the upper tail of the *t*-distribution with the indicated df. Cell B12 gives the critical value for a one-tailed test: That's the value that cuts off the proportion of the area in the upper tail equal to α.

Cell B13 doubles the proportion in B11. This cell holds the proportion of area from B11 added to the proportion of area that the negative value of the test statistic cuts off in the lower tail. Cell B13 shows the critical value for a two-tailed test: That's the positive value that cuts off $\alpha/2$ in the upper tail. The corresponding negative value (not shown) cuts off $\alpha/2$ in the lower tail.

Figure 11-12:
Results of
the paired
two-sample
t-Test data
analysis
tool.

	A	B	C	D
1	t-Test: Paired Two Sample for Means			
2				
3		Before	After	
4	Mean	192.4	189.5	
5	Variance	377.6	342.9444444	
6	Observations	10	10	
7	Pearson Correlation	0.986507688		
8	Hypothesized Mean Difference	0		
9	df	9		
10	t Stat	2.824139508		
11	P(T<=t) one-tail	0.009955992		
12	t Critical one-tail	1.833112923		
13	P(T<=t) two-tail	0.019911984		
14	t Critical two-tail	2.262157158		
15				

Testing Two Variances

The two-sample hypothesis testing I've described thus far pertains to means. It's also possible to test hypotheses about variances.

In this section I extend the one-variance manufacturing example I used in Chapter 10. FarKlempt Robotics, Inc. produces a part that has to be a certain length with a very small variability. It's considering two machines to produce this part, and wants to choose the one that results in the least variability.

FarKlempt takes a sample of parts from each machine, measures them, finds the variance for each sample, and performs a hypothesis test to see if one machine's variance is significantly greater than the other's.

The hypotheses are:

$H_0: \sigma_1^2 = \sigma_2^2$

$H_1: \sigma_1^2 \neq \sigma_2^2$

As always, an α is a must. As usual, I set it to .05.

When you test two variances, you don't subtract one from the other. Instead, you divide one by the other to calculate the test statistic. Sir Ronald Fisher is a famous statistician who worked out the mathematics and the family of distributions for working with variances in this way. The test statistic is named in his honor. It's called an *F*-ratio and the test is the *F*-test. The family of distributions for the test is called the *F*-distribution.

Without going into all the mathematics, I'll just tell you that, once again, df is the parameter that distinguishes one member of the family from another. What's different about this family is that two variance estimates are involved, so each member of the family is associated with two values of df, rather than one as in the *t*-test. Another difference between the *F*-distribution and the others you've seen is that the *F* cannot have a negative value. Figure 11-13 shows two members of the *F*-distribution family.

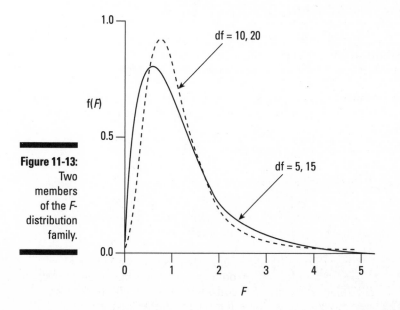

Figure 11-13: Two members of the *F*-distribution family.

The test statistic is:

$$F = \frac{\text{larger s}^2}{\text{smaller s}^2}$$

Suppose FarKlempt Robotics produces 10 parts with Machine 1 and finds a sample variance of .60 square inches. It produces 15 parts with Machine 2 and finds a sample variance of .44 square inches. Can it reject H_0?

Calculating the test statistic,

$$F = \frac{.60}{.44} = 1.36$$

The df's are 9 and 14: The variance estimate in the numerator of the F-ratio is based on ten cases, and the variance estimate in the denominator is based on 15 cases.

When the df's are 9 and 14 and it's a two-tailed test at α = .05, the critical value of F is 3.21. (In a moment, I'll show you an Excel function that finds that value for you.) The calculated value is less than the critical value, so the decision is to not reject H_0.

It makes a difference which df is in the numerator and which df is in the denominator. The F-distribution for df = 9 and df = 14 is different from the F-distribution for df = 14 and df = 9. For example, the critical value in the latter case is 3.98, not 3.21.

Using F in conjunction with t

One use of the F-distribution is in conjunction with the t-test for independent samples. Before you do the t-test, you use F to help decide whether to assume equal variances or unequal variances in the samples.

In the equal variances t-test example I showed you earlier, the standard deviations are 2.71 and 2.79. The variances are 7.34 and 7.78. The F-ratio of these variances is

$$F = \frac{7.78}{7.34} = 1.06$$

Each sample is based on 10 observations, so df=9 for each sample variance. An F-ratio of 1.06 cuts off the upper 47 percent of the F-distribution whose df are 9 and 9, so it's safe to use the equal variances version of the t-test for these data.

In the sidebar "A Point to Ponder" at the end of Chapter 10, I mention that on rare occasions a high α is a good thing. When H_0 is a desirable outcome and you'd rather not reject it, you stack the deck against rejection by setting α at a high level so that small differences cause you to reject H_0.

This is one of those rare occasions. It's more desirable to use the equal variances *t*-test, which typically provides more degrees of freedom than the unequal variances *t*-test. Setting a high value of α (.20 is a good one) for the *F*-test enables you to be confident when you assume equal variances.

FTEST

The worksheet function FTEST calculates an *F*-ratio on the data from two samples. It doesn't return the *F*-ratio. Instead, it provides the two-tailed probability of the calculated *F*-ratio under H_0. This means that the answer is the proportion of area to the right of the *F*-ratio and to the left of the reciprocal of the *F*-ratio (1 divided by the *F*-ratio).

Figure 11-14 presents the data for the FarKlempt machines example I just summarized for you. It shows the sample variances as well as the data. The figure also shows the Function Arguments dialog box for FTEST.

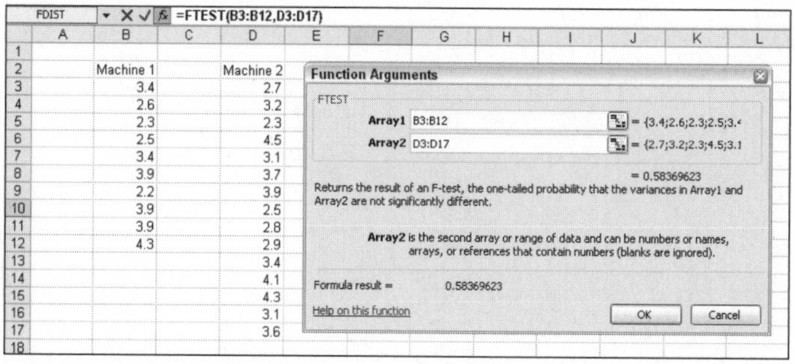

Figure 11-14:
Working
with FTEST.

Follow these steps:

1. **Type the data for each sample into a separate data array.**

   For this example, the data for the Machine 1 sample are in column B and the data for the Machine 2 sample are in column D.

2. **Select a cell.**

3. **Click the Insert Function button to open the Insert Function dialog box.**

4. **In the Insert Function dialog box, select FTEST and click OK to open the Function Arguments dialog box for FTEST.**

5. **In the Array1 box, type the sequence of cells that holds the data for the sample with the larger variance.**

   In this example, the Machine 1 data are in D3:B12.

6. **In the Array2 box, type the sequence of cells that holds the data for the other sample.**

   The Machine 2 data are in D3:D17. When you complete the entry in this box, the answer appears in the dialog box.

7. **Click OK to put the answer in the selected cell.**

   The value in the dialog box in Figure 11-14 is greater than .05, so the decision is to not reject H_0. Figure 11-15 shows the area that the answer represents.

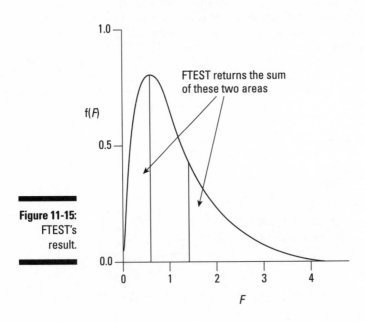

Figure 11-15:
FTEST's
result.

FDIST

You use the worksheet function FDIST to decide whether or not your calcu- lated *F*-ratio is in the region of rejection. You supply a value for *F*, a value for each df, and whether the test is one-tailed or two-tailed. FDIST returns the probability of obtaining an *F*-ratio at least as high as yours if H_0 is true. If that probability is less than your α, you reject H_0.

Here, I apply FDIST to the example I just used. The *F*-ratio is 1.36, with 9 and 14 df.

The steps are:

1. **Select a cell.**

2. **Click the Insert Function button to open the Insert Function dialog box.**

3. **In the Insert Function dialog box, select FDIST and click OK to open the Function Arguments dialog box for FDIST. (See Figure 11-16.)**

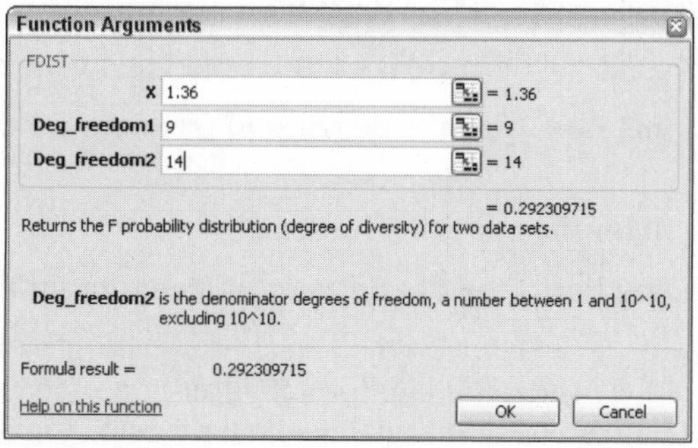

Figure 11-16: The Function Arguments dialog box for FDIST.

4. **In the X box, type your calculated *F*.**

   For this example, the calculated *F* is 1.36.

5. **In the Deg_freedom1 box, type the degrees of freedom for the variance estimate in the numerator of the *F*.**

   The degrees of freedom for the numerator in this example is 9 (10 scores - 1).

6. **In the Deg_freedom2 box, type the degrees of freedom for the variance estimate in the denominator of the *F*.**

   The degrees of freedom for the denominator in this example is 14 (15 scores - 1). After you type this number, the answer appears in the dialog box.

7. **Click OK to close the dialog box and put the answer in the selected cell.**

   The value in the dialog box in Figure 11-16 is greater than .05, so the decision is to not reject H_0.

FINV

Excel's FINV worksheet function finds the value in the *F*-distribution that cuts off a given proportion of the area in the upper (right-side) tail. You can use it to find the critical value of F. Here, I use it to find the critical value for the two-tailed test in the FarKlempt machines example.

1. **Select a cell.**

2. **Click the Insert Function button to open the Insert Function dialog box.**

3. **In the Insert Function dialog box, select FINV to open the Function Arguments dialog box for FINV.**

4. **In the Probability box, type the proportion of area in the upper tail.**

   In this example, that's .025 because it's a two-tailed test with $\alpha = .05$.

5. **In the Deg_freedom1 box, type the degrees of freedom for the numerator.**

   For this example, df for the numerator = 9.

6. **In the Deg_freedom2 box, type the degrees of freedom for the denominator.**

   For this example, df for the denominator = 9. After you type this value, the answer, 3.209 (and some additional decimal places), appears in the dialog box. (See Figure 11-17.)

7. **Click OK to put the answer into the selected cell.**

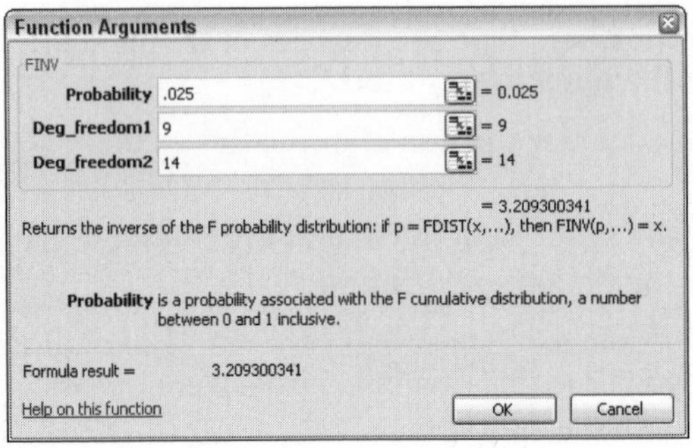

Figure 11-17:
The Function Arguments dialog box for FINV.

Data analysis tool: F-Test Two-Sample for Variances

Excel provides a data analysis tool for carrying out an F-test on two sample variances. I apply it here to the sample variances example I've been using. Figure 11-18 shows the data along with the dialog box for F-Test: Two-Sample for Variances.

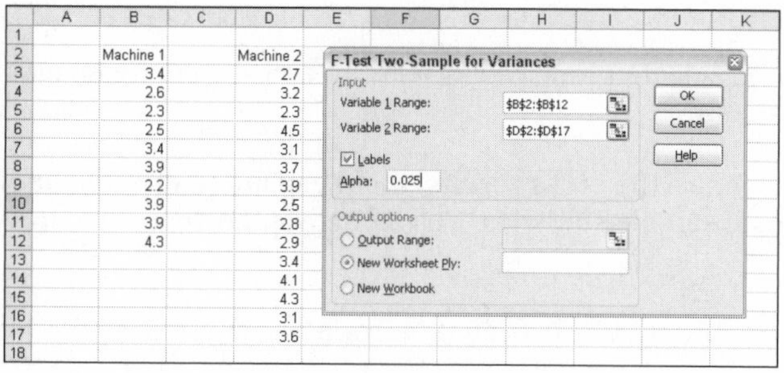

Figure 11-18:
The F-Test data analysis tool and data from two samples.

To use this tool, follow these steps:

1. **Type the data for each sample into a separate data array.**

   For this example, the data in the Machine 1 sample are in column B and the data for the Machine 2 sample are in column D.

2. **From the Tools menu, select Data Analysis to open the Data Analysis dialog box.**

3. **In the Data Analysis dialog box, scroll down the Analysis Tools list and select F-Test Two Sample For Variances. Click OK to open this tool's dialog box.**

   This is the dialog box in Figure 11-18.

4. **In the Variable 1 Range box, type the cell range that holds the data for the first sample.**

   For the example, the Machine 1 data are in $B$2:$B$12. (Note the dollar signs [$] for absolute referencing.)

5. **In the Variable 2 Range box, type the cell range that holds the data for the second sample.**

   The Machine 2 data are in $D$2:$D$17.

6. **If the cell ranges include column headings, click the Labels option to select it.**

   I included the headings in the ranges, so I checked the box.

7. **The Alpha box has 0.05 as a default. Change that value for a different α.**

   The Alpha box provides a one-tailed alpha. I want a two-tailed test, so I changed this value to .025.

8. **In the Output Options, select an option to indicate where you want the results.**

   I selected New Worksheet Ply to put the results on a new page in the worksheet.

9. **Click OK.**

   Because I selected New Worksheet Ply, a newly created page opens with the results.

Figure 11-19 shows the tool's results after I expanded the columns. Rows 4 through 6 hold sample statistics. Cell B7 shows the degrees of freedom.

Figure 11-19:
Results of
the F-Test
data
analysis
tool.

	A	B	C	D
1	F-Test Two-Sample for Variances			
2				
3		Machine 1	Machine 2	
4	Mean	3.24	3.34	
5	Variance	0.60044444	0.441143	
6	Observations	10	15	
7	df	9	14	
8	F	1.36111111		
9	P(F<=f) one-tail	0.29184811		
10	F Critical one-tail	3.20930034		
11				

The remaining rows present *F*-related information. The calculated value of *F* is in B8. Cell B9 gives the proportion of area the calculated *F* cuts off in the upper tail of the *F*-distribution. This is the right-side area in Figure 11-15. Cell B10 gives the critical value for a one-tailed test: That's the value that cuts off the proportion of the area in the upper tail equal to the value in the Alpha box.

Chapter 12

Testing More Than Two Samples

Statistics would be limited if you could only make inferences about one or two samples. In this chapter, I discuss the procedures for testing hypotheses about three or more samples. I show what to do when samples are independent of one another and what to do when they're not. In both cases, I discuss what to do after you test the hypotheses.

I also introduce Excel data analysis tools that do the work for you. Although these tools aren't at the level you'd find in a dedicated statistical package, you can combine them with Excel's standard features to produce some sophisticated analyses.

Testing More Than Two

Imagine this situation: Your company asks you to evaluate three different methods for training its employees to do a particular job. You randomly assign 30 employees to one of the three methods. Your plan is to train them, test them, tabulate the results, and make some conclusions. Before you can finish the study, three people leave the company — one from the Method 1 group, and two from the Method 3 group.

Table 12-1 shows the data.

Table 12-1	Data from Three Training Methods		
	Method 1	Method 2	Method 3
	95	83	68
	91	89	75
	89	85	79
	90	89	74
	99	81	75
	88	89	81
	96	90	73
	98	82	77
	95	84	
		80	
Mean	93.44	85.20	75.25
Variance	16.28	14.18	15.64
Standard Deviation	4.03	3.77	3.96

Do the three methods provide different results, or are they so similar that you can't distinguish among them? To decide, you have to carry out a hypothesis test:

H_0: $\mu_1 = \mu_2 = \mu_3$

H_1: Not H_0

with $\alpha = .05$.

A thorny problem

Sounds pretty easy, particularly if you've read Chapter 11. Take the mean of the scores from Method 1, the mean of the scores from Method 2, and do a t-test to see if they're different. Follow the same procedure for Method 1 versus Method 3, and for Method 2 versus Method 3. If at least one of those t-tests shows a significant difference, reject H_0. Nothing to it, right?

Wrong. If your α is .05 for each *t*-test, you're setting yourself up for a Type I error with a probability higher than you planned on. The probability that at least one of the three *t*-test results in a significant difference is way above .05. In fact, it's .14, which is way beyond acceptable. (The mathematics behind calculating that number is a little involved, so I won't elaborate.)

With more than three samples, the situation gets even worse. Four groups require six *t*-tests, and the probability that at least one of them is significant is .26. Table 12-2 shows what happens with increasing numbers of samples.

Table 12-2	The Incredible Increasing Alpha	
Number of Samples	*Number of t-Tests*	*Pr (At Least One Significant t)*
3	3	.14
4	6	.26
5	10	.40
6	15	.54
7	21	.66
8	28	.76
9	36	.84
10	45	.90

Carrying out multiple *t*-tests is clearly not the answer. So what do you do?

A solution

It's necessary to take a different approach. The idea is to think in terms of variances rather than means.

I'd like you to think of variance in a slightly different way. The formula for estimating population variance, remember, is

$$s^2 = \frac{\sum(x - \bar{x})^2}{N - 1}$$

Because the variance is almost a mean of squared deviations from the mean, statisticians also refer to it as *Mean Square*. In a way, that's an unfortunate nickname: It leaves out deviation from the mean, but there you have it.

The numerator of the variance, excuse me, Mean Square, is the sum of squared deviations from the mean. This leads to another nickname, *Sum of Squares*. The denominator, as I said in Chapter 10, is *degrees of freedom* (df). So, the slightly different way to think of variance is

$$\text{Mean Square} = \frac{\text{Sum of Squares}}{\text{df}}$$

You can abbreviate this as

$$\text{MS} = \frac{\text{SS}}{\text{df}}$$

Now, on to solving the thorny problem. One important step is to find the Mean Squares hiding in the data. Another is to understand that you use these Mean Squares to estimate the variances of the populations that produced these samples. In this case, assume those variances are equal, so you're really estimating one variance. The final step is to understand that you use these estimates to test the hypotheses I showed you at the beginning of the chapter.

Three different Mean Squares are inside the data in Table 12-1. Start with the whole set of 27 scores, forgetting for the moment that they're divided into three groups. Suppose you want to use those 27 scores to calculate an estimate of the population variance. (A dicey idea, but humor me.) The mean of those 27 scores is 85. I'll call that mean the *grand mean* because it's the average of everything.

So the Mean Square would be

$$\frac{(95-85)^2 + (91-85)^2 + \ldots + (73-85)^2 + (77-85)^2}{(27-1)} = 68.08$$

The denominator has 26 (27-1) degrees of freedom. I refer to that variance as the *Total Variance*, or in the new way of thinking about this, the MS_{Total}. It's often abbreviated as MS_T.

Here's another variance to consider. In Chapter 11, I describe the *t*-test for two samples with equal variances. For that test, you put the two sample variances together to create a *pooled* estimate of the population variance. The data in Table 12-1 provide three sample variances for a pooled estimate: 16.28, 14.18, and 15.64. Assuming these numbers represent equal population variances, the pooled estimate is:

$$s_p^2 = \frac{(n_1 - 1)s_1^2 + (n_2 - 1)s_2^2 + (n_3 - 1)s_3^2}{(n_1 - 1) + (n_2 - 1) + (n_3 - 1)}$$

$$= \frac{(9-1)(16.28) + (10-1)(14.18) + (8-1)(15.64)}{(9-1) + (10-1) + (8-1)} = 15.31$$

Because this pooled estimate comes from the variance within the groups, it's called $\text{MS}_{\text{Within}}$, or MS_W.

One more Mean Square to go — the variance of the sample means around the grand mean. In this example, that means the variance in these numbers: 93.44,

85.20, and 75.25 — sort of. I say sort of because these are means, not scores. When you deal with means you have to take into account the number of scores that produce each mean. To do that you multiply each squared deviation by the number of scores in that sample.

So this variance is:

$$\frac{(9)(93.44-85)^2+(10)(85.20-85)^2+(8)(75.25-85)^2}{3-1}=701.34$$

The df for this variance is 2 (the number of samples - 1).

Statisticians, not known for their crispness of usage, refer to this as the variance *between* sample means. (*Among* is the correct word when you're talking about more than two items.) This variance is known as $MS_{Between}$, or MS_B.

So you now have three estimates of population variance: MS_T, MS_W, and MS_B. What do you do with them?

Remember that the original objective is to test a hypothesis about three means. According to H_0, any differences you see among the three sample means are due strictly to chance. The implication is that the variance among those means is the same as the variance of any three numbers selected at random from the population.

If you could somehow compare the variance among the means (that's MS_B, remember) with the population variance, you could see if that holds up. If only you had an estimate of the population variance that's independent of the differences among the groups, you'd be in business.

Ah . . . but you do have that estimate. You have MS_W, an estimate based on pooling the variances within the samples. Assuming those variances represent equal population variances, this is a pretty solid estimate. In this example, it's based on 24 degrees of freedom.

The reasoning now becomes: If MS_B is about the same as MS_W, you have evidence consistent with H_0. If MS_B is significantly larger than MS_W, you have evidence that's inconsistent with H_0. In effect, you transform these hypotheses

H_0: $\mu_1 = \mu_2 = \mu_3$

H_1: Not H_0

into these

H_0: $\sigma_B^2 \leq \sigma_W^2$

H_1: $\sigma_B^2 > \sigma_W^2$

Instead of multiple *t*-tests among sample means, you perform a test of the difference between two variances.

What is that test? In Chapter 11 I show you the test for hypotheses about two variances. It's called the *F*-test. To perform this test, you divide one variance by the other. You evaluate the result against a family of distributions called the *F*-distribution. Because two variances are involved, two values for degrees of freedom define each member of the family.

For this example, *F* has df = 2 (for the MS_B) and df = 24 (for the MS_W). Figure 12-1 shows what this member of the *F* family looks like. For our purposes, it's the distribution of possible *F* values if H_0 is true.

The test statistic for the example is:

$$F = \frac{701.34}{15.31} = 45.82$$

What proportion of area does this value cut off in the upper tail of the *F*-distribution? From Figure 12-1, you can see that this proportion is microscopic, as the values on the horizontal axis only go up to 5. (And the proportion of area beyond 5 is tiny.) It's way less than .05.

This means it's highly unlikely that differences among the means are due to chance. It means you reject H_0.

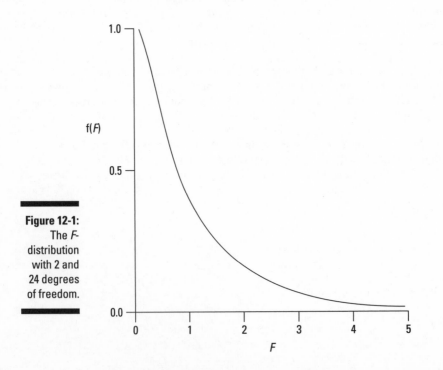

Figure 12-1:
The *F*-distribution with 2 and 24 degrees of freedom.

This whole procedure for testing more than two samples is called *the analysis of variance*, often abbreviated as ANOVA. In the context of an ANOVA, the denominator of an *F*-ratio has the generic name *error term*. The independent variable is sometimes called a *factor*. So this is a single-factor or (one-factor) ANOVA.

In this example, the factor is Teaching Method. Each instance of the independent variable is called a *level*. The independent variable in this example has three levels.

More complex studies have more than one factor, and each factor can have many levels.

Meaningful relationships

Take another look at the Mean Squares in this example, each with its Sum of Squares and degrees of freedom. Before, when I calculated each Mean Square for you, I didn't explicitly show you each Sum of Squares, but here I include them:

$$MS_B = \frac{SS_B}{df_B} = \frac{1402.68}{2} = 701.34$$

$$MS_W = \frac{SS_W}{df_W} = \frac{367.32}{24} = 15.31$$

$$MS_T = \frac{SS_T}{df_T} = \frac{1770}{26} = 68.08$$

Start with the degrees of freedom: $df_B = 2$, $df_W = 24$, and $df_T = 26$. Is it a coincidence that they add up? Hardly. It's always the case that

$df_B + df_W = df_T$

How about those Sums of Squares?

$1402.68 + 367.32 = 1770$

Again, this is no coincidence. In the analysis of variance, this always happens:

$SS_B + SS_W = SS_T$

In fact, statisticians who work with the analysis of variance speak of partitioning (read "breaking down into non-overlapping pieces") the SS_T into one portion for the SS_B and another for the SS_W, and partitioning the df_T into one amount for the df_B and another for the df_W.

After the F-test

The *F*-test enables you to decide whether or not to reject H_0. Once you decide to reject, then what? All you can say is that somewhere within the set of means, something is different from something else. The *F*-test doesn't specify what those "somethings" are.

Planned comparisons

In order to get more specific, you have to do some further tests. Not only that, you have to plan those tests in advance of carrying out the ANOVA.

What are those tests? Given what I said earlier, this may surprise you: *t*-tests. While this might sound inconsistent with the increased alpha of multiple *t*-tests, it's not. If an analysis of variance enables you to reject H_0, then it's OK to use *t*-tests to turn the magnifying glass on the data and find out where the differences are. And as I'm about to show you, the *t*-test you use is slightly different from the one I discuss in Chapter 11.

These post-ANOVA *t*-tests are called *planned comparisons*. Some refer to them as *a priori tests*. I illustrate by following through with the example. Suppose, before you gathered the data, you had reason to believe that Method 1 would result in higher scores than Method 2, and that Method 2 would result in higher scores than Method 3. In that case, you plan in advance to compare the means of those samples in the event your ANOVA-based decision is to reject H_0.

The formula for this kind of *t*-test is

$$t = \frac{\overline{x}_1 - \overline{x}_2}{\sqrt{MS_W \left[\frac{1}{n_1} + \frac{1}{n_2} \right]}}$$

It's a test of

H_0: $\mu_1 \le \mu_2$

H_1: $\mu_1 > \mu_2$

MS_W takes the place of the pooled estimate s_{p2} that I showed you in Chapter 11. In fact, when I introduced MS_W, I showed how it's just a pooled estimate that can incorporate variances from more than two samples. The df for this *t*-test is df_W, rather than $(n_1 - 1) + (n_2 - 1)$.

For this example, the Method 1 versus Method 2 comparison is:

$$t = \frac{\overline{x}_1 - \overline{x}_2}{\sqrt{MS_W \left[\frac{1}{n_1} + \frac{1}{n_2} \right]}} = \frac{93.44 - 85.2}{\sqrt{15.31 \left[\frac{1}{9} + \frac{1}{10} \right]}} = 4.59$$

With df = 24, this value of t cuts off a miniscule portion of area in the upper tail of the t-distribution. The decision is to reject H_0.

The planned comparison *t*-test formula I showed you matches up with the *t*-test for two samples. You can write the planned comparison *t*-test formula in a way that sets up additional possibilities. Start by writing the numerator

$$\bar{x}_1 - \bar{x}_2$$

a bit differently:

$$(+1)\,\bar{x}_1 + (-1)\,\bar{x}_2$$

The +1 and -1 are *comparison coefficients*. I refer to them, in a general way, as c_1 and c_2. In fact, c_3 and $\bar{x}_3$ can enter the comparison, even if you're just comparing $\bar{x}_1$ with $\bar{x}_2$:

$$(+1)\,\bar{x}_1 + (-1)\,\bar{x}_2 + (0)\,\bar{x}_3$$

The important thing is that the coefficients add up to zero.

Here's how the comparison coefficients figure into the planned comparison t-Test formula for a study that involves three samples:

$$t = \frac{c_1\,\bar{x}_1 + c_2\,\bar{x}_2 + c_3\,\bar{x}_3}{\sqrt{MS_W\left[\dfrac{c_1^2}{n_1} + \dfrac{c_2^2}{n_2} + \dfrac{c_3^2}{n_3}\right]}}$$

Applying this formula to Method 2 versus Method 3:

$$t = \frac{c_1\,\bar{x}_1 + c_2\,\bar{x}_2 + c_3\,\bar{x}_3}{\sqrt{MS_W\left[\dfrac{c_1^2}{n_1} + \dfrac{c_2^2}{n_2} + \dfrac{c_3^2}{n_3}\right]}} = \frac{(0)(93.44) + (+1)(85.2) + (-1)(75.25)}{\sqrt{15.31\left[\dfrac{0^2}{9} + \dfrac{1^2}{10} + \dfrac{1^2}{8}\right]}} = 5.36$$

The value for t indicates the results from Method 2 are significantly higher than the results from Method 3.

You can also plan a more complex comparison — say, Method 1 versus the average of Method 2 and Method 3. Begin with the numerator. That would be

$$\bar{x}_1 - \frac{\left(\bar{x}_2 + \bar{x}_3\right)}{2}$$

With comparison coefficients, you can write this as

$$(+1)\,\bar{x}_1 + \left(-\frac{1}{2}\right)\bar{x}_2 + \left(-\frac{1}{2}\right)\bar{x}_3$$

If you're more comfortable with whole numbers, you can write it as:

$$(+2)\,\bar{x}_1 + (-1)\,\bar{x}_2 + (-1)\,\bar{x}_3$$

Plugging these numbers into the formula gives you

$$t = \frac{c_1 \overline{X}_1 + c_2 \overline{X}_2 + c_3 \overline{X}_3}{\sqrt{MS_W \left[\frac{c_1^2}{n_1} + \frac{c_2^2}{n_2} + \frac{c_3^2}{n_3} \right]}} = \frac{(2)(93.44) + (-1)(85.2) + (-1)(75.25)}{\sqrt{15.31 \left[\frac{2^2}{9} + \frac{1^2}{10} + \frac{1^2}{8} \right]}} = 9.97$$

Again, strong evidence for rejecting H_0.

Unplanned comparisons

Things would get boring if your post-ANOVA testing is limited to comparisons you have to plan in advance. Sometimes you want to snoop around your data and see if anything interesting reveals itself. Sometimes something jumps out at you that you didn't anticipate.

When this happens, you can make comparisons you didn't plan on. These comparisons are called *a posteriori tests*, *post hoc tests*, or, simply, *unplanned comparisons*. Statisticians have come up with a wide variety of these tests, many of them with exotic names and many of them dependent on special sampling distributions.

The idea behind these tests is that you pay a price for not having planned them in advance. That price has to do with stacking the deck against rejecting H_0 for the particular comparison.

Of all the unplanned tests available, the one I like best is a creation of famed statistician Henry Scheffé. As opposed to esoteric formulas and distributions, you start with the test I already showed you and then add a couple of easy-to-do extras.

The first extra is to understand the relationship between t and F. I've showed you the F-test for three samples. You can also carry out an F-test for two samples. That F-test has df = 1 and df = $(n_1 - 1) + (n_2 - 1)$. The df for the t-test, of course, is $(n_1 - 1) + (n_2 - 1)$. Hmmm . . . seems like they should be related somehow.

They are. The relationship between the two-sample t and the two-sample F is

$t^2 = F$

Now I can tell you the steps for performing Scheffé's test:

1. **Calculate the planned comparison *t*-test.**

2. **Square the result to create F.**

3. **Find the critical value of F for df_B and df_W at $\alpha = .05$ (or whatever α you choose).**

4. **Multiply this critical F by the number of samples - 1. The result is your critical F for the unplanned comparison. I'll call this F'.**

5. **Compare the calculated F to F'. If the calculated F is greater, reject H_0 for this test. If it's not, don't reject H_0 for this test.**

Imagine that in the example, you didn't plan in advance to compare the mean of Method 1 with the mean of Method 3. (In a study involving only three samples that's hard to imagine, I grant you.) The t-test is:

$$t = \frac{c_1 \overline{x}_1 + c_2 \overline{x}_2 + c_3 \overline{x}_3}{\sqrt{MS_w \left[\frac{c_1^2}{n_1} + \frac{c_2^2}{n_2} + \frac{c_3^2}{n_3} \right]}} = \frac{(+1)(93.44) + (0)(85.2) + (-1)(75.25)}{\sqrt{15.31 \left[\frac{1^2}{9} + \frac{0^2}{10} + \frac{-1^2}{8} \right]}} = 9.57$$

Squaring this result gives

$$F = t^2 = (9.57)^2 = 91.61$$

For F with 2 and 24 df and $\alpha = .05$, the critical value is 3.403. (You can look that up in table in a statistics textbook or you can use the worksheet function FINV.) So

$$F' = (3-1)F = (2)(3.403) = 6.806$$

Because the calculated F, 91.61, is greater than F', the decision is to reject H_0. You have evidence that Method 1's results are different from Method 3's results.

Data analysis tool: Anova: Single Factor

The calculations for the ANOVA can get intense. Excel has a data analysis tool that does the heavy lifting. It's called Anova: Single Factor. Figure 12-2 shows this tool along with the data for the example I just went through.

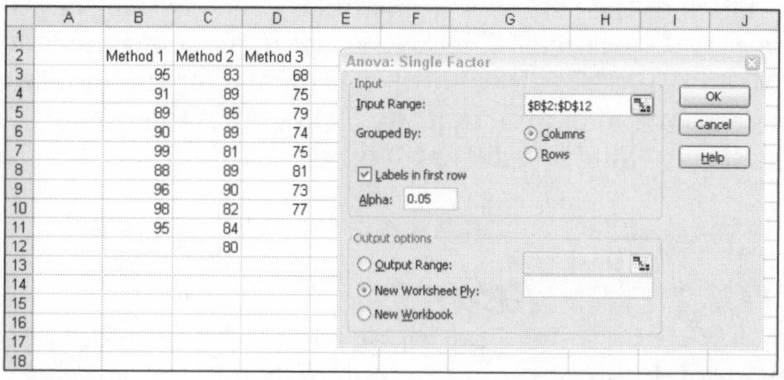

Figure 12-2:
The Anova:
Single
Factor data
analysis tool
dialog box.

The steps for using this tool are:

1. **Type the data for each sample into a separate data array.**

   For this example the data in the Method 1 sample are in column B, the data in the Method 2 sample are in Column D, and the data for the Method 3 sample are in column E.

2. **From the Tools menu, select Data Analysis to open the Data Analysis dialog box.**

3. **In the Data Analysis dialog box, scroll down the Analysis Tools list and select Anova: Single Factor. Click OK to open the Anova: Single Factor dialog box.**

   This is the dialog box in Figure 12-2.

4. **In the Input Range box, type the cell range that holds all the data.**

   For the example, the data are in $B$2:$D$12. (Note the dollar signs [$] for absolute referencing.)

5. **If the cell ranges include column headings, select the Labels in First Row option.**

   I included the headings in the ranges, so I checked the box.

6. **The Alpha box has 0.05 as a default.** I used the default value.

7. **In the Output options, select an option to indicate where you want the results.**

   I selected New Worksheet Ply to put the results on a new page in the worksheet.

8. **Click OK.**

   Because I selected New Worksheet Ply, a newly created page opens with the results.

Figure 12-3 shows the tool's output after I expanded the columns. The output features two tables, SUMMARY and ANOVA. The SUMMARY table provides summary statistics of the samples — the number in each group, the group sums, averages, and variances. The ANOVA table presents the Sums of Squares, df, Mean Squares, F, P-value, and critical F for the indicated df. The P-value is the proportion of area that the F cuts off in the upper tail of the F-distribution. If this value is less than .05, reject H_0.

Comparing the means

Excel's ANOVA tool does not provide a built-in facility for carrying out planned (or unplanned) comparisons among the means. With a little ingenuity, however, you can use the Excel worksheet function SUMPRODUCT to do those comparisons.

Figure 12-3:
Output from
the Anova:
Single
Factor
analysis
tool.

	A	B	C	D	E	F	G	H
1	Anova: Single Factor							
2								
3	SUMMARY							
4	*Groups*	*Count*	*Sum*	*Average*	*Variance*			
5	Method 1	9	841	93.44444	16.27778			
6	Method 2	10	852	85.2	14.17778			
7	Method 3	8	602	75.25	15.64286			
8								
9								
10	ANOVA							
11	*Source of Variation*	*SS*	*df*	*MS*	*F*	*P-value*	*F crit*	
12	Between Groups	1402.678	2	701.3389	45.82389	6.38E-09	3.402826	
13	Within Groups	367.3222	24	15.30509				
14								
15	Total	1770	26					
16								

The worksheet page with the ANOVA output is the launching pad for the planned comparisons. In this section, I take you through one planned comparison from the example — the mean of Method 1 versus the mean of Method 2.

Begin by creating columns that hold important information for the comparisons. Figure 12-4 shows what I mean. I put the comparison coefficients in column J, the squares of those coefficients in column K, and the reciprocal of each sample size ($1/n$) in column L.

Figure 12-4:
Carrying out
a planned
comparison.

K12 ▼ *fx* =SQRT(D13*(SUMPRODUCT(K5:K7,L5:L7)))

	A	B	C	D	E	F	G	H	I	J	K	L
1	Anova: Single Factor											
2												
3	SUMMARY											
4	*Groups*	*Count*	*Sum*	*Average*	*Variance*					*c*	*c^2*	*1/n*
5	Method 1	9	841	93.44444	16.27778					1	1	0.111111
6	Method 2	10	852	85.2	14.17778					-1	1	0.1
7	Method 3	8	602	75.25	15.64286					0	0	0.125
8												
9												
10	ANOVA									Comparison		
11	*Source of Variation*	*SS*	*df*	*MS*	*F*	*P-value*	*F crit*			numerator	8.244444444	
12	Between Groups	1402.678	2	701.3389	45.82389	6.38E-09	3.402826			denom	1.797519152	
13	Within Groups	367.3222	24	15.30509						t=	4.586568346	
14										P-value=	5.94147E-05	
15	Total	1770	26									
16												

A few rows below those cells, I put *t*-test-related information — the *t*-test numerator, the denominator, and the value of *t*. I used separate cells for the numerator and denominator to simplify the formulas. You can put them together in one big formula and just have a cell for *t*, but it's hard to keep track of everything.

SUMPRODUCT takes arrays of cells, multiplies the numbers in the corresponding cells, and sums the products. (This function is in the Math & Trig category, not the Statistical category.) I use SUMPRODUCT to multiply each coefficient by each sample mean and then add the products. I store that result in K11. That's the numerator for the planned comparison *t*-test. The formula for K11 is

=SUMPRODUCT(J5:J7,D5:D7)

The array J5:J7 holds the comparison coefficients, and D5:D7 holds the sample means.

K12 holds the denominator. I selected K12 so you could see its formula in the formula bar:

=SQRT(D13*(SUMPRODUCT(K5:K7,L5:L7)))

D13 has the MS_W. SUMPRODUCT multiplies the squared coefficients in K5:K7 by the reciprocals of the sample sizes in L5:L7 and sums the products. SQRT takes the square root of the whole thing.

K13 holds the value for t. That's just K11 divided by K12.

K14 presents the P-value for t — the proportion of area that t cuts off in the upper tail of the t-distribution with df = 24. The formula for that cell is

=TDIST(K13,C13,1)

The arguments are the calculated t (in K13), the degrees of freedom for MS_W (in C13), and the number of tails.

If you change the coefficients in J5:J7, you instantaneously create and complete another comparison.

In fact, I'll do that right now, and show you Scheffé's post hoc (unplanned) comparison. That one, in this example, compares the mean of Method 1 with the mean of Method 3. Figure 12-5 shows the extra information for this test, starting a couple of rows below the t-test.

Figure 12-5:
Carrying out
a post hoc
comparison.

	A	B	C	D	E	F	G	H	I	J	K	L
	K17	▼	f_x	=C12*G12								
1	Anova: Single Factor											
2												
3	SUMMARY											
4	Groups	Count	Sum	Average	Variance					c	c^2	1/n
5	Method 1	9	841	93.44444	16.27778					1	1	0.111111
6	Method 2	10	852	85.2	14.17778					0	0	0.1
7	Method 3	8	602	75.25	15.64286					-1	1	0.125
8												
9												
10	ANOVA									Comparison		
11	Source of Variation	SS	df	MS	F	P-value	F crit			numerator	18.19444444	
12	Between Groups	1402.678	2	701.3389	45.82389	6.38E-09	3.402826			denom	1.900974071	
13	Within Groups	367.3222	24	15.30509						t=	9.571116579	
14										P-value=	5.74498E-10	
15	Total	1770	26									
16										F=	91.60627256	
17										F'=	6.805652211	
18												

Cell K16 holds F, the square of the t value in K13. K17 has F', the product of C12 (df_B, which is the number of samples - 1) and G12 (the critical value of F for 2 and 24 degrees of freedom and α = .05). K16 is greater than K17, so reject H_0 for this comparison.

Another Kind of Hypothesis, Another Kind of Test

The ANOVA I just showed you works with independent samples. As you might remember from Chapter 11, sometimes you work with matched samples. For example, sometimes a person provides data in a number of different conditions. In this section, I introduce the ANOVA you use when you have more than two matched samples.

This type of ANOVA is called *repeated measures*. You'll see it called other names, too, like *randomized blocks* or *within subjects*.

Working with repeated measures ANOVA

To show how this works, I extend the example from Chapter 11. In that example, ten people participate in a weight-loss program. Table 12-3 shows their data over a three-month period.

Table 12-3		Data for the Weight-Loss Example			
Person	*Before*	*One Month*	*Two Months*	*Three Months*	*Mean*
1	198	194	191	188	192.75
2	201	203	200	196	200.00
3	210	200	192	188	197.50
4	185	183	180	178	181.50
5	204	200	195	191	197.50
6	156	153	150	145	151.00
7	167	166	167	166	166.50
8	197	197	195	192	195.25
9	220	215	209	205	212.25
10	186	184	179	175	181.00
Mean	192.4	189.5	185.8	182.4	187.525

Is the program effective? This question calls for a hypothesis test:

H_0: $\mu_{Before} = \mu_1 = \mu_2 = \mu_3$

H_1: Not H_0

Once again, I set $\alpha = .05$

As in the previous ANOVA, start with the variances in the data. The MS_T is the variance in all 40 scores from the grand mean, which is 187.525:

$$MS_T = \frac{(198 - 187.525)^2 + (201 - 187.525)^2 + \ldots + (175 - 187.525)^2}{40 - 1} = 318.20$$

The people participating in the weight-loss program also supply variance. Each one's overall mean (his or her average over the four measurements) varies from the grand mean. Because these data are in the rows, I call this MS_{Rows}:

$$MS_{Rows} = \frac{(192.75 - 187.525)^2 + (200 - 187.525)^2 + \ldots + (181 - 187.525)^2}{10 - 1} = 1292.41$$

The means of the columns also vary from the grand mean:

$$MS_{Columns} = \frac{(192.4 - 187.525)^2 + (189.5 - 187.525)^2 + (185.8 - 187.525)^2 + (182.4 - 187.525)^2}{4 - 1} = 189.69$$

One more source of variance is in the data. Think of it as the variance left over after you pull out the variance in the rows and the variance in the columns from the total variance. Actually, it's more correct to say it's based on the Sum of Squares left over when you subtract the SS_{Rows} and the $SS_{Columns}$ from the SS_T.

This variance is called MS_{Error}. As I said earlier, in the ANOVA the denominator of an F is called an *error term*. So the word "error" here gives you a hint that this MS is a denominator for an F.

To calculate MS_{Error}, you use the relationships among the Sums of Squares and among the df.

$$MS_{Error} = \frac{SS_{Error}}{df_{Error}} = \frac{SS_T - SS_{Rows} - SS_{Columns}}{df_T - df_{Rows} - df_{Columns}} = \frac{209.175}{27} = 7.75$$

Here's another way to calculate the df_{Error}:

df_{Error} = (number of rows -1)(number of columns -1)

To perform the hypothesis test, you calculate the F:

$$F = \frac{MS_{Column}}{MS_{Error}} = \frac{189.69}{7.75} = 24.49$$

With 3 and 27 degrees of freedom, the critical F for $\alpha = .05$ is 2.96. (Look it up, or use the Excel worksheet function FINV.) The calculated F is larger than the critical F, so the decision is to reject H_0.

What about an F involving MS_{Rows}? That one doesn't figure into H_0 for this example. If you find a significant F, all it shows is that people are different from one another with respect to weight and that doesn't tell you very much.

As is the case with the ANOVA I showed you before, you plan comparisons in order to zero in on the differences. You can use the same formula, except you substitute MS_{Error} for MS_W:

$$t = \frac{c_1 \overline{X}_1 + c_2 \overline{X}_2 + c_3 \overline{X}_3 + c_4 \overline{X}_4}{\sqrt{MS_{Error}\left[\dfrac{c_1^2}{n_1} + \dfrac{c_2^2}{n_2} + \dfrac{c_3^2}{n_3} + \dfrac{c_4^2}{n_4}\right]}}$$

The df for this test is df_{Error}.

For Scheffé's post hoc test, you also follow the same procedure as before and substitute MS_{Error} for MS_W. The only other change is to substitute $df_{Columns}$ for df_B and substitute df_{Error} for df_W when you find F'.

Getting trendy

In situations like the one in the weight-loss example, you have an independent variable that's quantitative — its levels are numbers (0 months, 1 month, 2 months, 3 months). Not only that, but in this case, the intervals are equal.

With that kind of an independent variable, it's often a good idea to look for trends in the data, rather than just plan comparisons among means. If you graph the means in the weight-loss example, they seem to approximate a line, as shown in Figure 12-6. *Trend analysis* is the statistical procedure that examines that pattern. The objective is to see if the pattern contributes to the significant differences among the means.

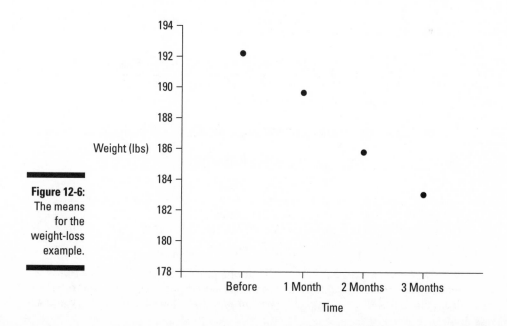

Figure 12-6: The means for the weight-loss example.

A trend can be linear, as it apparently is in this example, or nonlinear (in which the means fall on a curve). In this example, I only deal with linear trend.

To analyze a trend, you use comparison coefficients — those numbers you use in planned comparisons. You use them a bit differently from the way you did before.

Here, you use comparison coefficients to find a Sum of Squares for linear trend. I abbreviate that as SS_{Linear}. This is a portion of $SS_{Columns}$. In fact,

$$SS_{Linear} + SS_{Nonlinear} = SS_{Columns}$$

Also,

$$df_{Linear} + df_{Nonlinear} = df_{Columns}$$

After you calculate SS_{Linear}, you divide it by df_{Linear} to produce MS_{Linear}. This is extremely easy because $df_{Linear} = 1$. Divide MS_{Linear} by MS_{Error} and you have an F. If that F is higher than the critical value of F with df = 1 and df_{Error} at your α-level, then weight is decreasing in a linear way over the time period of the weight-loss program.

The comparison coefficients are different for different numbers of samples. For four samples, the coefficients are -3, -1, 1, and 3. To form the SS_{Linear} the formula is

$$SS_{Linear} = \frac{n\left(\sum c\bar{x}\right)^2}{\sum c^2}$$

In this formula, n is the number of people and c represents the coefficients. Applying the formula to this example,

$$SS_{Linear} = \frac{n\left(\sum c\bar{x}\right)^2}{\sum c^2} = \frac{10\left[(-3)(192.4) + (-1)(189.5) + (1)(185.8) + (3)(182.4)\right]^2}{(-3)^2 + (-1)^2 + (3)^2 + (1)^2}$$
$$= 567.845$$

This is such a large proportion of $SS_{Columns}$ that $SS_{Nonlinear}$ is really small:

$$SS_{Nonlinear} = SS_{Columns} - SS_{Linear} = 569.075 - 567.845 = 1.23$$

As I pointed out before, df = 1 so MS_{Linear} is conveniently the same as SS_{Linear}.

Finally,

$$F = \frac{MS_{Linear}}{MS_{Error}} = \frac{567.85}{7.75} = 73.30$$

A little more on trend

The coefficients I showed you represent one possible component of what underlies the differences among the four means in the example — the linear component. With four means, it's also possible to have other components. I lumped those other components together into a category I called nonlinear. Now I discuss them explicitly.

One possibility is that four means can differ from one another and form a trend that looks like a curve, as in the next figure.

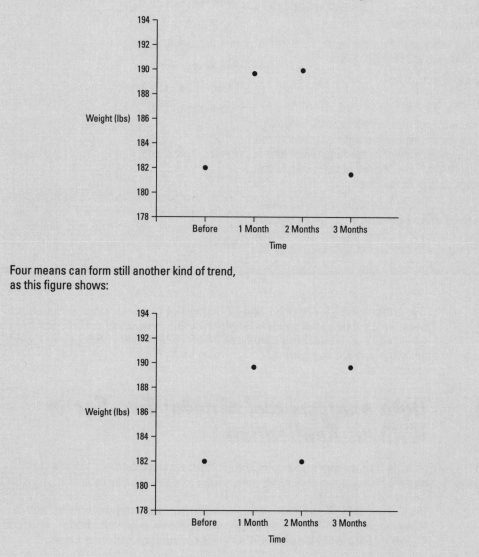

Four means can form still another kind of trend, as this figure shows:

(continued)

(continued)

The first kind, where the trend changes direction once, is called a *quadratic* component. In the first figure it increases, then it decreases. The second, where the trend changes direction twice, is called a *cubic* component. In the second figure it increases, decreases, and then increases again. In Figure 12-6, the trend is linear and doesn't change direction (it just keeps decreasing).

Quadratic and cubic components have coefficients, too, and here they are:

Quadratic: 1, -1, -1, 1

Cubic: -1, 3, -3, 1

You test for these components the same way you test for the linear component. A trend can be a combination of components: If you have a significant F, one or more of these trend components may be significant.

Linear, quadratic, and cubic are as far as you can go with four means. With five means, you can look for those three plus a *quartic* component (three direction changes), and with six you

can try and scope out all of the preceding plus a *quintic* component (four direction changes). What do the coefficients look like?

For five means, they're:

Linear: -2, -1, 0 , 1, 2

Quadratic: 2, -1, -2, -1, 2

Cubic: -1, 2, 0, -2, 1

Quartic: 1, -4, 6, -4, 1

And for six means:

Linear: -5, -3, -1 , 1, 3, 5

Quadratic: 5, -1, -4, -4. -1, 5

Cubic: -5, 7, 4, -4, -7, 5

Quartic: 1, -3, 2, 2, -3, 1

Quintic: -1, 5, -10, 10, -5, 1

I could go on with more means, coefficents, and exotic component names (hextic? septic?), but enough already. This should hold you for a while.

The critical value for *F* with 1 and 27 degrees of freedom and α = .05 is 4.21. Because the calculated value is larger than the critical value, statisticians would say the data show a *significant linear component*. This, of course, verifies what you see in Figure 12-6.

Data analysis tool: Anova: Two-Factor Without Replication

Huh? Is that a misprint? *Two-Factor???* Without Replication?? What's that all about?

Here's the story: If you're looking through the data analysis tools for something like *Anova: Single Factor Repeated Measures*, you won't find it. The tool you're looking for is there, but it's hiding out under a different name.

Figure 12-7 shows this tool's dialog box along with the data for the weight-loss example I just went through.

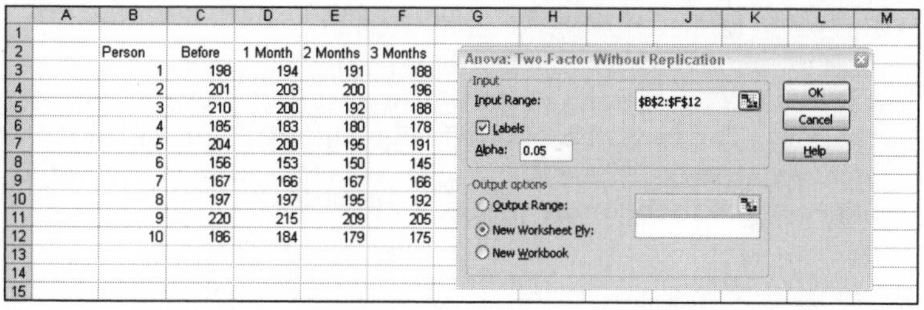

Figure 12-7:
The Anova:
Two-Factor
Without
Replication
data
analysis tool
dialog box.

The steps for using this tool are:

1. **Type the data for each sample into a separate data array. Put the label for each person in a data array.**

   For this example the labels for Person are in column B. The data in the Before sample are in column C, the data in the 1 Month sample are in column D, the data for the 2 Month sample are in column E, and the data for the 3 Month sample are in column F.

2. **From the Tools menu, select Data Analysis to open the Data Analysis dialog box.**

3. **In the Data Analysis dialog box, scroll down the Analysis Tools list and select Anova: Two-Factor Without Replication. Click OK to open the Anova: Two-Factor Without Replication dialog box.**

   This is the dialog box in Figure 12-7.

4. **In the Input Range box, type the cell range that holds all the data.**

   For the example, the data are in $B$2:$F$12. Note the dollar signs (\$) for absolute referencing. Note also — and this is important — the Person column is part of the data.

5. **If the cell ranges include column headings, select the Labels option.**

   I included the headings in the ranges, so I checked the box.

6. **The Alpha box has 0.05 as a default.** Change that value if you want a different α.

7. **In the Output Options, select an option to indicate where you want the results.**

   I selected New Worksheet Ply to put the results on a new page in the worksheet.

8. **Click OK.**

   Because I selected New Worksheet Ply, a newly created page opens with the results.

Figure 12-8 shows the tool's output after I expanded the columns. The output features two tables, SUMMARY and ANOVA.

The SUMMARY table is in two parts. The first part provides summary statistics for the rows. The second part provides summary statistics for the columns. Summary statistics include the number of scores in each row and in each column along with the sums, means, and variances.

The ANOVA table presents the Sums of Squares, df, Mean Squares, F, P-values, and critical F-ratios for the indicated df. The table features two values for F. One F is for the rows, the other for the columns. The P-value is the proportion of area that the F cuts off in the upper tail of the F-distribution. If this value is less than .05, reject H_0.

Although the ANOVA table includes an F for the rows, this doesn't concern you in this case, because H_0 is only about the columns in the data. Each row represents the data for one person. A high F just implies that people are different from one another, and that's not news.

	A	B	C	D	E	F	G
1	Anova: Two-Factor Without Replication						
2							
3	SUMMARY	Count	Sum	Average	Variance		
4	1	4	771	192.75	18.25		
5	2	4	800	200	8.666667		
6	3	4	790	197.5	94.33333		
7	4	4	726	181.5	9.666667		
8	5	4	790	197.5	32.33333		
9	6	4	604	151	22		
10	7	4	666	166.5	0.333333		
11	8	4	781	195.25	5.583333		
12	9	4	849	212.25	43.58333		
13	10	4	724	181	24.66667		
14							
15	Before	10	1924	192.4	377.6		
16	1 Month	10	1895	189.5	342.9444		
17	2 Months	10	1858	185.8	298.8444		
18	3 Months	10	1824	182.4	296.2667		
19							
20							
21	ANOVA						
22	Source of Variation	SS	df	MS	F	P-value	F crit
23	Rows	11631.73	9	1292.414	166.8229	2.71E-21	2.250131
24	Columns	569.075	3	189.6917	24.48512	7.3E-08	2.960351
25	Error	209.175	27	7.747222			
26							
27	Total	12409.98	39	318.2045			
28							

Figure 12-8: Output from the Anova: Two-Factor Without Replication data analysis tool.

Analyzing trend

Excel's Anova: Two-Factor Without Replication tool does not provide a way for performing a trend analysis. As with the planned comparisons, a little ingenuity takes you a long way. The Excel worksheet functions SUMPRODUCT and SUMSQ help with the calculations.

The worksheet page with the ANOVA output gives the information you need to get started. In this section, I take you through the analysis of linear trend.

I start by putting the comparison coefficients for linear trend into J15 through J18, as Figure 12-9 shows.

	J22		f_x =B15*SUMPRODUCT(J15:J18,D15:D18)^2							
	A	B	C	D	E	F	G	H	I	J
1	Anova: Two-Factor Without Replication									
2										
3	SUMMARY	Count	Sum	Average	Variance					
4	1	4	771	192.75	18.25					
5	2	4	800	200	8.666667					
6	3	4	790	197.5	94.33333					
7	4	4	726	181.5	9.666667					
8	5	4	790	197.5	32.33333					
9	6	4	604	151	22					
10	7	4	666	166.5	0.333333					
11	8	4	781	195.25	5.583333					
12	9	4	849	212.25	43.58333					
13	10	4	724	181	24.66667					
14										coefficients
15	Before	10	1924	192.4	377.6					-3
16	1 Month	10	1895	189.5	342.9444					-1
17	2 Months	10	1858	185.8	298.8444					1
18	3 Months	10	1824	182.4	296.2667					3
19										
20										
21	ANOVA									
22	Source of Variation	SS	df	MS	F	P-value	F crit		Numerator	11356.9
23	Rows	11631.73	9	1292.414	166.8229	2.7098E-21	2.250131		Denominator	20
24	Columns	569.075	3	189.6917	24.46512	7.3047E-08	2.960351		SS linear =	567.845
25	Linear	567.845	1	567.845	73.29659	3.5565E-09	4.210008			
26	Nonlinear	1.23	2	0.615	0.079383	0.3081547	3.354131			
27	Error	209.175	27	7.747222						
28										
29	Total	12409.98	39	318.2045						
30										

Figure 12-9: Carrying out a trend analysis.

In J22 through J24, I put information related to SS_{Linear} — the numerator, the denominator, and the value of the Sum of Squares. I use separate cells for the numerator and denominator to simplify the formulas.

As I pointed out before, SUMPRODUCT takes arrays of cells, multiplies the numbers in the corresponding cells, and sums the products. (This function is in the Math & Trig category, not the Statistical category.) I use SUMPRODUCT to multiply each coefficient by each sample mean and then add the products. I store that result in J22. That's the numerator for the SS_{Linear}. I select J22 so you can see its formula in the Formula Bar:

=B15*SUMPRODUCT(J15:J18,D15:D18)^2

The value in B15 is the number in each column. The array J15:J18 holds the comparison coefficients, and D15:D18 holds the column means.

J23 holds the denominator. Its formula is:

=SUMSQ(J15:J18)

SUMSQ squares the coefficients in J15:J18 and adds them.

J24 holds the value for SS_{Linear}. That's J22 divided by J23.

Figure 12-9 shows that in the ANOVA table I've inserted two rows above the row for Error. One row holds the SS, df, MS, F, P-Value and critical F for Linear, the other holds these values for Nonlinear. $SS_{Nonlinear}$ in B26 is B24-B25.

The F for Linear is D25 divided by D27. The formula for the P-Value in F25 is

=FDIST(E25,C25,C27)

The first argument, E25 is the F. The second and third arguments are the df.

The formula for the critical F in F25 is

=FINV(0.05,C25,C27)

The first argument is α, and the second and third are the df.

Chapter 13

Slightly More Complicated Testing

*I*n Chapter 11, I show you how to test hypotheses with two samples. In Chapter 12, I show you how to test hypotheses when you have more than two samples. The common thread through both chapters is that one independent variable (also called a *factor*) is involved.

Many times, you have to test the effects of more than one factor. In this chapter, I show how to analyze two factors within the same set of data. Several types of situations are possible, and I describe Excel data analysis tools that deal with each one.

Cracking the Combinations

FarKlempt Robotics, Inc. manufactures battery-powered robots. They want to test three rechargeable batteries for these robots on a set of three tasks — climbing, walking, and assembling. Which combination of battery and task results in the longest battery life?

They test a sample of nine robots. They randomly assign each robot one battery and one type of task. FarKlempt tracks the number of days each robot works before recharging. The data are in Table 13-1.

Table 13-1 FarKlempt Robots: Number of Days Before Recharging in Three Tasks with Three Batteries

Task	Battery 1	Battery 2	Battery 3	Average
Climbing	12	15	20	15.67
Walking	14	16	19	16.33
Assembling	11	14	18	14.33
Average	12.33	15.00	19.00	15.44

This calls for two hypothesis tests:

H_0: $\mu_{Battery1} = \mu_{Battery2} = \mu_{Battery3}$

H_1: Not H_0

and

H_0: $\mu_{Climbing} = \mu_{Walking} = \mu_{Assembling}$

H_1: Not H_0

In both tests, set $\alpha = .05$.

Breaking down the variances

The appropriate analysis for these tests is an analysis of variance (ANOVA). Each variable — Batteries and Tasks — is also called a *factor*. So this analysis is called a *two-factor ANOVA*.

To understand this ANOVA, consider the variances inside the data. First, focus on the variance in the whole set of nine numbers — MS_T. ("T" in the subscript stands for "Total.") The mean of those numbers is 15.44. Because it's the mean of all the numbers, it goes by the name *grand mean*.

This variance is

$$MS_T = \frac{(12 - 15.44)^2 + (15 - 15.44)^2 + \ldots + (18 - 15.44)^2}{9 - 1} = \frac{76.22}{8} = 9.53$$

The means of the three batteries (the column means) also vary from 15.44. That variance is

$$MS_{Batteries} = \frac{(3)(12.33 - 15.44)^2 + (3)(15.00 - 15.44)^2 + (3)(19.00 - 15.44)^2}{3 - 1}$$

$$= \frac{67.56}{2} = 33.78$$

Why does the 3 appear as a multiplier of each squared deviation? When you deal with means, you have to take into account the number of scores that produce each mean.

Similarly, the means of the tasks (the row means) vary from 15.44:

$$MS_{Tasks} = \frac{(3)(15.67 - 15.44)^2 + (3)(16.33 - 15.44)^2 + (3)(14.33 - 15.44)^2}{3 - 1}$$

$$= \frac{6.22}{2} = 3.11$$

One variance is left. It's called MS_{Error}. This is what remains when you subtract the $SS_{Batteries}$ and the SS_{Tasks} from the SS_T, and divide that by the df that remains when you subtract $df_{Batteries}$ and df_{Tasks} from df_T:

$$MS_{Error} = \frac{SS_T - SS_{Batteries} - SS_{Tasks}}{df_T - df_{Batteries} - df_{Tasks}} = \frac{2.44}{4} = 0.61$$

To test the hypotheses you calculate one F for the effects of the batteries and another for the effects of the tasks. For both, the denominator (the so-called "error term") is MS_{Error}:

$$F = \frac{MS_{Batteries}}{MS_{Error}} = \frac{33.77}{0.61} = 55.27$$

$$F = \frac{MS_{Tasks}}{MS_{Error}} = \frac{2.44}{0.61} = 5.09$$

Each F has 2 and 4 degrees of freedom. With $\alpha = .05$, the critical F in each case is 6.94. The decision is to reject H_0 for the batteries (they differ from one another to an extent greater than chance), but not for the tasks.

To zero in on the differences for the batteries, you carry out planned comparisons among the column means. (See Chapter 12 for the details.)

Data analysis tool: Anova: Two-Factor Without Replication

Excel's Anova: Two-Factor Without Replication tool carries out the analysis I just outlined. (I use this tool for another type of analysis in Chapter 12.) *Without Replication* means that only one robot is assigned to each battery-task combination. If you assign more than one to each combination, that's *replication*.

Figure 13-1 shows this tool's dialog box along with the data for the batteries-tasks example.

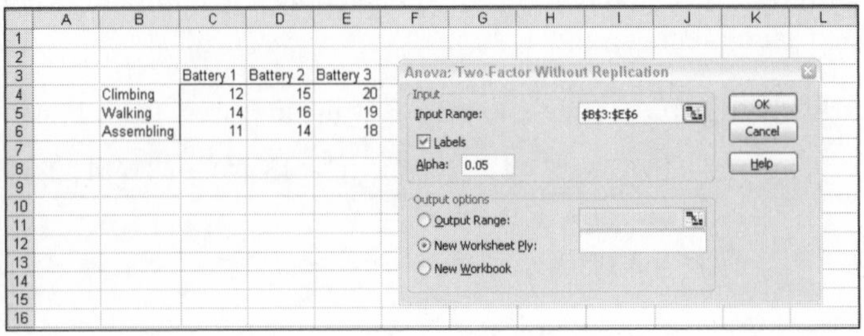

Figure 13-1:
The Anova:
Two-Factor
Without
Replication
data
analysis tool
dialog box
along with
the
batteries-
tasks data.

The steps for using this tool are:

1. **Type the data into the worksheet, and include labels for the rows and columns.**

   For this example the labels for the tasks are in cells B4, B5, and B6. The labels for the batteries are in cells C3, D3, and E3. The data are in cells C4 through E6.

2. **From the Tools menu, select Data Analysis to open the Data Analysis dialog box.**

3. **In the Data Analysis dialog box, scroll down the Analysis Tools list and select Anova: Two-Factor Without Replication. Click OK to open the select Anova: Two-Factor Without Replication dialog box.**

   This is the dialog box in Figure 13-1.

4. **In the Input Range box, type the cell range that holds all the data.**

   For the example, the data are in $B$2:$E$6. Note the dollar signs ($) for absolute referencing. Note also — and this is important — the row labels are part of the data.

5. **If the cell ranges include column headings, select the Labels option.**

   I included the headings in the ranges, so I checked the box.

6. **The Alpha box has 0.05 as a default.** Change that value if you want a different α.

7. **In the Output Options, select an option to indicate where you want the results.**

I selected the New Worksheet Ply option to put the results on a new page in the worksheet.

8. **Click OK.**

Because I selected New Worksheet Ply, a newly created page opens with the results.

Figure 13-2 shows the tool's output after I expanded the columns. The output features two tables, SUMMARY and ANOVA.

The SUMMARY table is in two parts. The first part provides summary statistics for the rows. The second part provides summary statistics for the columns. Summary statistics include the number of scores in each row and in each column along with the sums, means, and variances.

The ANOVA table presents the Sums of Squares, df, Mean Squares, F, P-values, and critical F for the indicated df. The table features two values for F. One F is for the rows, the other for the columns. The P-value is the proportion of area that the F cuts off in the upper tail of the F-distribution. If this value is less than .05, reject H_0.

In this example, the decisions are to reject H_0 for the batteries (the columns) and to not reject H_0 for the tasks (the rows).

Figure 13-2:
Output from the Anova: Two-Factor Without Replication data analysis tool.

	A	B	C	D	E	F	G
1	Anova: Two-Factor Without Replication						
2							
3	*SUMMARY*	*Count*	*Sum*	*Average*	*Variance*		
4	Climbing	3	47	15.66667	16.33333		
5	Walking	3	49	16.33333	6.333333		
6	Assembling	3	43	14.33333	12.33333		
7							
8	Battery 1	3	37	12.33333	2.333333		
9	Battery 2	3	45	15	1		
10	Battery 3	3	57	19	1		
11							
12							
13	ANOVA						
14	*Source of Variation*	*SS*	*df*	*MS*	*F*	*P-value*	*F crit*
15	Rows	6.222222	2	3.111111	5.090909	0.079553	6.944272
16	Columns	67.55556	2	33.77778	55.27273	0.001219	6.944272
17	Error	2.444444	4	0.611111			
18							
19	Total	76.22222	8	9.527778			
20							

Cracking the Combinations Again

The analysis I just showed you involves one score for each combination of the two factors. Assigning one individual to each combination is appropriate for robots and other manufactured objects, where you can assume that one object is pretty much the same as another.

When people are involved, however, it's a different story. Individual variation among humans is something you can't overlook. For this reason, it's necessary to assign a sample of people to a combination of factors — not just one person.

Rows and columns

I illustrate with an example. Imagine that a company has two methods of presenting its training information. One is via a person who gives an oral presentation, the other is via a written text. Imagine also that the information is presented in either a humorous way or in a technical way. I refer to the first factor as Presentation Method and to the second as Presentation Style.

Combining the two levels of Presentation Method with the two levels of Presentation Style gives four combinations. The company randomly assigns four people to each combination, for a total of 16 people. After providing the training, they test the 16 people on their comprehension of the material.

Figure 13-3 shows the combinations, the four comprehension scores within each combination, and summary statistics for the combinations, rows, and columns.

Presentation Style

	Humorous	Technical	
Spoken	Spoken and Humorous 54 55 62 68	Spoken and Technical 22 21 29 25	
Presentation Method	Mean = 57.25 Variance = 12.92	Mean = 24.25 Variance = 12.92	Mean = 40.75
Text	Text and Humorous 33 25 28 31	Test and Technical 66 65 71 72	
	Mean = 29.25 Variance = 12.25	Mean = 68.50 Variance = 12.33	Mean = 48.88
	Mean = 43.25	Mean = 46.38	Grand Mean = 44.81

Figure 13-3: Combining the levels of Presentation Method with the levels of Presentation Style.

Here are the hypotheses:

H_0: $\mu_{Spoken} = \mu_{Text}$

H_1: Not H_0

and

H_0: $\mu_{Humorous} = \mu_{Technical}$

H_1: Not H_0

Because the two presentation methods (Spoken and Text) are in the rows, I refer to Presentation Type as the *row factor*. The two presentation styles (Humorous and Technical) are in the columns, so Presentation Style is the *column factor*.

Interactions

When you have rows and columns of data, and you're testing hypotheses about the row factor and the column factor, you have an additional consideration. Namely, you have to be concerned about the row-column combinations. Do the combinations result in peculiar effects?

For the example I presented, it's possible that combining Spoken and Text with Humorous and Technical yields something unexpected. In fact, you can see that in the data in Figure 13-3: For Spoken presentation, the Humorous style produces a higher average than the Technical style. For Text presentation, the Humorous style produces a lower average than the Technical style.

A situation like that is called an *interaction*. In formal terms, an interaction occurs when the levels of one factor affect the levels of the other factor differently. The label for the interaction is row factor × column factor, so for this example that's Method × Type.

The hypotheses for this are:

H_0: Presentation Method does not interact with Presentation Style

H_1: Not H_0

The analysis

The statistical analysis, once again, is an analysis of variance (ANOVA). As is the case with the other ANOVAs I showed you, it depends on the variances in the data.

The first variance is the total variance, labeled MS_T. That's the variance of all 16 scores around their mean (the *grand mean*), which is 44.81:

$$MS_T = \frac{(54 - 44.81)^2 + (55 - 44.81)^2 + \ldots + (72 - 44.81)^2}{16 - 1} = \frac{5674.44}{15} = 378.30$$

The denominator tells you that df = 15 for MS_T.

The next variance comes from the row factor. That's MS_{Method}, and it's the variance of the row means around the grand mean:

$$MS_{Method} = \frac{(8)(40.75 - 44.81)^2 + (8)(48.88 - 44.81)^2}{2 - 1} = \frac{264.06}{1} = 264.06$$

The 8 multiplies each squared deviation because you have to take into account the number of scores that produce each row mean. The df for MS_{Method} is the number of rows - 1, which is 1.

Similarly, the variance for the column factor is

$$MS_{Style} = \frac{(8)(43.25 - 44.81)^2 + (8)(46.38 - 44.81)^2}{2 - 1} = \frac{39.06}{1} = 39.06$$

The df for MS_{Style} is 1 (the number of columns - 1).

Another variance is the pooled estimate based on the variances within the four row-column combinations. It's called the MS_{Within}, or MS_W. (For details on MS_w and pooled estimates, see Chapter 12.). For this example,

$$MS_W = \frac{(4-1)(12.92) + (4-1)(12.92) + (4-1)(12.25) + (4-1)(12.33)}{(4-1) + (4-1) + (4-1) + (4-1)}$$

$$= \frac{151.25}{12} = 12.60$$

This one is the error term (the denominator) for each *F* you calculate. Its denominator tells you that df = 12 for this MS.

The last variance comes from the interaction between the row factor and the column factor. In this example, it's labeled $MS_{Method \times Type}$. You can calculate this a couple of ways. The easiest way is to take advantage of this general relationship:

$$SS_{Row \times Column} = SS_T - SS_{Row\ Factor} - SS_{Column\ Factor} - SS_W$$

And this one:

$$df_{Row \times Column} = df_T - df_{Row\ Factor} - df_{Column\ Factor} - df_W$$

Another way to calculate this is

$$df_{Row \times Column} = (\text{number of rows } -1)(\text{number of columns } - 1)$$

The MS is

$$MS_{Row \times Column} = \frac{SS_{Row \times Column}}{df_{Row \times Column}}$$

For this example,

$$MS_{Method \times Style} = \frac{SS_{Method \times Style}}{df_{Method \times Style}} = \frac{5764.44 - 264.06 - 39.06 - 151.25}{15 - 12 - 1 - 1}$$

$$= \frac{5220.06}{1} = 5220.06$$

To test the hypotheses, you calculate three Fs:

$$F = \frac{MS_{Style}}{MS_W} = \frac{39.06}{12.60} = 3.10$$

$$F = \frac{MS_{Method}}{MS_W} = \frac{264.06}{12.60} = 20.95$$

$$F = \frac{MS_{Method \times Style}}{MS_W} = \frac{5220.06}{12.60} = 414.15$$

For df = 1 and 12, the critical F at $\alpha = .05$ is 4.75. (You can use the Excel function FINV to verify.). The decision is to reject H_0 for the Presentation Method and for the Method X Style interaction, and to not reject H_0 for the Presentation Style.

Data analysis tool: Anova: Two-Factor With Replication

Excel provides a data analysis tool that handles everything. This one is called Anova: Two-Factor With Replication. "Replication" means you have more than one score in each row-column combination.

Figure 13-4 shows this tool's dialog box along with the data for the batteries-tasks example.

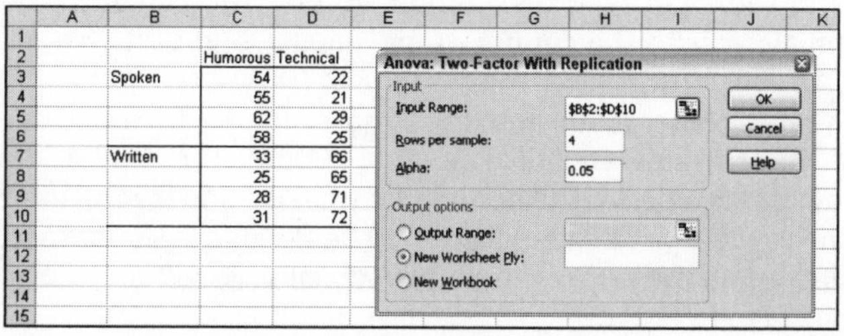

Figure 13-4:
The Anova:
Two-Factor
With
Replication
data
analysis tool
dialog box
along with
the type-
method
data.

The steps for using this tool are:

1. **Type the data into the worksheet, and include labels for the rows and columns.**

   For this example the labels for the presentation methods are in cells B3 and B7. The presentation types are in cells C2 and D2. The data are in cells C3 through D10.

2. **From the Tools menu, select Data Analysis to open the Data Analysis dialog box.**

3. **In the Data Analysis dialog box, scroll down the Analysis Tools list and select Anova: Two-Factor With Replication. Click OK to open the Anova: Two-Factor With Replication dialog box.**

   This is the dialog box in Figure 13-4.

4. **In the Input Range box, type the cell range that holds all the data.**

   For the example, the data are in $B$2:$D$10. Note the dollar signs ($) for absolute referencing. Note also — this is important — the labels for the row factor (presentation method) are part of the data range.

5. **The Alpha box has 0.05 as a default.** Change that value if you want a different α.

6. **In the Output Options, select an option to indicate where you want the results.**

   I selected New Worksheet Ply option to put the results on a new page in the worksheet.

7. **Click OK.**

   Because I selected the New Worksheet Ply option, a newly created page opens with the results.

Figure 13-5 shows the tool's output after I expanded the columns. The output features two tables, SUMMARY and ANOVA.

The SUMMARY table is in two parts. The first part provides summary statistics for the factor combinations and for the row factor. The second part provides summary statistics for the column factor. Summary statistics include the number of scores in each row-column combination, in each row, and in each column along with the counts, sums, means, and variances.

The ANOVA table presents the Sums of Squares, df, Mean Squares, F, P-values, and critical F for the indicated df. The table features three values for F. One F is for the row factor, one for the column factor, and one for the interaction. In the table, the row factor is called Sample. The P-value is the proportion of area that the F cuts off in the upper tail of the F-distribution. If this value is less than .05, reject H_0.

	A	B	C	D	E	F	G
1	Anova: Two-Factor With Replication						
2							
3	SUMMARY	Humorous	Technical	Total			
4	*Spoken*						
5	Count	4	4	8			
6	Sum	229	97	326			
7	Average	57.25	24.25	40.75			
8	Variance	12.91667	12.91667	322.2143			
9							
10	*Written*						
11	Count	4	4	8			
12	Sum	117	274	391			
13	Average	29.25	68.5	48.875			
14	Variance	12.25	12.33333	450.6964			
15							
16	*Total*						
17	Count	8	8				
18	Sum	346	371				
19	Average	43.25	46.375				
20	Variance	234.7857	570.2679				
21							
22							
23	ANOVA						
24	*Source of Variation*	SS	df	MS	F	P-value	F crit
25	Sample	264.0625	1	264.0625	20.95041	0.000636	4.747225
26	Columns	39.0625	1	39.0625	3.099174	0.103771	4.747225
27	Interaction	5220.063	1	5220.063	414.1537	1.14E-10	4.747225
28	Within	151.25	12	12.60417			
29							
30	Total	5674.438	15				
31							

Figure 13-5:
Output from
the Anova:
Two-Factor
Without
Replication
data
analysis
tool.

In this example, the decisions are to reject H_0 for the Presentation Method (the row factor labeled Sample in the table), to not reject H_0 for the Presentation Style (the column factor), and to reject H_0 for the interaction.

Chapter 14

Regression: Linear and Multiple

*O*ne of the main things you do when you work with statistics is make predictions. The idea is to take data on one or more variables, and use these data to predict a value of another variable. To do this, you have to understand how to summarize relationships among variables, and to test hypotheses about those relationships.

In this chapter, I introduce *regression*, a statistical way to do just that. Regression also enables you to use the details of relationships to make predictions. First, I show you how to analyze the relationship between one variable and another. Then I show you how to analyze the relationship between a variable and two others. These analyses involve a good bit of calculation, and Excel is more than equal to the task.

The Plot of Scatter

Sahutsket University is an exciting, dynamic institution. Every year, the school receives thousands of applications. One challenge the admissions office faces is this: Applicants want the office to predict what their GPAs (grade-point averages on a 4.0 scale) will be if they attend Sahutsket.

What's the best prediction? Without knowing anything about an applicant, and only knowing its own students' GPAs, the answer is clear: It's the average GPA at Sahutsket U. Regardless of who the applicant is, that's all the admissions office can say if its knowledge is limited.

With more knowledge about the students and about the applicants, a more accurate prediction becomes possible. For example, if Sahutsket keeps records on its students' SAT scores (Verbal and Math combined), the admissions office can match up each student's GPA with his or her SAT score and see if the two pieces of data are somehow related. If they are, an applicant can supply his or her SAT score, and the admissions office can use that score to help make a prediction.

Figure 14-1 shows the GPA-SAT matchup in a graphic way. Because the points are scattered, it's called a *scatterplot*. By convention, the vertical axis (the *y-axis*) represents what you're trying to predict. That's also called the *dependent variable* or the *y-variable*. In this case, that's GPA. Also by convention, the horizontal axis (the *x-axis*) represents what you're using to make your prediction. That's also called the *independent variable* or the *x-variable*. Here, that's SAT.

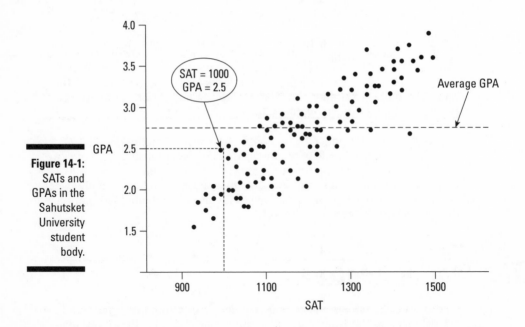

Figure 14-1:
SATs and GPAs in the Sahutsket University student body.

Each point in the graph represents an individual student's GPA and SAT. In a real scatterplot of a university student body, you'd see many more points than I show here. The general tendency of the set of points seems to be that high SAT scores are associated with high GPAs and low SAT scores are associated with low GPAs.

I singled out one of the points. It shows a Sahutsket student with an SAT score of 1000 and a GPA of 2.5. I also show the average GPA to give you a sense that knowing the GPA-SAT relationship provides an advantage over just knowing the mean.

How do you make that advantage work for you? You start by summarizing the relationship between SAT and GPA. The summary is a line through the points. How and where do you draw the line?

I'll get to that in a minute. First, I have to tell you about lines in general.

Graphing Lines

In the world of mathematics, a line is a way to picture a relationship between an independent variable (x) and a dependent variable (y). In this relationship

$$y = 4 + 2x$$

every time I supply a value for x, I can figure out the corresponding value for y. The equation says to take the x-value, multiply by 2 and then add 3.

If $x = 1$, for example, $y = 6$. If $x = 2$, $y = 8$. Table 14-1 shows a number of x-y pairs in this relationship, including the pair in which $x = 0$.

Table 14-1	x-y Pairs in $y = 4 + 2x$
x	**y**
0	4
1	6
2	8
3	10
4	12
5	14
6	16

Figure 14-2 shows these pairs as points on a set of x-y axes, along with a line through the points. Each time I list an x-y pair in parentheses, the x-value is first.

As the figure shows, the points fall neatly onto the line. The line *graphs* the equation $y = 4 + 2x$. In fact, whenever you have an equation like this, where x isn't squared or cubed or raised to any power higher than 1, you have what mathematicians call a *linear* equation. (If x is raised to a higher power than one, you connect the points with a curve, not a line.)

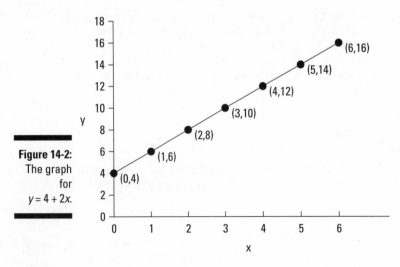

A couple of things to keep in mind about a line: You can describe a line in terms of how slanted it is, and where it runs into the y-axis.

The how-slanted-it-is part is called the *slope*. The slope tells you how much *y* changes when *x* changes by 1 unit. In the line in Figure 14-2, when *x* changes by 1 (from 4 to 5, for example) *y* changes by 2 (from 12 to 14).

The where-it-runs-into-the-y-axis part is called the *y-intercept* (or sometimes just the *intercept*). That's the value of *y* when *x* = 0. In Figure 14-2 the *y*-intercept is 4.

You can see these numbers in the equation. The slope is the number that multiplies *x* and the intercept is the number you add to *x*. In general,

$y = a + bx$

where *a* represents the intercept and *b* represents the slope.

The slope can be a positive number, a negative number, or zero. In Figure 14-2, the slope is positive. If the slope is negative, the line is slanted in a direction opposite to what you see in Figure 14-2. A negative slope means that *y decreases* as *x* increases. If the slope is zero, the line is parallel to the horizontal axis. If the slope is zero, *y* doesn't change as *x* changes.

The same applies to the intercept — it can be a positive number, a negative number, or zero. If the intercept is positive, the line cuts off the *y*-axis *above* the *x*-axis. If the intercept is negative, the line cuts off the *y*-axis *below* the *x*-axis. If the intercept is zero, it intersects with the *y*-axis and the *x*-axis at the point called the *origin*.

And now, back to what I was originally talking about.

Regression: What a Line!

Before I began telling you about lines, equations, slopes, and intercepts, I mentioned that a line is the best way to summarize the relationship in the scatterplot in Figure 14-1. It's possible to draw an infinite amount of straight lines through the scatterplot. Which one best summarizes the relationship?

Intuitively, the "best fitting" line ought to be the one that goes through the maximum number of points and isn't too far away from the others. For statisticians, that line has a special property: If you draw that line through the scatterplot, then draw distances (in the vertical direction) between the points and the line, and then square those distances and add them up, the sum of the squared distances is a minimum.

Statisticians call this line the *regression line*, and indicate it as

$$y' = a + bx$$

Each y' is on the line. It represents the best prediction of y for a given value of x.

To figure out exactly where this line is, you calculate its slope and its intercept. For a regression line, the slope and intercept are called *regression coefficients*.

The formulas for the regression coefficients are pretty straightforward. For the slope, the formula is

$$b = \frac{\sum (x - \bar{x})(y - \bar{y})}{\sum (x - \bar{x})^2}$$

The intercept formula is

$$a = \bar{y} - b\bar{x}$$

I illustrate with an example. To keep the numbers manageable and comprehensible, I use a small sample rather than the thousands of students you'd find in a scatterplot of an entire university student body. Table 14-2 shows a sample of data from 20 Sahutsket University students.

Table 14-2	SAT Scores and GPAs for 20 Sahutsket University Students	
Student	*SAT*	*GPA*
1	990	2.2
2	1150	3.2
3	1080	2.6

(continued)

Table 14-2 *(continued)*

Student	SAT	GPA
4	1100	3.3
5	1280	3.8
6	990	2.2
7	1110	3.2
8	920	2.0
9	1000	2.2
10	1200	3.6
11	1000	2.1
12	1150	2.8
13	1070	2.2
14	1120	2.1
15	1250	2.4
16	1020	2.2
17	1060	2.3
18	1550	3.9
19	1480	3.8
20	1010	2.0
Mean	1126.5	2.705
Variance	26171.32	0.46
Standard Deviation	161.78	0.82

For this set of data, the slope of the regression line is

$$b = \frac{(990 - 1126.5)(2.2 - 2.705) + (1150 - 1126.5)(3.2 - 2.705) + \ldots + (1010 - 1126.5)(2.0 - 2.705)}{(2.2 - 2.705)^2 + (3.2 - 2.705)^2 + \ldots + (2.0 - 2.705)^2} = 0.0034$$

The intercept is

$$a = \bar{y} - b\bar{x} = 2.705 - 0.0034(1126.5) = -1.1538$$

So the equation of the best fitting line through these 20 points is

$$y' = -1.1538 + 0.0034x$$

or in terms of GPAs and SATs

Predicted GPA = -1.1538 + 0.0034(*SAT*)

Using regression for forecasting

Based on this sample and this regression line, you can take an applicant's SAT score, say 1230, and predict the applicant's GPA:

Predicted GPA = -1.1538 + 0.0034(1230) = 3.028

Without this rule the only prediction is the mean GPA, 2.705.

Variation around the regression line

In Chapter 5, I described how the mean doesn't tell the whole story about a set of data. You have to show how the scores vary around the mean. For that reason, I introduced the variance and standard deviation.

You have a similar situation here. To get the full picture of the relationship in a scatterplot, you have to show how the scores vary around the regression line. Here, I introduce the *residual variance* and *standard error of estimate*, which are analogous to the variance and the standard deviation.

The residual variance is sort of an average of the squared deviations of the observed *y* values around the predicted *y*-values. Each deviation of a data point from a predicted point $(y - y')$ is called a *residual*, hence the name. The formula is

$$s_{yx}^{2} = \frac{\sum (y - y')^{2}}{N - 2}$$

I said "sort of" because the denominator is *N*-2, rather than *N*. The reason for the -2 is beyond our scope. As I've said before, the denominator of a variance estimate is *degrees of freedom* (df), and that concept comes in handy in a little while.

The standard error of estimate is

$$s_{yx} = \sqrt{s_{yx}^{2}} = \sqrt{\frac{\sum (y - y')^{2}}{N - 2}}$$

To show you how the residual error and the standard error of estimate play out for the data in the example, here's Table 14-3. This table extends Table 14-2 by showing the predicted GPA for each SAT:

Table 14-3	SAT Scores, GPAs, and Predicted GPAs for 20 Sahutsket University Students		
Student	SAT	GPA	Predicted GPA
1	990	2.2	2.24
2	1150	3.2	2.79
3	1080	2.6	2.55
4	1100	3.3	2.61
5	1280	3.8	3.23
6	990	2.2	2.24
7	1110	3.2	2.65
8	920	2.0	2.00
9	1000	2.2	2.27
10	1200	3.6	2.96
11	1000	2.1	2.27
12	1150	2.8	2.79
13	1070	2.2	2.51
14	1120	2.1	2.68
15	1250	2.4	3.13
16	1020	2.2	2.34
17	1060	2.3	2.48
18	1550	3.9	4.16
19	1480	3.8	3.92
20	1010	2.0	2.31
Mean	1126.5	2.705	
Variance	26171.32	0.46	
Standard Deviation	161.78	0.82	

As the table shows, sometimes the predicted GPA is pretty close, sometimes it's not. One predicted value (4.15) is impossible.

For these data, the residual variance is

$$s_{yx}{}^2 = \frac{\sum(y-y')^2}{N-2} = \frac{(2.2-2.24)^2+(3.2-2.79)^2+\ldots+(2.0-2.31)^2}{20-2} = \frac{2.91}{18} = .16$$

The standard error of estimate is

$$s_{yx} = \sqrt{s_{yx}{}^2} = \sqrt{.16} = .40$$

If the residual variance and the standard error of estimate are small, the regression line is a good fit to the data in the scatterplot. If the residual variance and the standard error of estimate are large, the regression line is a poor fit.

What's "small"? What's "large"? What's a "good" fit?

Keep reading.

Testing hypotheses about regression

The regression equation I've been working with

$$y' = a + bx$$

summarizes a relationship in a scatterplot of a sample. The regression coefficients *a* and *b* are sample statistics. You can use these statistics to test hypotheses about population parameters, and that's what I do in this section.

The regression line through the population that produces the sample (like the entire Sahutsket University student body, past and present) is the graph of an equation that consists of parameters, rather than statistics. By convention, remember, Greek letters stand for parameters, so the regression equation for the population is

$$y' = a + \beta x + \varepsilon$$

The first two Greek letters on the right are α (alpha) and β (beta), the equivalents of *a* and *b*. What about that last one? It looks something like the Greek equivalent of *e*. What's it doing there?

That last term is the Greek letter *epsilon*. It represents "error" in the population. In a way, error is an unfortunate term. It's a catchall for "things you don't know or things you have no control over." Error is reflected in the residuals — the deviations from the predictions. The more you learn about what you're measuring, the more you decrease the error.

You can't measure the error in the relationship between SAT and GPA, but it's lurking there. Someone might score low on the SAT, for example, and then go on to have a wonderful college career with a higher-than-predicted GPA. On a scatterplot, this person's SAT-GPA point looks like an error in prediction. As you learn more about that person, you might discover that he or she was sick on the day of the SAT, and that explains the "error."

You can test hypotheses about α, β, and ε, and that's what I do in the upcoming subsections.

Testing the fit

I begin with a test of how well the regression line fits the scatterplot. This is a test of ε, the error in the relationship.

The objective is to decide whether or not the line really does represent a relationship between the variables. It's possible that what looks like a relationship is just due to chance and the equation of the regression line doesn't mean anything (because the amount of error is overwhelming) — or it's possible that the variables are strongly related.

These possibilities are testable, and you set up hypotheses to test them:

H_0: No real relationship

H_1: Not H_0

While those hypotheses make nice light reading, they don't set up a statistical test. To set up the test, you have to consider the variances. To consider the variances, you start with the deviations. Figure 14-3 focuses on one point in a scatterplot and its deviation from the regression line (the residual) and from the mean of the y-variable. It also shows the deviation between the regression line and the mean.

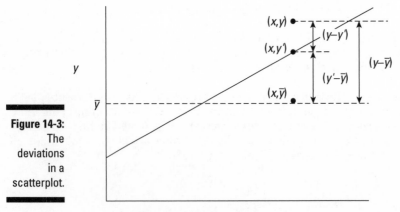

Figure 14-3:
The deviations in a scatterplot.

As the figure shows, the distance between the point and the regression line and the distance between the regression line and the mean add up to the distance between the point and the mean:

$$\left(y - y' \right) + \left(y' - \overline{y} \right) = \left(y - \overline{y} \right)$$

This sets the stage for some other important relationships.

Start by squaring each deviation. That gives you $\left(y - y' \right)^2$, $\left(y' - \overline{y} \right)^2$, and $\left(y - \overline{y} \right)^2$. If you add up each of the squared deviations, you have

- $\sum \left(y - y' \right)^2$

  I just showed you this one. That's the numerator for the residual variance. It represents the variability around the regression line — the "error" I spoke of earlier. In the terminology of Chapter 12, the numerator of a variance is called a Sum of Squares, or SS. So this is $SS_{Residual}$.

- $\sum \left(y' - \overline{y} \right)^2$

  This one is new. The deviation $\left(y' - \overline{y} \right)$ represents the gain in prediction due to using the regression line rather than the mean. The sum reflects this gain, and is called $SS_{Regression}$.

- $\sum \left(y - \overline{y} \right)^2$

  I showed you this one in Chapter 5 — although I used x instead of y. That's the numerator of the variance of y. In Chapter 12 terms, it's the numerator of *total variance*. This one is SS_{Total}.

This relationship holds among these three sums:

$$SS_{Residual} + SS_{Regression} = SS_{Total}$$

Each one is associated with a value for degrees of freedom — the denominator of a variance estimate. As I pointed out in the previous section, the denominator for $SS_{Residual}$ is N-2. The df for SS_{Total} is N-1 (see Chapter 5 and Chapter 12). As with the SS, the degrees of freedom add up:

$$df_{Regression} + df_{Residual} = df_{Total}$$

This leaves one degree of freedom for Regression.

Where is this all headed, and what does it have to do with hypothesis testing? Well, since you asked, you get variance estimates by dividing SS by df. Each variance estimate is called a *Mean Square*, abbreviated MS (see Chapter 12):

$$MS_{Regression} = \frac{SS_{Regression}}{df_{Regression}}$$

$$MS_{Residual} = \frac{SS_{Residual}}{df_{Residual}}$$

$$MS_{Total} = \frac{SS_{Total}}{df_{Total}}$$

Now for the hypothesis part. If H_0 is true and what looks like a relationship between x and y is really no big deal, the piece that represents the gain in prediction because of the regression line ($MS_{Regression}$) should be no greater than the variability around the regression line ($MS_{Residual}$). If H_0 is not true, and the gain in prediction is substantial, then $MS_{Regression}$ should be a lot bigger than $MS_{Residual}$.

So the hypotheses now set up as

H_0: $\sigma^2_{Regression} \le \sigma^2_{Residual}$

H_1: $\sigma^2_{Regression} > \sigma^2_{Residual}$

These are hypotheses you can test. How? To test a hypothesis about two variances, you use an F-test (see Chapter 11). The test statistic here is

$$F = \frac{MS_{Regression}}{MS_{Residual}}$$

To show you how it all works, I apply the formulas to the Sahutsket example. The $MS_{Residual}$ is the same as s_{yx2} from the preceding section, and that value is 0.16. The $MS_{Regression}$ is

$$MS_{Regression} = \frac{(2.24 - 2.705)^2 + (2.79 - 2.705)^2 + \ldots + (2.31 - 2.705)^2}{1} = 5.83$$

This sets up the F:

$$F = \frac{MS_{Regression}}{MS_{Residual}} = \frac{5.83}{0.16} = 36.03$$

With 1 and 18 df and $\alpha = .05$, the critical value of F is 4.41. (You can use the worksheet function FINV to verify.) The calculated F is greater than the critical F, so the decision is to reject H_0. That means the regression line provides a good fit to the data in the sample.

Testing the slope

Another question that arises in linear regression is whether or not the slope of the regression line is significantly different from zero. If it's not, the mean is just as good a predictor as the regression line.

The hypotheses for this test are:

H_0: $\beta \le 0$

H_1: $\beta > 0$

The statistical test is t, which I discuss in Chapters 9, 10, and 11 in connection with means. The t-test for the slope is

$$t = \frac{b - \beta}{s_b}$$

with df = N-2. The denominator estimates the standard error of the slope. This term sounds more complicated than it is. The formula is:

$$s_b = \frac{s_{yx}}{s_x\sqrt{N-1}}$$

where s_x is the standard deviation of the x-variable. For the data in the example

$$s_b = \frac{s_{yx}}{s_x\sqrt{(N-1)}} = \frac{0.402}{(161.776)\sqrt{(20-1)}} = .00057$$

The t-test is

$$t = \frac{b - \beta}{s_b} = \frac{.0034 - 0}{.00057} = 5.96$$

The actual value is 6.00. Rounding s_{yx} and s_b to a manageable number of decimal places before calculating results in 5.96. Either way, this is larger than the critical value of t for 18 df and α = .05 (2.10), so the decision is to reject H_0. This example, by the way, shows why it's important to test hypotheses. The slope, 0.0034 looks like a very small number (possibly because it's a very small number). Still, it's big enough to reject H_0 in this case.

Testing the intercept

For completeness, I include the hypothesis test for the intercept. I doubt you'll have much use for it, but it appears in the output of some of Excel's regression-related capabilities. I want you to understand all aspects of that output (which I tell you about in a little while), so here it is.

The hypotheses are

H_0: $\alpha = 0$

H_1: $\alpha \neq 0$

The test, once again, is a t-test. The formula is

$$t = \frac{a - \alpha}{s_a}$$

The denominator is the estimate of the standard error of the intercept. Without going into detail, the formula for s_a is

$$s_a = \frac{s_{yx}}{s_x\sqrt{\left[\dfrac{1}{N} + \dfrac{\overline{x}^2}{(N-1)s_x^2}\right]}}$$

where s_x is the standard deviation of the x-variable, s_x^2 is the variance of the x-variable, and $\overline{x}^2$ is the squared mean of the x-variable. Applying this formula to the data in the example,

$$s_a = \frac{s_{yx}}{s_x\sqrt{\left[\dfrac{1}{N} + \dfrac{\overline{x}^2}{(N-1)s_x^2}\right]}} = \frac{0.40}{161.78\sqrt{\dfrac{1}{20} + \dfrac{(1126.5)^2}{(20-1)(161.78)^2}}} = 0.649$$

The *t*-test is

$$t = \frac{a - \alpha}{s_a} = \frac{-1.15}{0.649} = -1.78$$

With 18 degrees of freedom, and the probability of a Type I error at .05, the critical *t* is 2.45 for a two-tailed test. It's a two-tailed test because H_1 is that the intercept doesn't equal zero. It doesn't specify whether the intercept is greater than zero or less than zero. Because the calculated value isn't more negative than the negative critical value, the decision is to not reject H_0.

Worksheet Functions for Regression

Excel is a big help for computation-intensive work like linear regression. An assortment of functions and data analysis tools makes life a lot easier. In this section, I concentrate on the worksheet functions and on two array functions.

Figure 14-4 shows the data I use to illustrate each function. The data are GPA and SAT scores for 20 students in the example I showed you earlier. As the figure shows, the SAT scores are in C3:C22 and the GPAs are in D3:D22. The SAT is the *x*-variable and GPA is the *y*-variable.

	A	B	C	D
1				
2		Student	SAT	GPA
3		1	990	2.2
4		2	1150	3.2
5		3	1080	2.6
6		4	1100	3.3
7		5	1280	3.8
8		6	990	2.2
9		7	1110	3.2
10		8	920	2.0
11		9	1000	2.2
12		10	1200	3.6
13		11	1000	2.1
14		12	1150	2.8
15		13	1070	2.2
16		14	1120	2.1
17		15	1250	2.4
18		16	1020	2.2
19		17	1060	2.3
20		18	1550	3.9
21		19	1480	3.8
22		20	1010	2.0
23				

Figure 14-4: Data for the regression-related worksheet functions.

SLOPE, INTERCEPT, and STEYX

These three functions work the same way, so I give a general description and provide details as necessary for each function.

1. **With the data entered, select a cell.**

2. **Click the Insert Function button to open the Insert Function dialog box.**

3. **In the Insert Function dialog box, select a regression function and click OK to open its Function Arguments dialog box.**

   - To calculate the slope of a regression line through the data, select SLOPE.

   - To calculate the intercept, select INTERCEPT.

   - To calculate the standard error of estimate, select STEYX.

Figures 14-5, 14-6, and 14-7 show the Function Arguments dialog box for these three functions.

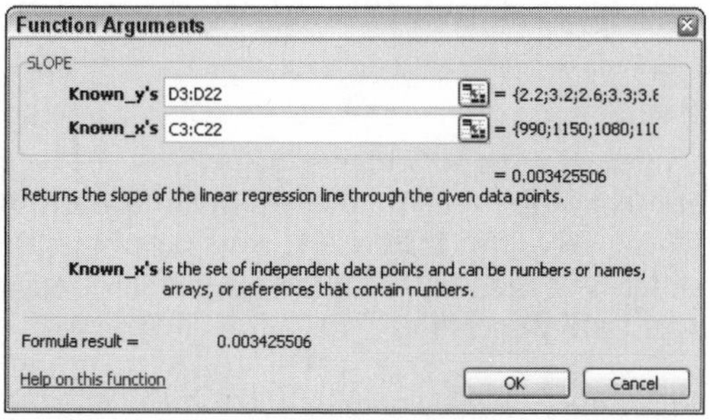

Figure 14-5:
The
Function
Arguments
dialog box
for SLOPE.

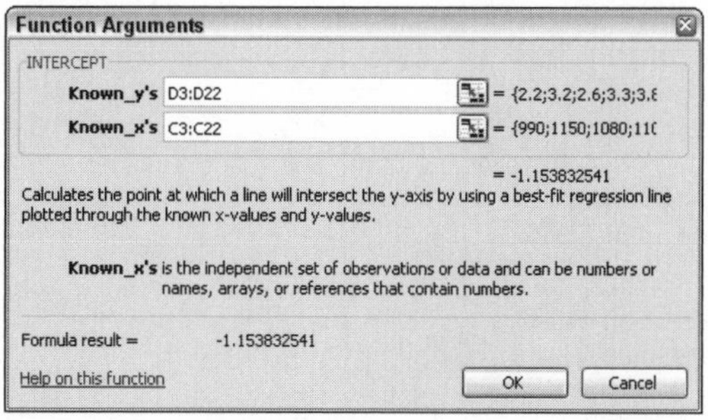

Figure 14-6:
The
Function
Arguments
dialog
box for
INTERCEPT.

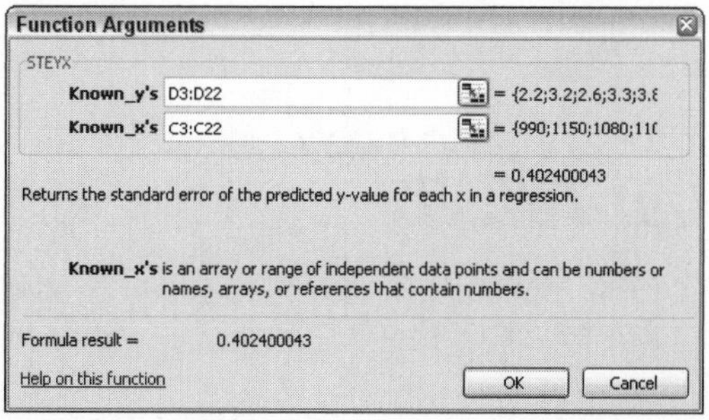

Figure 14-7:
The
Function
Arguments
dialog box
for STEYX.

4. **In the Known_y's box, enter the column that holds the scores for the *y*-variable.**

   For this example, that's C3:C22.

5. **In the Known_x's box, enter the column that holds the scores for the *x*-variable.**

   For this example, it's D3:D22. After you enter this column, the answer appears in the dialog box.

   - SLOPE's answer is .00342556 (Figure 14-5).
   - INTERCEPT's answer is -1.153832541 (Figure 14-6).
   - STEYX's answer is 0.402400043 (Figure 14-7).

6. **Click OK to put the answer into your selected cell.**

FORECAST

This one is a bit different from the preceding three. In addition to the columns for the *x*- and *y*-variables, for FORECAST you supply a value for *x* and the answer is a prediction based on the linear regression relationship between the *x*-variable and the *y*-variable.

Figure 14-8 shows the Function Arguments dialog box for FORECAST. To get to this dialog box follow the steps in the preceding section, except in Step 3 select FORECAST. In the X box, I entered 1290. For this SAT, the predicted GPA is 3.265070236.

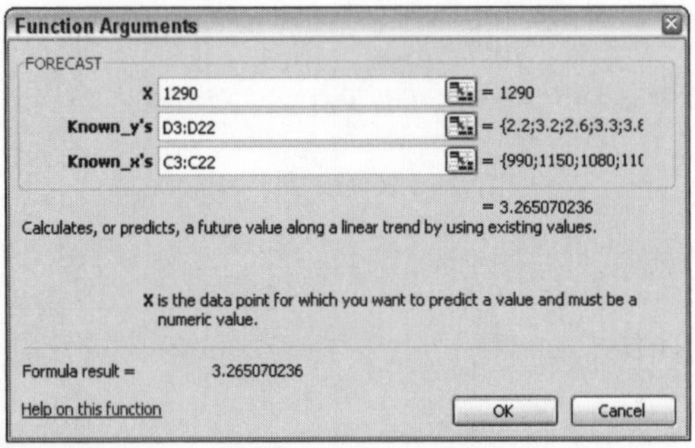

Figure 14-8: The Function Arguments dialog box for FORECAST.

Array function: TREND

TREND is a versatile function. You can use TREND to generate a set of predicted *y*-values for the *x*-values in the sample.

You can also supply a new set of *x*-values and generate a set of predicted *y*-values based on the linear relationship in your sample. It's like applying FORECAST repeatedly in one fell swoop.

In this section, I go through both uses.

Predicting y's for the x's in your sample

First, I use TREND to predict GPAs for the 20 students in the sample. Figure 14-9 shows TREND set up to do this. I included the Formula Bar in this screen shot so you can see what the formula looks like for this use of TREND.

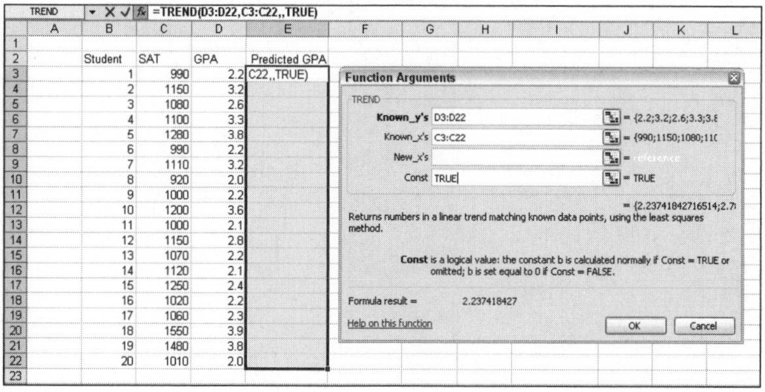

Figure 14-9:
The
Function
Arguments
dialog box
for TREND
along with
data.
TREND is
set up to
predict
GPAs for the
sample
SATs.

1. **With the data entered, select a column for TREND's answers.**

   I selected E3:E22. That puts the predicted GPAs right next to the sample GPAs.

2. **Click the Insert Function button to open the Insert Function dialog box.**

3. **In the Insert Function dialog box, select TREND and click OK to open the Function Arguments dialog box for TREND.**

4. **In the Known_y's box, enter the column that holds the scores for the *y*-variable.**

   For this example, that's D3:D22, the GPAs.

5. **In the Known_x's box, enter the column that holds the scores for the *x*-variable.**

   For this example, it's C3:C22, the SAT scores.

6. **Leave the New_x's box blank.**

7. **In the Const box, type TRUE (or leave it blank) to calculate the *y*-intercept. Type FALSE to set the *y*-intercept to zero.**

I typed TRUE. (I really don't know why you'd type FALSE.) A note of caution: In the dialog box, the instruction for the Const box refers to *b*. That's
the *y*-intercept. Earlier in the chapter, I used *a* to represent the *y*-intercept,
and *b* to represent the slope. No particular usage is standard for this.

8. **IMPORTANT: Do NOT click OK. Because this is an array function, press
   Ctrl+Shift+Enter to put TREND's answers into the selected column.**

   Figure 14-10 shows the answers in E3:E22.

	A	B	C	D	E
1					
2		Student	SAT	GPA	Predicted GPA
3		1	990	2.2	2.237418427
4		2	1150	3.2	2.785499392
5		3	1080	2.6	2.54571397
6		4	1100	3.3	2.61422409
7		5	1280	3.8	3.230815175
8		6	990	2.2	2.237418427
9		7	1110	3.2	2.648479151
10		8	920	2.0	1.997633005
11		9	1000	2.2	2.271673487
12		10	1200	3.6	2.956774693
13		11	1000	2.1	2.271673487
14		12	1150	2.8	2.785499392
15		13	1070	2.2	2.511458909
16		14	1120	2.1	2.682734211
17		15	1250	2.4	3.128049994
18		16	1020	2.2	2.340183608
19		17	1060	2.3	2.477203849
20		18	1550	3.9	4.155701803
21		19	1480	3.8	3.915916381
22		20	1010	2.0	2.305928548
23					

Figure 14-10:
The results
of TREND:
Predicted
GPAs for the
sample
SATs.

Predicting a new set of y's for a new set of x's

Here, I use TREND to predict GPAs for a new set of SATs. Figure 14-11 shows
TREND set up for this. The figure also shows the selected column for the
results. Once again, I include the Formula Bar to show you the formula for
this use of the function.

1. **With the data entered, select a column for TREND's answers.**

   I selected G8:G11.

2. **Click the Insert Function button to open the Insert Function dialog box.**

3. **In the Insert Function dialog box, select TREND and click OK to open
   the Function Arguments dialog box for TREND.**

4. **In the Known_y's box, enter the column that holds the scores for the
   y-variable.**

   For this example, that's D3:D22, the GPAs.

5. **In the Known_x's box, enter the column that holds the scores for the
   x-variable.**

   For this example, it's C3:C22, the SAT scores.

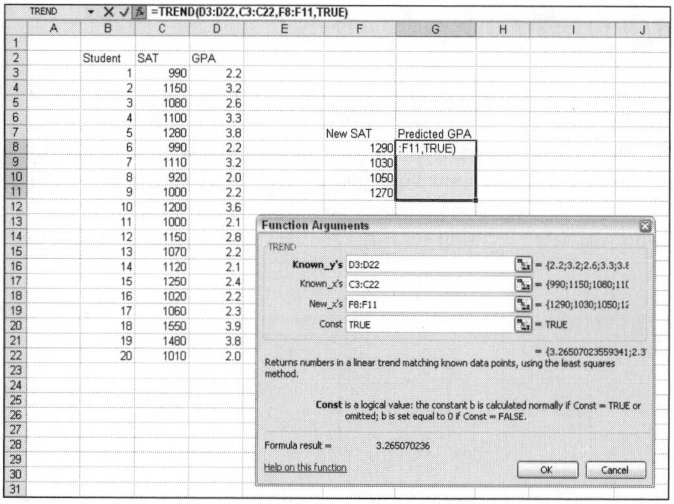

Figure 14-11:
The
Function
Arguments
dialog box
for TREND
along with
data.
TREND is
set up to
predict
GPAs for a
new set of
SATs.

6. **In the New_x's box enter the column that holds the new scores for the x-variable.**

   Here, that column is F8:F11, the New SAT scores.

7. **In the Const box, type TRUE (or leave it blank) to calculate the y-intercept. Type FALSE to set the y-intercept to zero.**

   I typed TRUE. (Again, I really don't know why you'd type FALSE.)

8. **IMPORTANT: Do NOT click OK. Because this is an array function, press Ctrl+Shift+Enter to put TREND's answers into the selected column.**

   Figure 14-12 shows the answers in G8:G11.

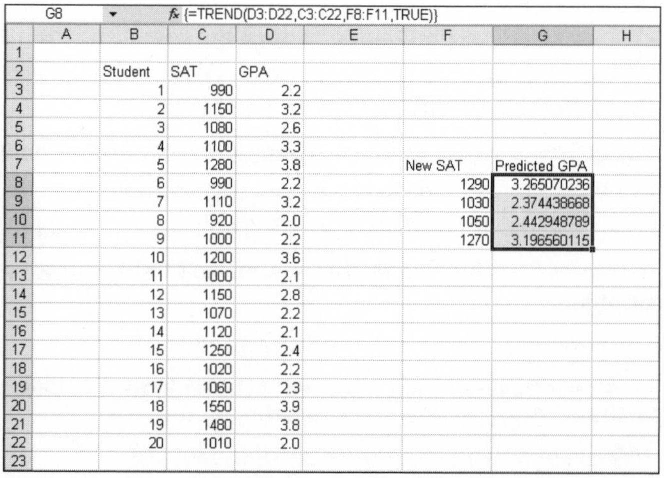

Figure 14-12:
The results
of TREND:
Predicted
GPAs for a
new set of
SATs.

Array function: LINEST

LINEST combines SLOPE, INTERCEPT, and STEYX, and throws in a few extras. Figure 14-13 shows the Function Arguments dialog box for LINEST, along with the data and the selected array for the answers. Notice that it's a five-row-by-two-column array. For linear regression, that's what the selected array has to be. How would you know the exact row-column dimensions of the array if I didn't tell you? Well . . . you wouldn't.

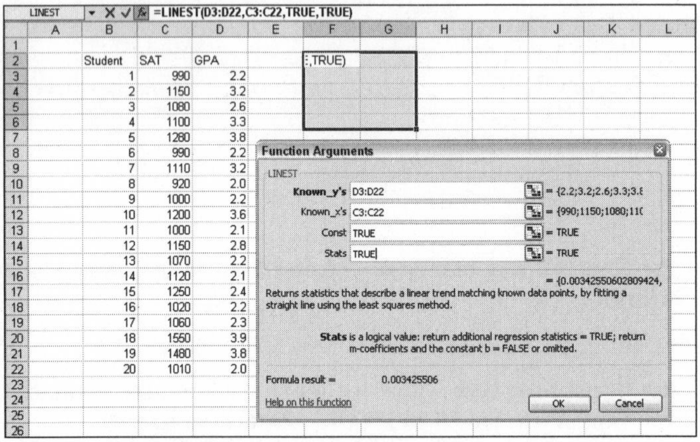

Figure 14-13: The Function Arguments dialog box for LINEST along with the data and the selected array for the results.

Here are the steps for using LINEST:

1. **With the data entered, select a five-row-by-two-column array of cells for LINEST's results.**

   I selected F2:G6.

2. **Click the Insert Function button to open the Insert Function dialog box.**

3. **In the Insert Function dialog box, select LINEST and click OK to open the Function Arguments dialog box for LINEST.**

4. **In the Known_y's box, enter the column that holds the scores for the *y*-variable.**

   For this example, that's D3:D22, the GPAs.

5. **In the Known_x's box, enter the column that holds the scores for the *x*-variable.**

   For this example, it's C3:C22, the SAT scores.

6. **In the Const box, type TRUE (or leave it blank) to calculate the *y*-intercept. Type FALSE to set the *y*-intercept to zero.**

   I typed TRUE.

7. **In the Stats box, type** TRUE **to return regression statistics in addition to the slope and the intercept. Type** FALSE **(or leave it blank) to return just the slope and the intercept.**

In the dialog box, *b* refers to intercept and *m-coefficient* refers to slope. As I said earlier, no set of symbols is standard for this.

8. **IMPORTANT: Do NOT click OK. Because this is an array function, press Ctrl+Shift+Enter to put LINEST's answers into the selected array.**

Figure 14-14 shows LINEST's results. They're not labeled in any way, so I added the labels for you in the worksheet. The left column gives you the slope, standard error of the slope, something called "R Square," *F*, and the $SS_{regression}$. What's R Square? That's another measure of the strength of the relationship between SAT and GPA in the sample. I discuss it in detail in Chapter 15.

The right column provides the intercept, standard error of the intercept, standard error of estimate, degrees of freedom, and $SS_{residual}$.

	A	B	C	D	E	F	G	H	I
1									
2		Student	SAT	GPA					
3		1	990	2.2		Slope	0.0034255	-1.153833	Intercept
4		2	1150	3.2		Standard error of slope	0.0005706	0.649102	Standard error of intercept
5		3	1080	2.6		R Square	0.6668765	0.4024	Standard error of estimate
6		4	1100	3.3		F	36.03401	18	df
7		5	1280	3.8		SSregression	5.8348357	2.9146643	SSresidual
8		6	990	2.2					
9		7	1110	3.2					
10		8	920	2.0					
11		9	1000	2.2					
12		10	1200	3.6					
13		11	1000	2.1					
14		12	1150	2.8					
15		13	1070	2.2					
16		14	1120	2.1					
17		15	1250	2.4					
18		16	1020	2.2					
19		17	1060	2.3					
20		18	1550	3.9					
21		19	1480	3.8					
22		20	1010	2.0					
23									

Figure 14-14: LINEST's results in the selected array.

Data Analysis Tool: Regression

Excel's Regression data analysis tool does everything LINEST does (and more) and labels the output for you, too. Figure 14-15 shows the Regression tool's dialog box along with the data for the SAT-GPA example.

The steps for using this tool are:

1. **Type the data into the worksheet, and include labels for the columns.**

2. **From the Tools menu, select Data Analysis to open the Data Analysis dialog box.**

3. **In the Data Analysis dialog box, scroll down the Analysis Tools list and select Regression. Click OK to open the Regression dialog box.**

This is the dialog box in Figure 14-15.

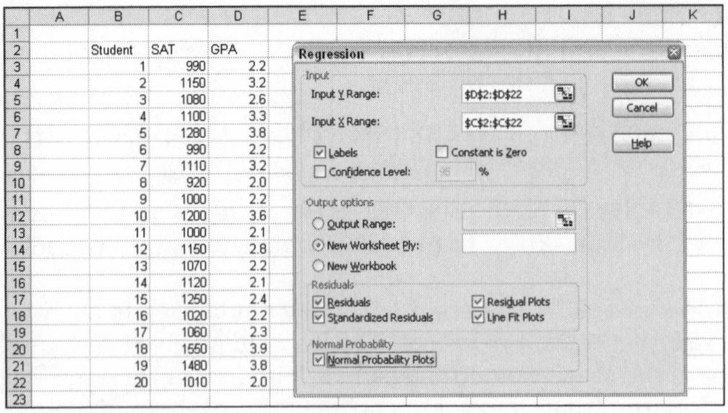

Figure 14-15:
The
Regression
Data
Analysis
Tool dialog
box along
with the
SAT-GPA
data.

4. **In the Input Y Range box, enter the cell range that holds the data for the *y*-variable.**

   For the example, the GPAs (including the label) are in $D$2:$D$22. Note the dollar signs ($) for absolute referencing.

5. **In the Input X Range box, enter the cell range that holds the data for the *x*-variable.**

   The SATs (including the label) are in $C$2:$C$22.

6. **If the cell ranges include column headings, select the Labels option.**

   I included the headings in the ranges, so I selected the option.

7. **The Alpha box has 0.05 as a default.** Change that value if you want a different alpha.

8. **In the Output Options section, select an option to indicate where you want the results.**

   I selected New Worksheet Ply to put the results on a new page in the worksheet.

9. **The Residuals area provides four options for viewing the deviations between the data points and the predicted points. Select as many as you like or none of them.**

   I selected all four. I'll explain them when I show you the output.

10. **Select the Normal Probability Plot option if you want to produce a graph of the percentiles of the *y*-variable.**

    I selected this one so I could show it to you in the output.

11. **Click OK.**

    Because I selected New Worksheet Ply, a newly created page opens with the results.

Tabled output

Figure 14-16 shows the upper half of the tool's tabled output after I expanded the columns. The title is SUMMARY OUTPUT. This part of the output features one table for Regression Statistics, another for ANOVA, and one for the regression coefficients.

Figure 14-16:
The upper
half of the
Regression
data
analysis
tool's tabled
output.

	A	B	C	D	E	F	G	H	I
1	SUMMARY OUTPUT								
2									
3	*Regression Statistics*								
4	Multiple R	0.81662505							
5	R Square	0.666876472							
6	Adjusted R Square	0.648369609							
7	Standard Error	0.402400043							
8	Observations	20							
9									
10	ANOVA								
11		*df*	*SS*	*MS*	*F*	*Significance F*			
12	Regression	1	5.834835693	5.834835693	36.03401	1.12048E-05			
13	Residual	18	2.914664307	0.161925795					
14	Total	19	8.7495						
15									
16		*Coefficients*	*Standard Error*	*t Stat*	*P-value*	*Lower 95%*	*Upper 95%*	*Lower 95.0%*	*Upper 95.0%*
17	Intercept	-1.153832541	0.649101962	-1.777582888	0.092372	-2.517545157	0.2098801	-2.51754516	0.20988008
18	SAT	0.003425506	0.000570648	6.002833489	1.12E-05	0.002226619	0.0046244	0.002226619	0.00462439
19									

The first three rows of the Regression Statistics table present information related to R^2, a measure of the strength of the SAT-GPA relationship in the sample (discussed in Chapter 15). The fourth row shows the standard error of estimate and the fifth gives the number of individuals in the sample.

The ANOVA table shows the results of testing

H_0: $\alpha_{Regression} \leq \alpha_{Residual}$

H_1: $\alpha_{Regression} > \alpha_{Residual}$

If the value in the F-significance column is less than .05 (or whatever alpha level you're using), reject H_0. In this example, it's less than .05.

Just below the ANOVA table is a table that gives the information on the regression coefficients. Excel doesn't name it, but I refer to it as the coefficients table. The Coefficients column provides the values for the intercept and the slope. The slope is labeled with the name of the *x*-variable. The Standard Error column presents the standard error of the intercept and the standard error of the slope. The remaining columns provide the results for the *t*-tests of the intercept and the slope. The P-value column lets you know whether or not to reject H_0 for each test. If the value is less than your alpha, reject H_0. In this example, the decision is to reject H_0 for the slope but not for the intercept.

Figure 14-17 shows the lower half of the Regression tool's tabled output.

21						
22	RESIDUAL OUTPUT				PROBABILITY OUTPUT	
23						
24	Observation	Predicted GPA	Residuals	Standard Residuals	Percentile	GPA
25	1	2.237418427	-0.037418427	-0.095536221	2.5	2
26	2	2.785499392	0.414500608	1.058297332	7.5	2
27	3	2.54571397	0.05428603	0.138602356	12.5	2.1
28	4	2.61422409	0.68577591	1.750913753	17.5	2.1
29	5	3.230815175	0.569184825	1.453234976	22.5	2.2
30	6	2.237418427	-0.037418427	-0.095536221	27.5	2.2
31	7	2.648479151	0.551520849	1.408135554	32.5	2.2
32	8	1.997633005	0.002366995	0.006043379	37.5	2.2
33	9	2.271673487	-0.071673487	-0.182995776	42.5	2.2
34	10	2.956774693	0.643225307	1.642274131	47.5	2.3
35	11	2.271673487	-0.171673487	-0.43831442	52.5	2.4
36	12	2.785499392	0.014500608	0.037022757	57.5	2.6
37	13	2.511458909	-0.311458909	-0.795212664	62.5	2.8
38	14	2.682734211	-0.582734211	-1.487829085	67.5	3.2
39	15	3.128049994	-0.728049994	-1.858847373	72.5	3.2
40	16	2.340183608	-0.140183608	-0.357914887	77.5	3.3
41	17	2.477203849	-0.177203849	-0.452434465	82.5	3.6
42	18	4.155701803	-0.255701803	-0.652854376	87.5	3.8
43	19	3.915916381	-0.115916381	-0.295956132	92.5	3.8
44	20	2.305928548	-0.305928548	-0.78109262	97.5	3.9
45						

Figure 14-17: The lower half of the Regression data analysis tool's tabled output.

Here, you find the RESIDUAL OUTPUT and the PROBABILITY OUTPUT. The RESIDUAL OUTPUT is a table that shows the predicted value and the residual $(y - y')$ for each individual in the sample. It also shows the *standard residual* for each observation, which is

$$\text{standard residual} = \frac{\text{residual} - \text{average residual}}{S_{yx}}$$

The tabled data on residuals and standard residuals are useful for analyzing the variability around the regression line. You can scan these data for outliers, for example, and see if outliers are associated with particular values of the *x*-variable. (If they are, it might mean that something weird is going on in your sample.)

The PROBABILITY OUTPUT is a table of the percentiles in the *y*-variable data in the sample. (Yes, PERCENTILE OUTPUT would be a better name.)

Graphic output

Figures 14-18, 14-19, and 14-20 show the Regression tool's graphic output. The Normal Probability Plot in Figure 14-18 is a graphic version of the PROBABILITY OUTPUT table. The SAT Residual Plot in Figure 14-19 shows the residuals graphed against the *x*-variable: For each SAT score in the sample, this plot shows the corresponding residual. Figure 14-20 shows the SAT Line Fit Plot — a look at the scatterplot and the predicted *y*-values.

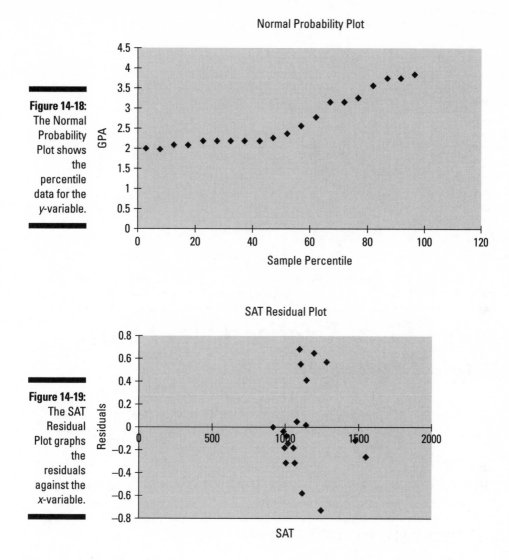

Figure 14-18:
The Normal Probability Plot shows the percentile data for the *y*-variable.

Figure 14-19:
The SAT Residual Plot graphs the residuals against the *x*-variable.

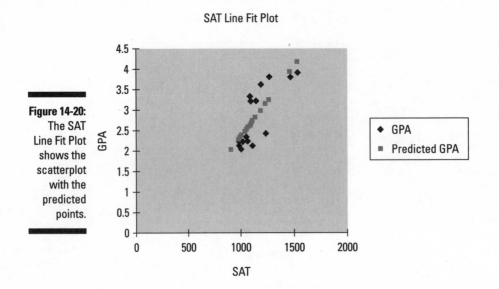

Juggling Many Relationships at Once: Multiple Regression

Linear regression is a great tool for making predictions. Once you know the slope and the intercept of the line that relates two variables, you can take a new *x*-value and predict a new *y*-value. In the example I've been working through, you take an SAT score and predict a GPA for a Sahutsket University student.

What if you knew more than just the SAT score for each student? What if you had the student's high school average (on a 100 scale), and you could use that information, too? If you could combine SAT with HS average, you might have a more accurate predictor than just SAT alone.

When you work with more than one independent variable, you're in the realm of *multiple regression*. As in linear regression, you find regression coefficients for the best-fitting line through a scatterplot. Once again, "best-fitting" means that the sum of the squared distances from the data points to the line is a minimum.

With two independent variables, however, you can't show a scatterplot in two dimensions. You need three dimensions, and that becomes difficult to draw. Instead, I'll just show you the equation of the regression line:

$$y' = a + b_1 x_1 + b_2 x_2$$

For the SAT-GPA example, that translates to

Predicted GPA = a + b_1(SAT) + b_2(High School Average)

You can test hypotheses about the overall fit, and about all three of the regression coefficients.

I won't go through all the formulas for finding the coefficients, because that gets *really* complicated. Instead, I'll go right to the Excel capabilities.

A couple of things to bear in mind before I proceed:

- ✔ **You can have any number of *x*-variables.** I just use two in the upcoming example.

- ✔ **Expect the coefficient for SAT to change from linear regression to multiple regression.** Expect the intercept to change, too.

- ✔ **Expect the standard error of estimate to decrease from linear regression to multiple regression.** Because multiple regression uses more information than linear regression, it typically reduces the error.

Excel Tools for Multiple Regression

The good news about Excel's multiple regression tools is that they're some of the same ones I just told you about for linear regression — you just use them in a slightly different way.

The bad news is . . . well . . . uh . . . I can't think of any bad news!

TREND revisited

I begin with TREND. I already showed you how to use this function to predict values based on one *x*-variable. Change what you type into the dialog box, and it predicts values based on more than one.

Figure 14-21 shows the TREND dialog box and data for 20 students. In the data, I've added a column for each student's high school average. The figure also shows the selected column for TREND's predictions. I include the Formula Bar in this screen shot so you can see what the formula looks like.

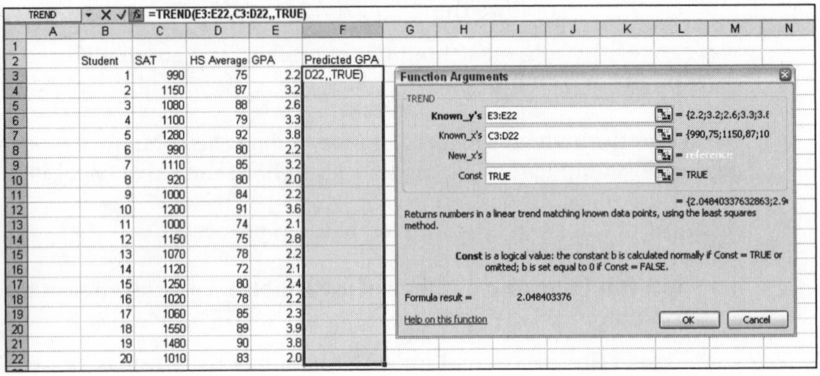

Figure 14-21:
The
Function
Arguments
dialog box
for TREND
along with
data.
TREND is
set up to
predict
GPAs for the
sample
SATs and
high school
averages.

1. **With the data entered, select a column for TREND's answers.**

   I selected F3:F22. That puts the predicted GPAs right next to the sample GPAs.

2. **Click the Insert Function button to open the Insert Function dialog box.**

3. **In the Insert Function dialog box, select TREND and click OK to open the Function Arguments dialog box for TREND.**

4. **In the Known_y's box, enter the column that holds the scores for the *y*-variable.**

   For this example, that's E3:E22, the GPAs.

5. **In the Known_x's box, enter the columns that hold the scores for the *x*-variables.**

   For this example, it's C3:D22, the columns that hold the SAT scores and the high school averages.

6. **Leave the New_x's box blank.**

7. **In the Const box, type TRUE (or leave it blank) to calculate the *y*-intercept. Type FALSE to set the *y*-intercept to zero.**

   I typed TRUE. (I really don't know why you'd type FALSE.) A note of caution: In the dialog box, the instruction for the Const box refers to *b*. That's the *y*-intercept. Earlier in the chapter, I use *a* to represent the *y*-intercept and *b* to represent the slope. No particular usage is standard for this. Also, the dialog box makes it sound like this function just works for linear regression. As you're about to see, it works for multiple regression, too.

8. **IMPORTANT: Do NOT click OK. Because this is an array function, press Ctrl+Shift+Enter to put TREND's answers into the selected column.**

Figure 14-22 shows the answers in F3:F22.

	A	B	C	D	E	F	G
1							
2		Student	SAT	HS Average	GPA	Predicted GPA	
3		1	990	75	2.2	2.048403376	
4		2	1150	87	3.2	2.967217927	
5		3	1080	88	2.6	2.831485598	
6		4	1100	79	3.3	2.499039035	
7		5	1280	92	3.8	3.511405481	
8		6	990	80	2.2	2.261402606	
9		7	1110	85	3.2	2.780114135	
10		8	920	80	2.0	2.083070431	
11		9	1000	84	2.2	2.457278015	
12		10	1200	91	3.6	3.264997435	
13		11	1000	74	2.1	2.031279555	
14		12	1150	75	2.8	2.456019776	
15		13	1070	78	2.2	2.380011114	
16		14	1120	72	2.1	2.251792163	
17		15	1250	80	2.4	2.923779255	
18		16	1020	78	2.2	2.252630989	
19		17	1060	85	2.3	2.65273401	
20		18	1550	89	3.9	4.071458617	
21		19	1480	90	3.8	3.935726288	
22		20	1010	83	2.0	2.440154194	
23							

So TREND predicts the values, and I haven't even shown you how to find the coefficients yet!

LINEST revisited

To find the multiple regression coefficients, I turn again to LINEST.

In Figure 14-23, I show the data and the dialog box for LINEST, along with the data and the selected array for the answers. The selected array is five rows by three columns. It's always five rows. The number of columns is equal to the number of regression coefficients. For linear regression, it's two — the slope and the intercept. For this case of multiple regression, it's three.

Here are the steps for using LINEST for multiple regression with three coefficients:

1. **With the data entered, select a five-row-by-three-column array of cells for LINEST's results.**

   I selected H3:J7.

2. **Click the Insert Function button to open the Insert Function dialog box.**

3. **In the Insert Function dialog box, select LINEST and click OK to open the Function Arguments dialog box for LINEST.**

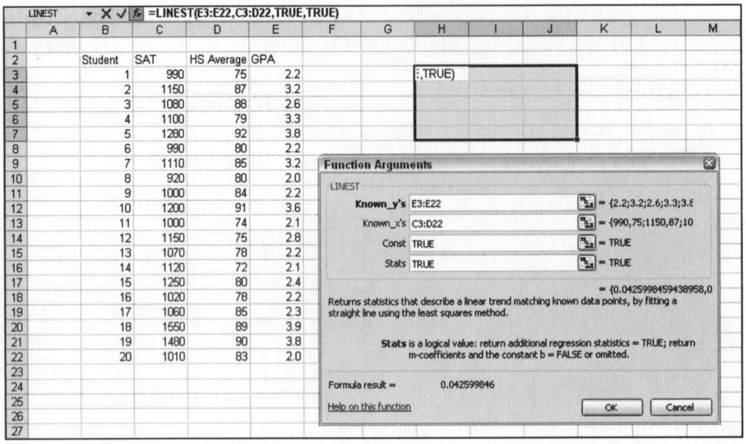

Figure 14-23:
The
Function
Arguments
dialog box
for LINEST
along with
the data and
the selected
array for the
results of a
multiple
regression.

4. **In the Known_y's box, enter the column that holds the scores for the** *y***-variable.**

   For this example, that's E3:E22, the GPAs.

5. **In the Known_x's box, enter the columns that hold the scores for the** *x***-variables.**

   For this example, it's C3:D22, the SAT scores and the high school averages.

6. **In the Const box, type TRUE (or leave it blank) to calculate the** *y***-intercept. Type FALSE to set the** *y***-intercept to zero.**

   I typed TRUE.

7. **In the Stats box, type TRUE to return regression statistics in addition to the slope and the intercept. Type FALSE (or leave it blank) to return just the slope and the intercept.**

   I typed TRUE. The dialog box refers to the intercept as *b* and to the other coefficients as *m-coefficients*. I use *a* to represent the slope and *b* to refer to the other coefficients. No set of symbols is standard.

8. **IMPORTANT: Do NOT click OK. Because this is an array function, press Ctrl+Shift+Enter to put LINEST's answers into the selected array.**

Figure 14-24 shows LINEST's results. They're not labeled in any way, so I added the labels for you in the worksheet. I also drew a box around part of the results to clarify what goes with what.

The entries that stand out are the ugly #N/A symbols in the last three rows of the rightmost column. These indicate that LINEST doesn't put anything into these cells.

The top two rows of the array provide the values and standard errors for the coefficients. I drew the box around those rows to separate them from the three remaining rows, which present information in a different way. Before I get to those rows, I'll just tell you that the top row gives you the information for writing the regression equation:

$$y' = -3.67 + .0025x_1 + .043x_2$$

In terms of SAT, GPA, and high school average, it's:

Predicted GPA = -3.67 + .0025(SAT) + .043(High School Average)

The third row has R Square (a measure of the strength of the relationship between GPA and the other two variables, which I cover in Chapter 15) and the standard error of estimate. Compare the standard error of estimate for the multiple regression with the standard error for the linear regression, and you'll see that the multiple one is smaller. (Never mind. I'll do it for you. It's 0.40 for the linear and 0.35 for the multiple.)

The fourth row shows the *F*-ratio that tests the hypothesis about whether or not the line is a good fit to the scatterplot, and the df for the denominator of the *F*. The df for the numerator (not shown) is the number of coefficients minus 1. You can use FINV to verify that this *F* with df = 2 and 17 is significant.

The last row gives you $SS_{Regression}$ and $SS_{Residual}$.

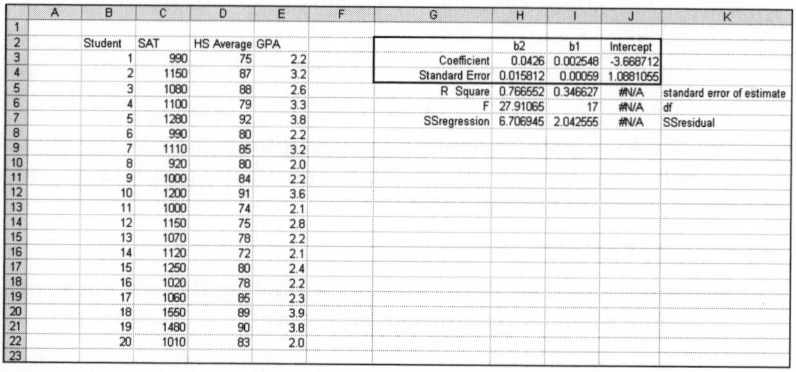

Figure 14-24: LINEST's multiple results in the selected array.

	A	B	C	D	E	F	G	H	I	J	K
1											
2		Student	SAT	HS Average	GPA			b2	b1	Intercept	
3		1	990	75	2.2		Coefficient	0.0426	0.002548	-3.668712	
4		2	1150	87	3.2		Standard Error	0.015812	0.00059	1.0881055	
5		3	1080	88	2.6		R Square	0.766552	0.346627	#N/A	standard error of estimate
6		4	1100	79	3.3		F	27.91065	17	#N/A	df
7		5	1280	92	3.8		SSregression	6.706945	2.042555	#N/A	SSresidual
8		6	990	80	2.2						
9		7	1110	85	3.2						
10		8	920	80	2.0						
11		9	1000	84	2.2						
12		10	1200	91	3.6						
13		11	1000	74	2.1						
14		12	1150	75	2.8						
15		13	1070	78	2.2						
16		14	1120	72	2.1						
17		15	1250	80	2.4						
18		16	1020	78	2.2						
19		17	1060	85	2.3						
20		18	1550	89	3.9						
21		19	1480	90	3.8						
22		20	1010	83	2.0						
23											

Regression data analysis tool revisited

In the same way you use TREND and LINEST for multiple regression, you use the Regression data analysis tool. Specify the appropriate array for the *x*-variables, and you're off and running.

Here are the steps:

1. **Type the data into the worksheet, and include labels for the columns.**

2. **From the Tools menu, select Data Analysis to open the Data Analysis dialog box.**

3. **In the Data Analysis dialog box, scroll down the Analysis Tools list and select Regression. Click OK to open the Regression dialog box.**

   This is the dialog box in Figure 14-15.

4. **In the Input Y Range box, enter the cell range that holds the data for the *y*-variable.**

   Using the example I just used for LINEST, the GPAs (including the label) are in $E$2:$E$22. Note the dollar signs ($) for absolute referencing.

5. **In the Input X Range box, enter the cell range that holds the data for the *x*-variable.**

   The SATs and the high school averages (including the labels) are in $C$2:$D$22.

6. **If the cell ranges include column headings, select the Labels option.**

   I included the headings in the ranges, so I selected the option.

7. **The Alpha box has 0.05 as a default.**

   I used the default value.

8. **In the Output Options, select an option to indicate where you want the results.**

   I selected New Worksheet Ply to put the results on a new page in the worksheet.

9. **The Residuals area provides four options for viewing the deviations between the data points and the predicted points. Select as many as you like or none of them.**

   I selected all four.

10. **Select the Normal Probability Plot option if you want to produce a graph of the percentiles of the *y*-variable.**

    I selected this one.

11. **Click OK.**

Go back to the section "Data analysis tool: Regression" for the details of what's in the output. It's the same as before, with a couple of changes and additions because of the new variable. Figure 14-25 shows the Regression Statistics, the ANOVA table, and the coefficients table.

Figure 14-25:
Part of
the output
from the
Regression
data
analysis
tool: The
Regression
Statistics,
the ANOVA
table
and the
Coefficients
table.

	A	B	C	D	E	F	G	H	I
1	SUMMARY OUTPUT								
2									
3	*Regression Statistics*								
4	Multiple R	0.875529451							
5	R Square	0.76655182							
6	Adjusted R Square	0.739087328							
7	Standard Error	0.346627012							
8	Observations	20							
9									
10	ANOVA								
11		*df*	*SS*	*MS*	*F*	*Significance F*			
12	Regression	2	6.706945148	3.353473	27.91065	4.26206E-06			
13	Residual	17	2.042554852	0.12015					
14	Total	19	8.7495						
15									
16		*Coefficients*	*Standard Error*	*t Stat*	*P-value*	*Lower 95%*	*Upper 95%*	*Lower 95.0%*	*Upper 95.0%*
17	Intercept	-3.66871154	1.088105488	-3.37165	0.003623	-5.964413427	-1.373009653	-5.964413427	-1.373009653
18	SAT	0.002547602	0.000589753	4.319781	0.000465	0.001303333	0.003791872	0.001303333	0.003791872
19	HS Average	0.042599846	0.015811932	2.694158	0.015361	0.009239586	0.075960106	0.009239586	0.075960106
20									

The ANOVA table shows the new df (2, 17, and 19 for Regression, Residual, and Total, respectively). The coefficients table adds information for the HS Average. It shows the values of all the coefficients, as well as standard errors, and *t*-test information for hypothesis testing.

If you go through the example, you'll see the table of residuals in the output. (I don't show them in Figure 14-25.) Compare the absolute values of the residuals from the linear regression with the absolute values of the residuals from the multiple regression, you'll see the multiple ones are smaller, on average.

The graphic output has some additions, too: A scatterplot of HS Average and GPA that also shows predicted GPAs, and a plot of residuals and HS Average.

Chapter 15

Correlation: The Rise and Fall of Relationships

....

In This Chapter

▶ What correlation is all about

▶ How correlation connects to regression

▶ Conclusions from correlations

▶ Analyzing items

....

*I*n Chapter 14, I show you the ins and outs of regression, a tool for summarizing relationships between (and among) variables. In this chapter, I introduce you to the ups and downs of correlation, another tool for looking at relationships.

I use the example of SAT and GPA from Chapter 14, and show how to think about the data in a slightly different way. The new concepts connect to what I showed you in the preceding chapter, and you'll see how that works. I also show you how to test hypotheses about relationships and how to use Excel functions and data analysis tools for correlation.

Scatterplots Again

A *scatterplot* is a graphic way of showing a relationship between two variables. Figure 15-1 is a scatterplot that represents the GPAs and SAT scores of 20 students at the fictional Sahutsket University. The GPAs are on a 4.0 scale and the SATs are combined Verbal and Math.

Each point represents one student. A point's location in the horizontal direction represents the student's SAT. That same point's location in the vertical direction represents the student's GPA.

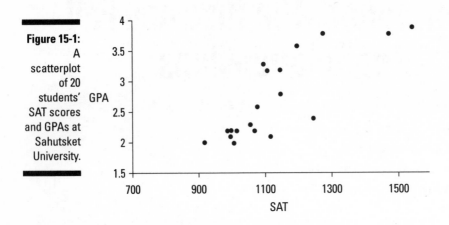

Figure 15-1:
A scatterplot of 20 students' SAT scores and GPAs at Sahutsket University.

Understanding Correlation

In Chapter 14, I refer to the SAT as the *independent variable* and to the GPA as the *dependent variable*. The objective in Chapter 14 is to use SAT to predict GPA. Here's a very important point: Although I use scores on one variable to *predict* scores on the other, I do *not* mean that the score on one variable *causes* a score on the other. "Relationship" doesn't necessarily mean "causality."

Correlation is a statistical way of looking at a relationship. When two things are correlated, it means that they vary together. *Positive* correlation means that high scores on one are associated with high scores on the other, and that low scores on one are associated with low scores on the other. The scatterplot in Figure 15-1 is an example of positive correlation.

Negative correlation, on the other hand, means that high scores on the first thing are associated with *low* scores on the second. Negative correlation also means that low scores on the first are associated with high scores on the second. An example is the correlation between body weight and the time spent on a weight-loss program. If the program is effective, the higher the amount of time spent on the program, the lower the body weight. Also, the lower the amount of time spent on the program, the higher the body weight.

Table 15-1, a repeat of Table 14-2, shows the data from the scatterplot.

Table 15-1	SAT Scores and GPAs for 20 Sahutsket University Students	
Student	*SAT*	*GPA*
1	990	2.2
2	1150	3.2
3	1080	2.6
4	1100	3.3
5	1280	3.8
6	990	2.2
7	1110	3.2
8	920	2.0
9	1000	2.2
10	1200	3.6
11	1000	2.1
12	1150	2.8
13	1070	2.2
14	1120	2.1
15	1250	2.4
16	1020	2.2
17	1060	2.3
18	1550	3.9
19	1480	3.8
20	1010	2.0
Mean	1126.5	2.705
Variance	26171.32	0.46
Standard Deviation	161.78	0.82

In keeping with the way I used SAT and GPA in Chapter 14, SAT is the *x*-variable and GPA is the *y*-variable.

The formula for calculating the correlation between the two is

$$r = \frac{\left[\frac{1}{N-1}\right]\sum(x-\bar{x})(y-\bar{y})}{s_x s_y}$$

The term on the left, r, is called the *correlation coefficient*. It's also called *Pearson's product-moment correlation coefficient* after its creator Karl Pearson.

The two terms in the denominator on the right are the standard deviation of the x-variable and the standard deviation of the y-variable. The term in the numerator is called the *covariance*. So, another way to write this formula is

$$r = \frac{cov(x,y)}{s_x s_y}$$

The covariance represents x and y varying together. Dividing the covariance by the product of the two standard deviations imposes some limits. The lower limit of the correlation coefficient is -1.00, and the upper limit is +1.00.

A correlation coefficient of -1.00 represents perfect negative correlation (low x-scores associated with high y-scores, and high x-scores associated with low y-scores.) A correlation of +1.00 represents perfect positive correlation (low x-scores associated with low y-scores and high x-scores associated with high y-scores.) A correlation of 0.00 means that the two variables are not related.

Applying the formula to the data in Table 15-1,

$$r = \frac{\left[\frac{1}{N-1}\right]\sum(x-\bar{x})(y-\bar{y})}{s_x s_y}$$

$$= \frac{\left[\frac{1}{20-1}\right]\left[(990-1126.5)(2.2-2.705)+...+(1010-1126.5)(2.0-2.705)\right]}{(161.78)(0.82)}$$

$$= .817$$

What, exactly, does this number mean? I'm about to tell you.

Correlation and Regression

Figure 15-2 shows the scatterplot with the line that "best fits" the points. It's possible to draw an infinite number of lines through these points. Which one is best?

To be "best," a line has to meet a specific standard: If you draw the distances in the vertical direction between the points and the line, and you square those distances, and then you add those squared distances, the best-fitting line is the one that makes the sum of those squared distances as small as possible. This line is called the *regression line*.

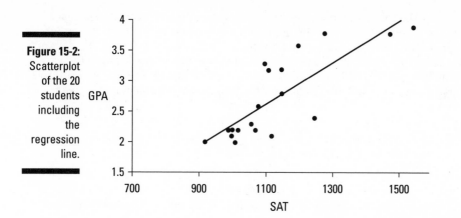

Figure 15-2:
Scatterplot
of the 20
students
including
the
regression
line.

The regression line's purpose in life is to enable you to make predictions. As I mention in Chapter 14, without a regression line your best predicted value of the y-variable is the mean of the y's. A regression line takes the x-variable into account and delivers a more precise prediction. Each point on the regression line represents a predicted value for y. In the symbology of regression, each predicted value is a y'.

Why do I tell you all this? Because correlation is closely related to regression. Figure 15-3 focuses on one point in the scatterplot and its distance to the regression line and to the mean. (This is a repeat of Figure 14-3.)

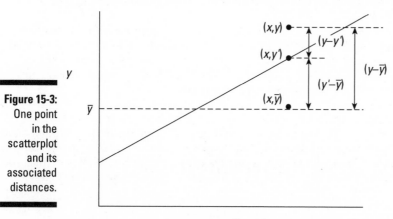

Figure 15-3:
One point
in the
scatterplot
and its
associated
distances.

Notice the three distances laid out in the figure. The distance labeled $(y-y')$ is the difference between the point and the regression line's prediction for where the point should be. (In Chapter 14, I call that a *residual*.) The distance labeled $(y-\bar{y})$ is the difference between the point and the mean of the y's. The distance labeled $(y'-\bar{y})$ is the gain in prediction ability that you get from using the regression line to predict the point rather than using the mean.

Figure 15-3 shows the three distances are related like this:

$$(y-y') + (y'-\bar{y}) = (y-\bar{y})$$

As I point out in Chapter 14, you can square all the residuals and add them, square all the deviations of the predicted points from the mean and add them, and square all the deviations of the actual points from the mean and add them, too.

It turns out that these sums of squares are related in the same way as the deviations I just showed you:

$$SS_{Residual} + SS_{Regression} = SS_{Total}$$

If $SS_{Regression}$ is large in comparison to $SS_{Residual}$, that indicates the relationship between the x-variable and the y-variable is a strong one. It means that throughout the scatterplot, the variability around the regression line is small.

On the other hand, if $SS_{Regression}$ is small in comparison to $SS_{Residual}$, that means the relationship between the x-variable and the y-variable is weak. In this case, the variability around the regression line is large throughout the scatterplot.

One way to test $SS_{Regression}$ against $SS_{Residual}$ is to divide each by its degrees of freedom (1 for $SS_{Regression}$ and N-2 for $SS_{Residual}$) to form variance estimates (also known as Mean Squares, or MS), and then divide one by the other to calculate an F. If $MS_{Regression}$ is significantly larger than $MS_{Residual}$, you have evidence that the x-y relationship is strong. (See Chapter 14 for details.)

Here's the clincher as far as correlation is concerned: Another way to assess the size of $SS_{Regression}$ is to compare it with SS_{Total}. Divide the first by the second. If the ratio is large, this tells you that the x-y relationship is strong. This ratio has a name. It's called the *coefficient of determination*. Its symbol is r^2. Take the square root of this coefficient, and you have ... the correlation coefficient!

$$r = \pm\sqrt{r^2} = \pm\sqrt{\frac{SS_{Regression}}{SS_{Total}}}$$

The plus-or-minus sign ($\pm$) means that r is either the positive or negative square root, depending on whether the slope of the regression line is positive or negative.

So, if you calculate a correlation coefficient and you quickly want to know what its value signifies, just square it. The answer — the coefficient of determination — lets you know the proportion of the SS_{Total} that's tied up in the relationship between the x-variable and the y-variable. If it's a large proportion, the correlation coefficient signifies a strong relationship. If it's a small proportion, the correlation coefficient signifies a weak relationship.

In the GPA-SAT example, the correlation coefficient is .817. The coefficient of determination is

$$r^2 = (.817)^2 = .667$$

In this sample of 20 students, the $SS_{Regression}$ is 66.7 percent of the SS_{Total}. Sounds like a large proportion, but what's large? What's small? Those questions scream out for hypotheses tests.

Testing Hypotheses About Correlation

In this section, I show you how to answer important questions about correlation. Like any other kind of hypothesis testing, the idea is to use sample statistics to make inferences about population parameters. Here, the sample statistic is r, the correlation coefficient. By convention, the population parameter is ρ (rho), the Greek equivalent of r. (Yes, it does look like our letter p, but it really is the Greek equivalent of r.)

Two kinds of questions are important in connection with correlation: (1) Is a correlation coefficient greater than zero? (2) Are two correlation coefficients different from one another?

Is a correlation coefficient greater than zero?

Returning once again to the Sahutsket SAT-GPA example, you can use the sample r to test hypotheses about the population ρ — the correlation coefficient for all students at Sahutsket University.

Assuming we know in advance (before we gather any sample data) that any correlation between SAT and GPA should be positive, the hypotheses are:

$H_0: \rho \leq 0$

$H_1: \rho > 0$

I set $\alpha = .05$.

The appropriate statistical test is a t-test. The formula is:

$$t = \frac{r - \rho}{s_r}$$

This test has N-2 df.

For the example, the values in the numerator are set: r is .817 and ρ (in H_0) is zero. What about the denominator? I won't burden you with the details. I'll just tell you that it's

$$\sqrt{\frac{1 - r^2}{N - 2}}$$

With a little algebra, the formula for the t-test simplifies to

$$t = \frac{r\sqrt{N - 2}}{\sqrt{1 - r^2}}$$

For the example,

$$t = \frac{r\sqrt{N - 2}}{\sqrt{1 - r^2}} = \frac{.817\sqrt{20 - 2}}{\sqrt{1 - .817^2}} = 6.011$$

with df = 18 and $\alpha = .05$ (one-tailed), the critical value of t is 2.10 (use the worksheet function TINV to check). Because the calculated value is greater than the critical value, the decision is to reject H_0.

Do two correlation coefficients differ?

In a sample of 24 students at Farshimmelt College, the correlation between SAT and GPA is .752. Is this different from the correlation (.817) at Sahutsket University? If I have no way of assuming that one correlation should be higher than the other, the hypotheses are:

H_0: $\rho_{\text{Sahutsket}} = \rho_{\text{Farshimmelt}}$

H_1: $\rho_{\text{Sahutsket}} \neq \rho_{\text{Farshimmelt}}$

Again, $\alpha = .05$.

For highly technical reasons, you can't set up a t-test for this one. In fact, you can't even work with .817 and .752, the two correlation coefficients.

Instead, what you do is *transform* each correlation coefficient into something else and then work with the two something elses in a formula that gives you — believe it or not — a z-test.

The transformation is called *Fisher's r to z transformation*. Fisher is the statistician who's remembered as the "F" in the *F*-test. He transformed the *r* into a z by doing this:

$$z_r = \frac{1}{2}\left[\log_e(1+r) - \log_e(1-r)\right]$$

If you know what $\log_e$ means, fine. If not, don't worry about it. (I explain it in Chapter 20.) Excel takes care of all this for you, as you'll see in a moment.

Anyway, for this example

$$z_{.817} = \frac{1}{2}\left[\log_e(1+.817) - \log_e(1-.817)\right] = 1.1477$$

$$z_{.752} = \frac{1}{2}\left[\log_e(1+.752) - \log(1-.752)\right] = 0.9775$$

once you transform *r* to z, the formula is

$$Z = \frac{z_1 - z_2}{\sigma_{z_1 - z_2}}$$

The denominator turns out to be easier than you might think. It's:

$$\sigma_{z_1 - z_2} = \sqrt{\frac{1}{N_1 - 3} + \frac{1}{N_2 - 3}}$$

For this example,

$$\sigma_{z_1 - z_2} = \sqrt{\frac{1}{N_1 - 3} + \frac{1}{N_2 - 3}} = \sqrt{\frac{1}{20 - 3} + \frac{1}{24 - 3}} = .326$$

the whole formula is

$$Z = \frac{z_1 - z_2}{\sigma_{z_1 - z_2}} = \frac{1.1477 - .9775}{.326} = .522$$

The next step is to compare the calculated value to a standard normal distribution. For a two-tailed test with $\alpha = .05$ the critical values in a standard normal distribution are 1.96 in the upper tail and -1.96 in the lower tail. The calculated value falls in between those two, so the decision is to not reject H_0.

Worksheet Functions for Correlation

Excel provides two worksheet functions for calculating correlation — and they do exactly the same thing in exactly the same way! Why Excel offers both CORREL and PEARSON I do not know, but there you have it. Those are the two main correlation functions.

The others are RSQ and COVAR. RSQ calculates the coefficient of determination (the square of the correlation coefficient), and COVAR calculates the covariance, sort of.

CORREL and PEARSON

Figure 15-4 shows the data for the Sahutsket SAT-GPA example along with the Function Arguments dialog box for CORREL.

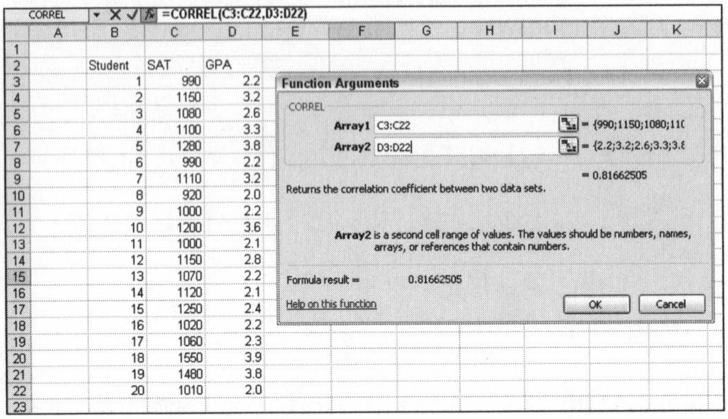

Figure 15-4:
The Function Arguments dialog box for CORREL along with data.

To use this function, the steps are:

1. **Type the data into cell arrays.**

   I've entered the SAT data into C3:C22 and the GPA data into D3:D22.

2. **Select a cell for CORREL's answer.**

   I selected F15.

3. **Click the Insert Function button to open the Insert Function dialog box.**

4. **In the Insert Function dialog box, select CORREL and click OK to open its Function Arguments dialog box.**

5. **In the Array1 box, type the column that holds the scores for one of the variables.**

   I entered C3:C22 for the SAT scores.

6. **In the Array2 box, type the column that holds the scores for the other variable.**

   I entered D3:D22 for the GPAs. The answer, 0.81662505, appears in the dialog box.

7. **Click OK to put the answer into your selected cell.**

   Selecting PEARSON instead of CORREL gives you exactly the same answer, and you use it exactly the same way.

Item analysis: a useful application of correlation

Instructors often want to know how performance on a particular exam question is related to overall performance on the exam. Ideally, someone who knows the material answers the question correctly, someone who doesn't answers it incorrectly. If everyone answers it correctly — or if no one does — it's a useless question. This evaluation is called *item analysis.*

Suppose it's possible to answer the exam question either correctly or incorrectly, and it's possible to score from 0 to 100 on the exam. Arbitrarily, you can assign a score of 0 for an incorrect answer to the question, and 1 for a correct answer, and then calculate a correlation coefficient where each pair of scores is either 0 or 1 for the question and a number from 0 to 100 for the exam. The score on the exam question is called a *dichotomous variable,* and this type of correlation is called *point biserial correlation.*

If the point biserial correlation is high for an exam question, it's a good idea to retain that question. If the correlation is low, the question probably serves no purpose.

Because one of the variables can only be 0 or 1, the formula for the biserial correlation coefficient is a bit different from the formula for the regular correlation coefficient. If you use Excel for the calculations, however, that doesn't matter. Just use CORREL (or PEARSON) in the way I outlined.

RSQ

If you have to quickly calculate the coefficient of determination (r^2), RSQ is the function for you. I see no particular need for this function because it's easy enough to use CORREL and then square the answer. To access the Function Arguments dialog box for RSQ, follow the steps for CORREL (or PEARSON), but select RSQ in the Insert Function dialog box.

Here's what the Excel Formula Bar looks like after you fill in RSQ's Function Arguments dialog box for this example:

=RSQ(D3:D22,C3:C22)

In terms of the dialog box, the only difference between this one and CORREL (and PEARSON) is that the boxes you fill in are called Known_y's and Known_x's rather than Array1 and Array2.

COVAR

This is another function for which I see no burning need. A minute ago I said COVAR calculates covariance — sort of. I said that because the covariance I introduced earlier (as the numerator of the correlation coefficient) is

$$\text{covariance}=\left[\frac{1}{N-1}\right]\sum(x-\bar{x})(y-\bar{y})$$

COVAR, however, calculates

$$\text{covariance}=\left[\frac{1}{N}\right]\sum(x-\bar{x})(y-\bar{y})$$

You use this function the same way you use CORREL. To access the Function Arguments dialog box for COVAR, follow the steps for CORREL (or PEARSON), but select COVAR in the Insert Function dialog box.

After you fill in COVAR's Function Arguments dialog box for this example, the formula in the Formula Bar is

=COVAR(C3:C22,D3:D22)

If you want to use this function to calculate *r*, you divide the answer by the product of STDEVP(C3:C22) and STDEVP(D3:D22). I don't know why you'd bother with all this when you can just use CORREL.

Data Analysis Tool: Correlation

If you have to calculate a single correlation coefficient, you'll find that Excel's Correlation data analysis tool does the same thing CORREL does although the output is in tabular form. This tool becomes useful if you have to calculate multiple correlations on a set of data.

For example, Figure 15-5 shows SAT, High School Average, and GPA for 20 Sahutsket University students along with the dialog box for the Correlation data analysis tool.

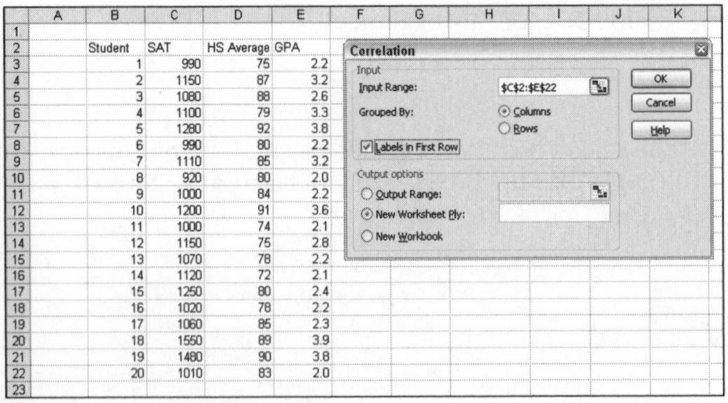

Student	SAT	HS Average	GPA
1	990	75	2.2
2	1150	87	3.2
3	1080	88	2.6
4	1100	79	3.3
5	1280	92	3.8
6	990	80	2.2
7	1110	85	3.2
8	920	80	2.0
9	1000	84	2.2
10	1200	91	3.6
11	1000	74	2.1
12	1150	75	2.8
13	1070	78	2.2
14	1120	72	2.1
15	1250	80	2.4
16	1020	78	2.2
17	1060	85	2.3
18	1550	89	3.9
19	1480	90	3.8
20	1010	83	2.0

Figure 15-5: The Correlation Data Analysis Tool dialog box along with data for SAT, high school average, and GPA.

The steps for using this tool are:

1. **Type the data into the worksheet, and include labels for the columns.**

   In this example, the data (including labels) are in C2:E22.

2. **From the Tools menu, select Data Analysis to open the Data Analysis dialog box.**

3. **In the Data Analysis dialog box, scroll down the Analysis Tools list and select Correlation. Click OK to open the Correlation dialog box.**

   This is the dialog box in Figure 15-5.

4. **In the Input Range box, enter the cell range that holds the data for the y-variable.**

   I entered $C$2:$E$22. Note the dollar signs ($) for absolute referencing.

5. **If the cell ranges include column headings, select the Labels in First Row option.**

   I included the headings in the ranges, so I selected this option.

6. **In the Output Options section, select an option to indicate where you want the results.**

   I selected the New Worksheet Ply option to put the results on a new page in the worksheet.

7. **Click OK.**

   Because I selected the New Worksheet Ply option, a newly created page opens with the results.

Tabled output

Figure 15-6 shows the tool's tabled output after I expanded the columns. The table is a *correlation matrix*.

Figure 15-6:
The Correlation data analysis tool's tabled output.

	A	B	C	D
1		SAT	HS Average	GPA
2	SAT	1		
3	HS Average	0.552527	1	
4	GPA	0.816625	0.714353653	1
5				

Each cell in the matrix represents the correlation of the variable in the row with the variable in the column. Cell B3 presents the correlation of SAT with High School Average, for example. Each cell in the main diagonal contains 1. This is because each main diagonal cell represents the correlation of a variable with itself.

It's only necessary to fill in half the matrix. The cells above the main diagonal would contain the same values as the cells below the main diagonal.

What does this table tell you, exactly? Read on. . . .

Multiple correlation

The correlation coefficients in this matrix combine to produce a *multiple correlation coefficient*. This is a number that summarizes the relationship between the dependent variable — GPA in this example — and the two independent variables (SAT and High School Average).

To show you how these correlation coefficients combine, I abbreviate GPA as G, SAT as S, and High School Average as H. So r_{GS} is the correlation coefficient for GPA and SAT, r_{GH} is the correlation coefficient for GPA and High School Average, and r_{SH} is the correlation coefficient for SAT and High School Average.

Here's the formula that puts them all together:

$$R_{G.SH} = \sqrt{\frac{r_{GS}^2 + r_{GH}^2 - 2r_{GS}\,r_{GH}\,r_{SH}}{1 - r_{GS}^2}}$$

The uppercase R on the left indicates that this is a multiple correlation coefficient, as opposed to the lowercase r that indicates a correlation between two variables. The subscript $G.SH$ means that the multiple correlation is between GPA and the combination of SAT and High School Average.

This is the calculation that produces Multiple R in the Regression Statistics section of the Regression data analysis tool's results. (See Chapter 14.)

For this example,

$$R_{G.SH} = \sqrt{\frac{(.816625)^2 + (.714354)^2 - 2(.816625)(.714354)(.552527)}{1 - (.816625)^2}} = .875529$$

Because I used the same data to show you multiple regression in Chapter 14, this value (with some additional decimal places) is in Figure 14-25, in cell B4.

If you square this number, you get the *multiple coefficient of determination*. In Chapter 14, I told you about R Square, and that's what this is. It's another item in the Regression Statistics that the Regression data analysis tool calculates. You also find it in LINEST's results, although it's not labeled.

Adjusting R²

Here's some more information about R² as it relates to Excel. In addition to R² — or as Excel likes to write it, R Square — the Regression data analysis tool calculates *Adjusted R Square*. In Figure 14-25, it's in cell B6. Why is it necessary to "adjust" R Square?

In multiple regression, adding independent variables (like High School Average) sometimes makes the regression equation less accurate. The multiple coefficient of determination, R Square, doesn't reflect this. Its denominator is SS_{Total} (for the dependent variable) and

that never changes. The numerator can only increase or stay the same. So any decline in accuracy doesn't result in a lower R Square.

Taking degrees of freedom into account fixes the flaw. Every time you add an independent variable, you change the degrees of freedom and that makes all the difference. Just so you know, here's the adjustment:

$$Adjusted\ R^2 = 1 - \left(1 - R^2\right)\left[\frac{(N-1)}{(N-k-1)}\right]$$

The k in the denominator is the number of independent variables.

For this example, that result is:

$$R^2{}_{G.SH} = (.875529)^2 = .766552$$

You can go back and see this number in Figure 14-24 in cell H5 (the LINEST results). You can also see it in Figure 14-25, cell B5 (the Regression data analysis tool report).

Partial correlation

GPA and SAT are associated with High School Average (in the example). Each one's association with High School Average might somehow hide the true correlation between them.

What would their correlation be if you could remove that association? Another way to say this: What would be the GPA-SAT correlation if you could hold High School Average constant?

One way to hold High School Average constant is to find the GPA-SAT correlation for a sample of students who have one High School Average — 87, for example. In a sample like that, the correlation of each variable with High School Average is zero. This usually isn't feasible in the real world, however.

Another way is to find the *partial correlation* between GPA and SAT. This is a statistical way of removing each variable's association with High School Average in your sample. You use the correlation coefficients in the correlation matrix to do this:

$$r_{GS.H} = \frac{r_{GS} - r_{GH}\, r_{SH}}{\sqrt{1 - r^2_{\,GH}}\,\sqrt{1 - r^2_{\,SH}}}$$

Once again, G stands for GPA, S for SAT, and H for High School Average. The subscript $GS.H$ means that the correlation is between GPA and SAT with High School Average "partialled out."

For this example,

$$r_{GS.H} = \frac{.816625 - (.714353)(.552527)}{\sqrt{1 - (.714353)^2}\,\sqrt{1 - (.552527)^2}} = .547005$$

Semipartial correlation

It's also possible to remove the correlation with High School Average from SAT without removing it from GPA. This is called *semipartial correlation*. The formula for this one also uses the correlation coefficients from the correlation matrix:

$$r_{G(S.H)} = \frac{r_{GS} - r_{GH}\, r_{SH}}{\sqrt{1 - r^2_{\,SH}}}$$

The subscript $G(S.H)$ means the correlation is between GPA and SAT with High School Average "partialled out" of SAT only.

Applying this formula to the example,

$$r_{G(S.H)} = \frac{.816625 - (.714353)(.552527)}{\sqrt{1 - (.552527)^2}} = .315714$$

Some statistics textbooks refer to semipartial correlation as *part correlation*.

Data Analysis Tool: Covariance

You use the Covariance data analysis tool the same way you use the Correlation data analysis tool. I won't go through the steps again. Instead, I'll just show you the tabled output in Figure 15-7. The data are from Figure 15-5.

The table is a *covariance matrix*. Each cell in the matrix shows the covariance of the variable in the row with the variable in the column (again, using N rather than N-1). Cell C4 shows the covariance of GPA with High School Average. The main diagonal in this matrix presents the variance of each variable (which is equivalent to the covariance of a variable with itself). In this case, the variance is what you compute if you use VARP.

Again, it's only necessary to fill half the matrix. Cells above the main diagonal would hold the same values as the cells below the main diagonal.

As is the case with COVAR, I don't see why you'd use this tool. I just include it for completeness.

Figure 15-7:
The
Covariance
data
analysis
tool's tabled
output for
SAT, high
school
average,
and GPA.

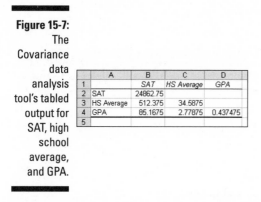

	A	B	C	D
1		SAT	HS Average	GPA
2	SAT	24862.75		
3	HS Average	512.375	34.5875	
4	GPA	85.1675	2.77875	0.437475
5				

Testing Hypotheses About Correlation

Excel has no worksheet function for testing hypotheses about *r*. As I pointed out before, you perform a *t*-test whose formula is:

$$t = \frac{r\sqrt{N-2}}{\sqrt{1-r^2}}$$

With 0.817 stored in cell H12, I used this formula to calculate *t*:

=H12*SQRT(20-2)/SQRT(1-H12^2)

I then used the answer (6.011 and some additional decimal places) as input to TDIST (along with 18 df and 1 tail) to find that the one-tailed probability of the result is way less than .05.

Worksheet Functions: FISHER, FISHERINV

Excel handles the rather complex transformations that enable you to test hypotheses about the difference between two correlation coefficients. FISHER transforms *r* to z. FISHERINV does the reverse. Just to refresh your memory, you use the transformed values in the formula

$$Z = \frac{z_1 - z_2}{\sigma_{z_1 - z_2}}$$

in which the denominator is

$$\sigma_{z_1 - z_2} = \sqrt{\frac{1}{N_1 - 3} + \frac{1}{N_2 - 3}}$$

In the example I discussed earlier (Sahutsket versus Farshimmelt), the correlation coefficients were .817 and .752, and I did a two-tailed test. The first step is to transform each correlation. I'll go through the steps for using FISHER to transform .817:

1. **Select a cell for FISHER's answer.**

   I selected B3 for the transformed value.

2. **Click the Insert Function button to open the Insert Function dialog box.**

3. **In the Insert Function dialog box, select FISHER and click OK to open its Function Arguments dialog box.**

   The FISHER Function Arguments dialog box appears in Figure 15-8.

Function Arguments

FISHER

X .817   = 0.817

= 1.147727958

Returns the Fisher transformation.

X is the value for which you want the transformation, a number between -1 and 1, excluding -1 and 1.

Formula result =   1.147727958

Help on this function   OK   Cancel

4. **In the X box, type the correlation coefficient.**

   I entered .817. The answer, 1.147728, appears in the dialog box.

5. **Click OK to put the answer into your selected cell.**

I selected B4 to store the transformation of .752. Next, I used this formula to calculate Z:

=(B3-B4)/SQRT((1/(20-3))+1/(24-3))

Finally, I used NORMSINV to find the critical value of z for rejecting H_0 with a two-tailed α of .05. Because the result of the formula (0.521633) is less than that critical value (1.96), the decision is to not reject H_0.

Part IV
Working with Probability

In this part . . .

Statistical analysis and decision-making rest on a foundation of probability. Throughout the book, I give a smattering of probability ideas — just enough to get you through the statistics. Part IV gives a more in-depth treatment, and covers related Excel features. You find out about discrete and continuous random variables, counting rules, conditional probability, and probability distributions. In this part, I also discuss specific probability distributions that are appropriate for specific purposes. Part IV ends with an exploration of modeling, tests of how well a model fits data, and how Excel deals with modeling and testing.

Chapter 16

Introducing Probability

. . . .

In This Chapter

▶ Defining probability

▶ Working with probability

▶ Dealing with random variables and their distributions

▶ Focusing on the binomial distribution

. . . .

*T*hroughout this book, I toss around the concept of probability because it's the basis of hypothesis testing and inferential statistics. Most of the time, I represent probability as the proportion of area under part of a distribution. For example, the probability of a Type I error (a/k/a α) is the area in a tail of the standard normal distribution or the t distribution.

In this chapter, I explore probability in greater detail, including random variables, permutations, and combinations. I examine probability's fundamentals and applications, zero in on a couple of specific probability distributions, and I discuss probability-related Excel worksheet functions.

What Is Probability?

Most of us have an intuitive idea of what probability is all about. Toss a fair coin, and you have a 50-50 chance it comes up "Head." Toss a fair die (one of a pair of dice), and you have a one-in-six chance it comes up "2."

If you want to be more formal in your definition, you most likely say something about all the possible things that can happen, and the proportion of those things you care about. Two things can happen when you toss a coin, and if you only care about one of them (Head), the probability of that event happening is one out of two. Six things can happen when you toss a die, and if you only care about one of them (2), the probability of that event happening is one out of six.

Experiments, trials, events, and sample spaces

Statisticians and others who work with probability refer to a process like tossing a coin or throwing a die as an *experiment*. Each time you go through the process you complete a *trial*.

This might not fit your personal definition of an experiment (or of a trial, for that matter), but for a statistician, an experiment is any process that produces one of at least two distinct results (like a Head or a Tail).

Another piece of the definition of an experiment: You can't predict the result with certainty. Each distinct result is called an *elementary outcome*. Put a bunch of elementary outcomes together and you have an *event*. For example, with a die, the elementary outcomes 2, 4, and 6 make up the event "even number."

Put all the possible elementary outcomes together and you've got yourself a *sample space*. The numbers 1, 2, 3, 4, 5, and 6 make up the sample space for a die. "Head" and "Tail" make up the sample space for a coin.

Sample spaces and probability

How does all this play into probability? If each elementary outcome in a sample space is equally likely, the probability of an event is

$$pr\,(\text{Event}) = \frac{\text{Number of Elementary Outcomes in the Event}}{\text{Number of Elementary Outcomes in the Sample Space}}$$

So the probability of tossing a die and getting an even number is

$$pr\,(\text{Even Number}) = \frac{\text{Number of Even–Numbered Elementary Outcomes}}{\text{Number of Possible Outcomes of a Die}} = \frac{3}{6} = .5$$

If the elementary outcomes are not equally likely, you find the probability of an event in a different way. First, you have to have some way of assigning a probability to each one. Then you add up the probabilities of the elementary outcomes that make up the event.

A couple of things to bear in mind about outcome probabilities: Each probability has to be between zero and one. All the probabilities of elementary outcomes in a sample space have to add up to 1.00.

How do you assign those probabilities? Sometimes you have advance information — such as knowing that a coin is biased toward coming up Head 60 percent of the time. Sometimes you just have to think through the situation to figure out the probability of an outcome.

Here's a quick example of "thinking through." Suppose a die is biased so that the probability of an outcome is proportional to the numerical label of the outcome: A 6 comes up six times as often as a 1, a 5 comes up five times as often as a 1, and so on. What is the probability of each outcome? All the probabilities have to add up to 1.00, and all the numbers on a die add up to 21 (1+2+3+4+5+6 = 21), so the probabilities are: $pr(1) = 1/21$, $pr(2) = 2/21$, ..., $pr(6) = 6/21$.

Compound Events

Some rules for dealing with *compound events* help you "think through." A compound event consists of more than one event. It's possible to combine events by either *union* or *intersection* (or both).

Union and intersection

On a toss of a fair die, what's the probability of getting a 1 or a 4? Mathematicians have a symbol for *or*. It looks like this ∪ and it's called *union*. Using this symbol, the probability of a 1 or a 4 is $pr(1 \cup 4)$.

In approaching this kind of probability, it's helpful to keep track of the elementary outcomes. One elementary outcome is in each event, so the event "1 or 4" has two elementary outcomes. With a sample space of six outcomes, the probability is 2/6 or 1/3. Another way to calculate this is

$$pr(1 \cup 4) = pr(1) + pr(4) = (1/6) + (1/6) = 2/6 = 1/3$$

Here's a slightly more involved one: What's the probability of getting a number between 1 and 3 or a number between 2 and 4?

Just adding the elementary outcomes in each event won't get it done this time. Three outcomes are in the event "between 1 and 3" and three are in the event "between 2 and 4." The probability can't be 3 + 3 divided by the six outcomes in the sample space because that's 1.00, leaving nothing for $pr(5)$ and $pr(6)$. For the same reason, you can't just add the probabilities.

The challenge arises in the overlap of the two events. The elementary outcomes in "between 1 and 3" are 1, 2, and 3. The elementary outcomes in "between 2 and 4" are 2, 3, 4. Two outcomes overlap: 2 and 3. In order to not count them twice, the trick is to subtract them from the total.

A couple of things will make life easier as I proceed. I abbreviate "between 1 and 3" as *A* and "between 2 and 4" as *B*. Also, I use the mathematical symbol for "overlap." The symbol is ∩ and it's called *intersection*.

Using the symbols, the probability of "between 1 and 3" or "between 2 and 4" is

$$pr(A \cup B) = \frac{\text{Number of outcomes in } A + \text{Number of outcomes in } B - \text{Number of outcomes in } (A \cap B)}{\text{Number of outcomes in the sample space}}$$

$$pr(A \cup B) = \frac{3+3-2}{6} = \frac{4}{6} = \frac{2}{3}$$

You can also work with the probabilities:

$$pr(A \cup B) = \frac{3}{6} + \frac{3}{6} - \frac{2}{6} = \frac{4}{6} = \frac{2}{3}$$

The general formula is:

$$pr(A \cup B) = pr(A) + pr(B) - pr(A \cap B)$$

Why was it OK to just add the probabilities together in the earlier example? Because $pr(1 \cap 4)$ is zero: It's impossible to get a 1 and a 4 in the same toss of a die. Whenever $pr(A \cap B) = 0$, A and B are said to be *mutually exclusive*.

Intersection again

Imagine throwing a coin and rolling a die at the same time. These two experiments are *independent* because the result of one has no influence on the result of the other.

What's the probability of getting a Head and a 4? You use the intersection symbol and write this as $pr(\text{Head} \cap 4)$:

$$pr(Head \cap 4) = \frac{\text{Number of Elementary Outcomes in Head} \cap 4}{\text{Number of Elementary Outcomes in the Sample Space}}$$

Start with the sample space. Table 16-1 lists all the elementary outcomes.

Table 16-1	The Elementary Outcomes in the Sample Space for Throwing a Coin and Rolling a Die
Head, 1	Tail, 1
Head, 2	Tail, 2
Head, 3	Tail, 3
Head, 4	Tail, 4
Head, 5	Tail, 5
Head, 6	Tail, 6

As the table shows, 12 outcomes are possible. How many outcomes are in the event "Head and 4"? Just one. So

$$pr(Head \cap 4) = \frac{\text{Number of Elementary Outcomes in } Head \cap 4}{\text{Number of Elementary Outcomes in the Sample Space}} = \frac{1}{12}$$

You can also work with the probabilities:

$$pr(Head \cap 4) = pr(Head) \times pr(4) = \frac{1}{2} \times \frac{1}{6} = \frac{1}{12}$$

In general, if A and B are independent,

$$pr(A \cap B) = pr(A) \times pr(B)$$

Conditional Probability

In some circumstances, you narrow the sample space. For example, suppose I toss a die, and I tell you the result is greater than 2. What's the probability that it's a 5?

Ordinarily, the probability of a 5 would be 1/6. In this case, however, the sample space isn't 1, 2, 3, 4, 5, and 6. When you know the result is greater than 3, the sample space becomes 3, 4, 5, and 6. The probability of a 5 is now 1/4.

This is an example of *conditional probability.* It's "conditional" because I've given a "condition" — the toss resulted in a number greater than 2. The notation for this is

pr(5 | Greater than 2)

The vertical line is a shorthand for the word *given*, and you read that notation as "the probability of a 5 given Greater than 2."

Working with the probabilities

In general, if you have two events A and B,

$$pr(A|B) = \frac{pr(A \cap B)}{pr(B)}$$

as long as $pr(B)$ isn't zero.

For the intersection in the numerator in the right, this is *not* a case where you just multiply probabilities together. In fact, if you could do that, you don't have a conditional probability because that means A and B are independent. If they're independent, one event can't be conditional on the other.

You have to think through the probability of the intersection. In a die, how many outcomes are in the event "$5 \cap$ Greater than 2"? Just one, so $pr(5 \cap$ Greater than 2) is 1/6, and

$$pr\left(5|\text{Greater than 2}\right) = \frac{pr\left(5|\text{Greater than 2}\right)}{pr\left(\text{Greater than 2}\right)} = \frac{\frac{1}{6}}{\frac{4}{6}} = \frac{1}{4}$$

The foundation of hypothesis testing

All the hypothesis testing I've gone through in previous chapters involves conditional probability. When you calculate a sample statistic, compute a statistical test, and then compare the test statistic against a critical value, you're looking for a conditional probability. Specifically, you're trying to find

pr(obtained test statistic or a more extreme value | H_0 is true)

If that conditional probability is low (less than .05 in all the examples I show you in hypothesis-testing chapters), you reject H_0.

Large Sample Spaces

When dealing with probability, it's important to understand the sample space. In the examples I show you, the sample spaces are small. With a coin or a die, it's easy to list all the elementary outcomes.

The world, of course, isn't that simple. In fact, probability problems that live in statistics textbooks aren't even that sample. Most of the time, sample spaces are large and it's not convenient to list every elementary outcome.

Take, for example, rolling a die twice. How many elementary outcomes are in the sample space consisting of both tosses? You can sit down and list them, but it's better to reason it out: Six possibilities for the first toss, and each of those six can pair up with six possibilities on the second. So the sample space has $6 \times 6 = 36$ possible elementary outcomes. (This is similar to the coin-and-die sample space in Table 16-1, where the sample space consists of $2 \times 6 = 12$ elementary outcomes. With 12 outcomes, it was easy to list them all in a table. With 36 outcomes, it starts to get . . . well . . . dicey.)

Events often require some thought, too. What's the probability of rolling a die twice and totaling five? You have to count the number of ways the two tosses can total five, and then divide by the number of elementary outcomes in the sample space (36). You total a five by getting any of these pairs of tosses: 1 and 4, 2 and 3, 3 and 2, or 4 and 1. That's four ways and they don't overlap (excuse me, intersect), so

$$pr\,(5) = \frac{\text{Number of Ways of Rolling a 5}}{\text{Number of Possible Outcomes of Two Tosses}} = \frac{4}{36} = .11$$

Listing all the elementary outcomes for the sample space is often a nightmare. Fortunately, shortcuts are available, as I show in the upcoming subsections. Because each shortcut quickly helps you count a number of items, another name for a shortcut is a *counting rule*.

Believe it or not, I just slipped one counting rule past you. A couple of paragraphs ago, I said that in two tosses of a die you have a sample space of 6×6 = 36 possible outcomes. This is the *product rule:* If N_1 outcomes are possible on the first trial of an experiment, and N_2 outcomes on the second trial, the number of possible outcomes is N_1N_2. Each possible outcome on the first trial can associate with all possible outcomes on the second. What about three trials? That's $N_1N_2N_3$.

Now for a couple more counting rules.

Permutations

Suppose you have to arrange five objects into a sequence. How many ways can you do that? For the first position in the sequence, you have five choices. Once you make that choice, you have four choices for the second position. Then you have three choices for the third, two for the fourth, and one for the fifth. The number of ways is (5)(4)(3)(2)(1) = 120.

In general, the number of sequences of N objects is $N(N\text{-}1)(N\text{-}2)\ldots(2)(1)$. This kind of computation occurs fairly frequently in probability applications, and it has its own notation, $N!$ You don't read this by screaming out "N" in a loud voice. Instead, it's *N factorial.* By definition, 1! = 1, and 0! = 1.

Now for the good stuff. If you have to order the 26 letters of the alphabet, the number of possible sequences is 26!, a huge number. But suppose the task is to create five-letter sequences so that no letter repeats in the sequence. How many ways can you do that? You have 26 choices for the first letter, 25 for the second, 24 for the third, 23 for the fourth, 22 for the fifth, and that's it. So that's (26)(25(24)(23)(22). Here's how that product is related to 26!:

$$\frac{26!}{21!}$$

Each sequence is called a *permutation.* In general, if you take permutations of N things r at a time, the notation is $_NP_r$ (the P stands for permutation). The formula is

$$_NP_r = \frac{N!}{(N-r)!}$$

Combinations

In the example I just showed you, these sequences are different from one another: *abcde*, *adbce*, *dbcae*, and on and on and on. In fact, you could come up with 5! = 120 of these different sequences just for the letters a, b, c, d, and e.

Suppose I add the restriction that one of these sequences is no different from another, and all I'm concerned about is having sets of five nonrepeating letters in no particular order. Each set is called a *combination*. For this example, the number of combinations is the number of permutations divided by 5!:

$$\frac{26!}{5!\,(21!)}$$

In general, the notation for combinations of *N* things taken *r* at a time is $_NC_r$ (the *C* stands for "combination"). The formula is

$$_NC_r = \frac{N!}{r!(N-r)!}$$

Worksheet Functions

Three Excel functions help you with factorials, permutations, and combinations. Excel categorizes one of them as a Statistical function, but, surprisingly, not the other two.

FACT

FACT, which computes factorials, is one of the functions not categorized as Statistical. Instead, you find it under Math & Trig Functions. It's easy to use. Supply it with a number, and it returns the factorial:

1. **Select a cell for FACT's answer.**

2. **Click the Insert Function button to open the Insert Function dialog box.**

3. **In the Insert Function dialog box, select FACT and click OK to open its Function Arguments dialog box.**

4. **In the Number box, type the number whose factorial you want to compute.**

   The answer appears in the dialog box. If you typed 5 in Step 3, for example, 120 appears.

5. **Click OK to put the answer into the selected cell.**

PERMUT

You'll find this one listed under Statistical Functions. As its name suggests, PERMUT enables you to calculate $_NP_r$. Here's how to use it to find $_{26}P_5$, the number of five-letter sequences (no repeating letters) you can create from the 26 letters of the alphabet. In a permutation, remember, *abcde* is considered different from *bcdae*.

1. **Select a cell for PERMUT's answer.**

2. **Click the Insert Function button to open the Insert Function dialog box.**

3. **In the Insert Function dialog box, select PERMUT and click OK to open its Function Arguments dialog box (Figure 16-1).**

Figure 16-1:
The
Function
Arguments
dialog
box for
PERMUT.

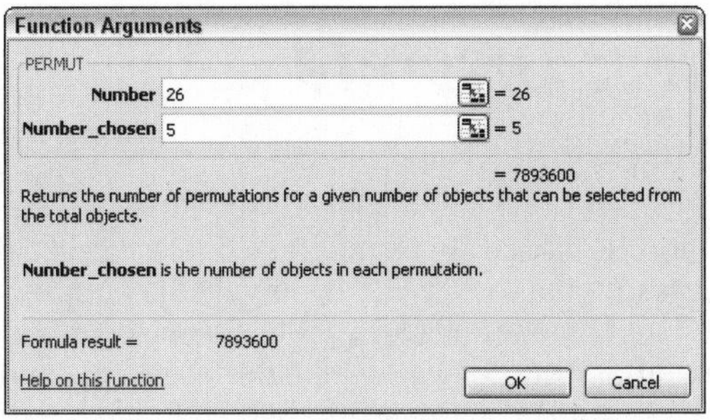

Function Arguments

PERMUT

Number 26 = 26
Number_chosen 5 = 5

= 7893600

Returns the number of permutations for a given number of objects that can be selected from the total objects.

Number_chosen is the number of objects in each permutation.

Formula result = 7893600

Help on this function   OK   Cancel

4. **In the Number box, type the *N* in $_NP_r$.**

For this example, *N* is 26.

5. **In the Number_selected box, type the *r* in $_NP_r$.**

That would be 5.

The answer appears in the dialog box.

For this example, the answer is 7893600.

6. **Click OK to put the answer into the selected cell.**

COMBIN

COMBIN works pretty much the same way as PERMUT. Excel groups COMBIN under Math & Trig Functions.

Here's how you use it to find $_{26}C_5$, the number of ways to construct a five-letter sequence (no repeating letters) from the 26 letters of the alphabet. In a combination, *abcde* is considered equivalent to *bcdae*.

1. **Select a cell for COMBIN's answer.**

2. **Click the Insert Function button to open the Insert Function dialog box.**

3. **In the Insert Function dialog box, select COMBIN and click OK to open its Function Arguments dialog box.**

4. **In the Number box, type the N in $_NC_r$.**

   Once again, N is 26.

5. **In the Number_chosen box, type the r in $_NC_r$.**

   And again, r is 5. The answer appears in the dialog box. For this example, the answer is 65870.

6. **Click OK to put the answer into the selected cell.**

Random Variables: Discrete and Continuous

Return to tosses of a fair die, where six elementary outcomes are possible. If I use x to refer to the result of a toss, x can be any whole number from 1 to 6. Because x can take on a set of values, it's a variable. Because x's possible values correspond to the elementary outcomes of an experiment (meaning you can't predict its values with absolute certainty) x is called a *random variable*.

Random variables come in two varieties. One variety is *discrete*, of which die-tossing is a good example. A discrete random variable can only take on what mathematicians like to call a *countable* number of values — like the numbers 1 through 6. Values between the whole numbers 1 through 6 (like 1.25 or 3.1416) are impossible for a random variable that corresponds to the outcomes of die tosses.

The other kind of random variable is *continuous*. A continuous random variable can take on an infinite number of values. Temperature is an example. Depending on the precision of a thermometer, it's possible to have temperatures like 34.516 degrees.

Probability Distributions and Density Functions

Back again to die tossing. Each value of the random variable x (1-6, remember) has a probability. If the die is fair, each probability is 1/6. Pair each value of a discrete random variable like x with its probability, and you have a *probability distribution*.

Probability distributions are easy enough to represent in graphs. Figure 16-2 shows the probability distribution for *x*.

A random variable has a mean, a variance, and a standard deviation. Calculating these parameters is pretty straightforward. In the world of random variables, the mean is called the *expected value*, and the expected value of random variable *x* is abbreviated as *E(x)*. Here's how you calculate it:

$$E(x) = \sum x \big(pr(x) \big)$$

For the probability distribution in Figure 16-2, that's

$$E(x) = \sum x \big(pr(x) \big) = 1\left(\frac{1}{6}\right) + 2\left(\frac{1}{6}\right) + 3\left(\frac{1}{6}\right) + 4\left(\frac{1}{6}\right) + 5\left(\frac{1}{6}\right) + 6\left(\frac{1}{6}\right) = 3.5$$

The variance of a random variable is often abbreviated as *V(x)*, and the formula is

$$V(x) = \sum x^2 \big(pr(x) \big) - \big[E(x) \big]^2$$

Working with the probability distribution in Figure 16-2 once again,

$$V(x) = 1^2\left(\frac{1}{6}\right) + 2^2\left(\frac{1}{6}\right) + 3^2\left(\frac{1}{6}\right) + 4^2\left(\frac{1}{6}\right) + 5^2\left(\frac{1}{6}\right) + 6^2\left(\frac{1}{6}\right) - 3.5^2 = 2.917$$

The standard deviation is the square root of the variance, which in this case is 1.708.

For continuous random variables, things get a little trickier. You can't pair a value with a probability because you can't really pin down a value. Instead, you associate a continuous random variable with a mathematical rule (an equation) that generates *probability density,* and the distribution is called a *probability density function.* To calculate the mean and variance of a continuous random variable, you need calculus.

In Chapter 8, I show you a probability density function — the standard normal distribution. I reproduce it here as Figure 16-3.

In the figure, *f(x)* represents the probability density. Because probability density can involve some heavyweight mathematical concepts, I won't go into it. As I mention in Chapter 8, think of probability density as something that turns the area under the curve into probability.

While you can't speak of the probability of a specific value of a continuous random variable, you can work with the probability of an interval. To find the probability that the random variable takes on a value within an interval, you find the proportion of the total area under the curve that's inside that interval. Figure 16-3 shows this. The probability that *x* is between 0 and 1σ is .3413.

For the rest of this chapter, I deal just with discrete random variables. A specific one is up next.

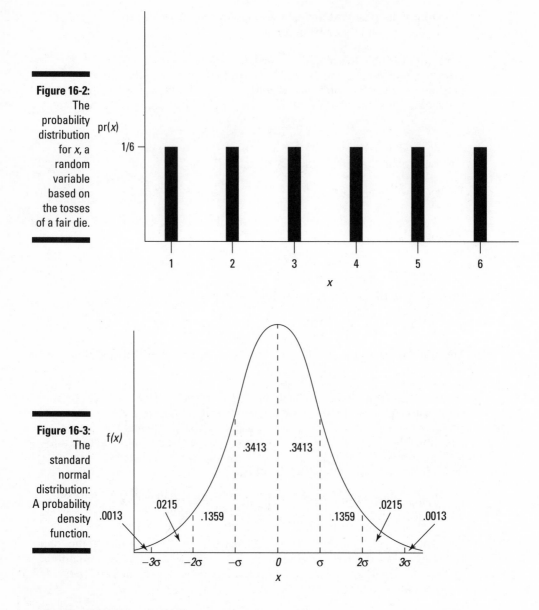

Figure 16-2: The probability distribution for *x*, a random variable based on the tosses of a fair die.

Figure 16-3: The standard normal distribution: A probability density function.

The Binomial Distribution

Imagine an experiment that has these five characteristics:

✔ The experiment consists of *N* identical trials.

A trial could be a toss of a die, or a toss of a coin.

✔ Each trial results in one of two elementary outcomes.

It's standard to call one outcome a *success* and the other a *failure*. For die tossing, a success might be a toss that comes up 3, in which case a failure is any other outcome.

✔ The probability of a success remains the same from trial to trial.

It's pretty standard to use p to represent the probability of a success and *1-p* (or q) to represent the probability of a failure.

✔ The trials are independent.

✔ The discrete random variable x is the number of successes in the N trials.

This type of experiment is called a *binomial experiment*. The probability distribution for x follows this rule:

$$pr(x) = \frac{N!}{x!(N-x)!}p^x(1-p)^{N-x}$$

On the extreme right, $p^x(1-p)^{N-x}$ is the probability of one combination of x successes in N trials. The term to its immediate left is $_NC_x$, the number of possible combinations of x successes in N trials.

This is called the *binomial distribution*. You use it to find probabilities like the probability you'll get four 3s in ten tosses of a die:

$$pr(4) = \frac{10!}{4!(6)!}\left(\frac{1}{6}\right)^4\left(\frac{5}{6}\right)^6 = .054$$

The *negative binomial distribution* is closely related. In this distribution, the random variable is the number of trials before the xth success. For example, you use the negative binomial to find the probability of five tosses that result in anything but a 3 before the fourth time you roll a 3.

For this to happen, in the eight tosses before the fourth 3, you have to get five non-3s and three successes (tosses when a 3 comes up). Then, the next toss results in a 3. The probability of a combination of four successes and five failures is $p^4(1-p)^5$. The number of ways you can have a combination of five failures and 4-1 successes is $_{5+4-1}C_{4-1}$. So the probability is

$$pr(5 \text{ failures before the 4th success}) = \frac{(5+4-1)!}{4-1!(5)!}\left(\frac{1}{6}\right)^4\left(\frac{5}{6}\right)^5 = .017$$

In general, the negative binomial distribution (sometimes called the *Pascal distribution*) is

$$pr(f \text{ failures before the } x\text{th success}) = \frac{(f+x-1)!}{(x-1)!(f)!}p^x(1-p)^f$$

Worksheet Functions

These distributions are computation-intensive, so I'll get to the worksheet functions right away.

BINOMDIST

BINOMDIST is Excel's worksheet function for the binomial distribution. As an example, I use BINOMDIST to calculate the probability of getting four 3s in ten tosses of a fair die:

1. **Select a cell for BINOMDIST's answer.**

2. **Click the Insert Function button to open the Insert Function dialog box.**

3. **In the Insert Function dialog box, select BINOMDIST and click OK to open its Function Arguments dialog box (see Figure 16-4).**

4. **In the Number_s box, type the number of successes.**

   For this example, the number of successes is 4.

5. **In the Trials box, type the number of trials.**

   The number of trials is 10.

6. **In the Probability_s box, type the probability of a success.**

   I entered 1/6, the probability of a 3 on a toss of a fair die.

7. **In the Cumulative box, type** FALSE **for the probability of exactly the number of successes typed in the Number_s box. Type** TRUE **for the probability of getting that number of successes or fewer.**

   I typed FALSE.

   The answer appears in the dialog box.

8. **Click OK to put the answer into the selected cell.**

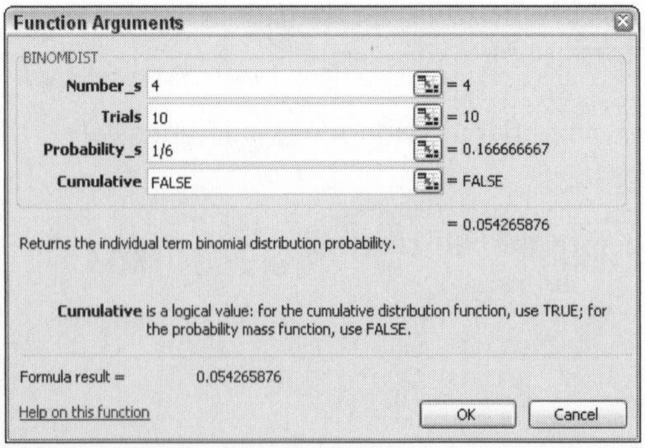

Figure 16-4:
The Function Arguments dialog box for BINOMDIST.

Function Arguments		
BINOMDIST		
Number_s	4	= 4
Trials	10	= 10
Probability_s	1/6	= 0.166666667
Cumulative	FALSE	= FALSE
		= 0.054265876

Returns the individual term binomial distribution probability.

Cumulative is a logical value: for the cumulative distribution function, use TRUE; for the probability mass function, use FALSE.

Formula result = 0.054265876

Help on this function   OK   Cancel

To give you a better idea of what the binomial distribution looks like, I use BINOMDIST (with FALSE entered in the Cumulative box) to find $pr(0)$ through $pr(10)$, and then I use the Excel Chart Wizard (see Chapter 3) to graph the results. Figure 16-5 shows the data and the graph.

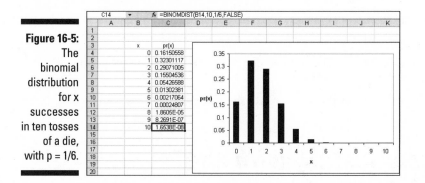

Figure 16-5:
The binomial distribution for x successes in ten tosses of a die, with p = 1/6.

Incidentally, if you type TRUE in the Cumulative box, the result is .984 (and some more decimal places), which is $pr(0) + pr(1) + pr(2) + pr(3) + pr(4)$.

NEGBINOMDIST

As its name suggests, NEGBINOMDIST handles the negative binomial distribution. I use it here to work out the example I gave you earlier — the probability of getting five failures (tosses that result in anything but a 3) before the fourth success (the fourth 3).

1. **Select a cell for NEGBINOMDIST's answer.**

2. **Click the Insert Function button to open the Insert Function dialog box.**

3. **In the Insert Function dialog box, select NEGBINOMDIST and click OK to open its Function Arguments dialog box (see Figure 16-6).**

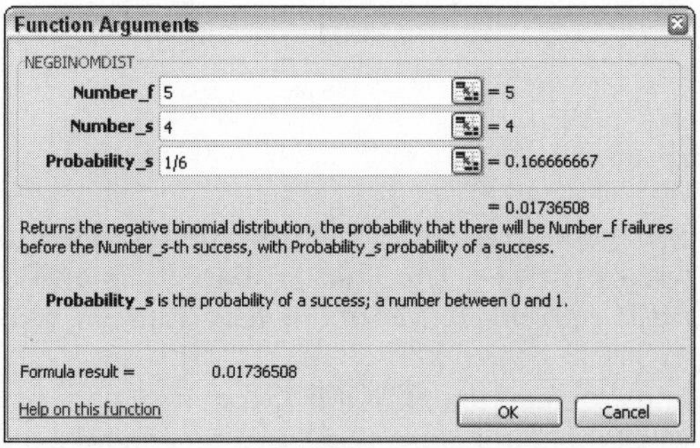

Figure 16-6:
The Function Arguments dialog box for NEG-BINOMDIST.

4. **In the Number_f box, type the number of failures.**

   The number of failures is 5 for this example.

5. **In the Number_s box, type the number of successes.**

   I typed 4.

6. **In the Probability_s box, type the probability of a success.**

   I typed 1/6.

   The answer appears in the dialog box. The answer is 0.017 and some additional decimal places.

7. **Click OK to put the answer into the selected cell.**

Hypothesis Testing with the Binomial Distribution

Hypothesis tests sometimes involve the binomial distribution. Typically, you have some idea about the probability of a success, and you put that idea into a null hypothesis. Then, you perform N trials and record the number of successes. Finally, you compute the probability of getting that many successes or a more extreme amount if your H_0 is true. If the probability is low, reject H_0.

When you test in this way, you're using sample statistics to make an inference about a population parameter. Here, that parameter is the probability of a success in the population of trials. By convention, Greek letters represent parameters. Statisticians use π (pi), the Greek equivalent of p, to stand for the probability of a success in the population.

Continuing with the die-tossing example, suppose you have a die and you want to test whether or not it's fair. You suspect that if it's not, it's biased toward 3. Define a toss that results in 3 as a success. You toss it ten times. Four tosses are successes. Casting all this into hypothesis-testing terms:

H_0: $\pi \leq 1/6$

$H_{1:}$ $\pi > 1/6$

As I usually do, I set $\alpha = .05$

To test these hypotheses, you have to find the probability of getting at least four successes in ten tosses with $p = 1/6$. That probability is $pr(4) + pr(5) + pr(6) + pr(7) + pr(8) + pr(9) + pr(10)$. If the total is less than .05, reject H_0.

That's a lot of calculating. You can use BINOMDIST to take care of it all (as I did when I set up the worksheet in Figure 16-5), or you can take a different route. You can find a critical value for the number of successes, and if the number of successes is greater than the critical value, reject H_0.

How do you find the critical value? You can use a convenient worksheet function that I'm about to show you.

CRITBINOM

This function is tailor-made for binomial-based hypothesis testing. Give CRITBINOM the number of trials, the probability of a success, and a criterion cumulative probability. CRITBINOM returns the smallest value of x (the number of successes) for which the cumulative probability is greater than or equal to the criterion.

Here are the steps for the hypothesis testing example I just showed you:

1. **Select a cell for CRITBINOM's answer.**

2. **Click the Insert Function button to open the Insert Function dialog box.**

3. **In the Insert Function dialog box, select CRITBINOM and click OK to open its Function Arguments dialog box (see Figure 16-7).**

4. **In the Trials box, type the number of trials.**

   The number of trials is 10.

5. **In the Probability_s box, type the probability of a success.**

   I typed 1/6, the value of π according to H_0.

6. **In the Alpha box, type the cumulative probability you want to exceed.**

   I typed .95 because I want to find the critical value that cuts off the upper 5 percent of the binomial distribution.

   The critical value appears in the dialog box. The critical value is 4.

7. **Click OK to put the answer into the selected cell.**

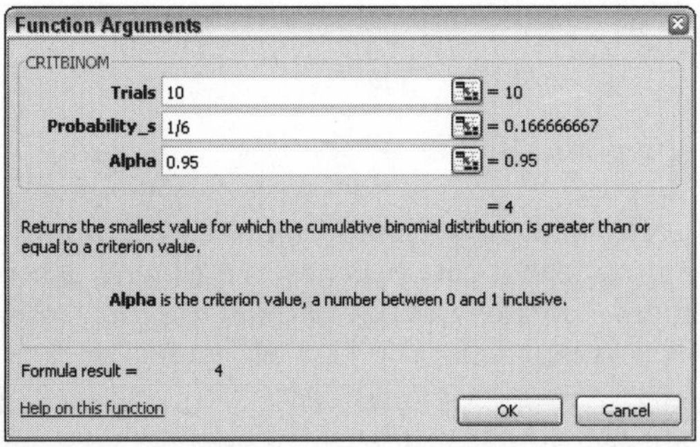

Figure 16-7:
The
Function
Arguments
dialog
box for
CRITBINOM.

Function Arguments

CRITBINOM

Trials   10   = 10
Probability_s   1/6   = 0.166666667
Alpha   0.95   = 0.95
                                               = 4

Returns the smallest value for which the cumulative binomial distribution is greater than or equal to a criterion value.

Alpha is the criterion value, a number between 0 and 1 inclusive.

Formula result =   4

Help on this function   OK   Cancel

As it happens, the critical value is the number of successes in the sample. The decision is to reject H_0.

More on hypothesis testing

In some situations, the binomial distribution approximates the standard normal distribution. When this happens, you use the statistics of the normal distribution to answer questions about the binomial distribution.

Those statistics involve z-scores, which means you have to know the mean and the standard deviation of the binomial. Fortunately, they're easy to compute. If N is the number of trials, and π is the probability of a success, the mean is

$$\mu = N\pi$$

the variance is

$$\sigma^2 = N\pi\left(1 - \pi\right)$$

and the standard deviation is

$$\sigma = \sqrt{N\pi\left(1 - \pi\right)}$$

The binomial approximation to the normal is appropriate when $N\pi \geq 5$ and $N(1 - \pi) \geq 5$.

When you test a hypothesis, you're making an inference about π, and you have to start with an estimate. You run N trials and get x successes. The estimate is

$$P = \frac{x}{N}$$

In order to create a z-score, you need one more piece of information — the standard error of P. This sounds harder than it is, because this standard error is just

$$\sigma_P = \sqrt{\frac{\pi\left(1 - \pi\right)}{N}}$$

Now you're ready for a hypothesis test.

Here's an example: The CEO of the FarKlempt Robotics Corporation believes that 50 percent of FarKlempt robots are purchased for home use. A sample of 1,000 FarKlempt customers indicates that 550 of them use their robots at home. Is this significantly different from what the CEO believes? The hypotheses:

$H_0: \pi = .50$

$H_1: \pi \neq .50$

I set $\alpha = .05$

$N\pi = 500$, and $N(1 - \pi) = 500$, so the normal approximation is appropriate.

First, calculate P:

$$P = \frac{x}{N} = \frac{550}{1000} = .55$$

Now, create a z-score

$$z = \frac{P - \pi}{\sqrt{\dfrac{\pi(1 - \pi)}{N}}} = \frac{.55 - .50}{\sqrt{\dfrac{(.50)(1 - .50)}{1000}}} = \frac{.05}{\sqrt{\dfrac{.25}{1000}}} = 3.162$$

With $\alpha = .05$, is 3.162 a large enough z-score to reject H_0? An easy way to find out is to use the worksheet function NORMSDIST (Chapter 8). If you do, you'll find that this z-score cuts off less than .01 of the area in the upper tail of the standard normal distribution. The decision is to reject H_0.

The Hypergeometric Distribution

Here's another distribution that deals with successes and failures.

I start with an example. In a set of 16 lightbulbs, 9 are good and 7 are defective. If you randomly select 6 lightbulbs out of these 16, what's the probability that 3 of the 6 are good? Consider selecting a good lightbulb as a "success."

When you finish selecting, your set of selections is a combination of 3 of the 9 good lightbulbs together with a combination of 3 of the 7 defective lightbulbs. The probability of getting 3 good bulbs is a . . . well . . . combination of counting rules:

$$pr(3) = \frac{(_9C_3)(_7C_3)}{_{16}C_6} = \frac{(84)(35)}{8008} = .37$$

Each outcome of the selection of the good lightbulbs can associate with all outcomes of the selection of the defective lightbulbs, so the product rule is appropriate for the numerator. The denominator (the sample space) is the number of possible combinations of 6 items in a group of 16.

This is an example of the *hypergeometric distribution.* In general, with a small population that consists of N_1 successes and N_2 failures, the probability of x successes in a sample of m items is

$$pr(x) = \frac{\left(_{N_1}C_x\right)\left(_{N_2}C_{m-x}\right)}{_{N_1+N_2}C_m}$$

The random variable x is said to be a *hypergeometrically distributed random variable.*

HYPERGEOMDIST

This function calculates everything for you when you deal with the hypergeometric distribution. Here's how to use it to go through the example I just showed you:

1. **Select a cell for HYPERGEOMDIST's answer.**

2. **Click the Insert Function button to open the Insert Function dialog box.**

3. **In the Insert Function dialog box, select HYPERGEOMDIST and click OK to open its Function Arguments dialog box (see Figure 16-8).**

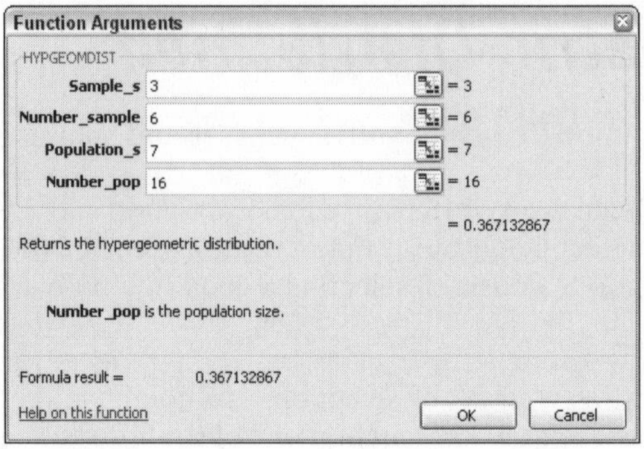

Figure 16-8: The Function Arguments dialog box for HYPER-GEOMDIST.

4. **In the Sample_s box, type the number of successes in the sample.**

   The number of successes in the sample is 3 for this example.

5. **In the Number_sample box, type the number of items in the sample.**

   I typed 6, the sample size for this example.

6. **In the Population_s box, type the number of successes in the population.**

   I typed 7, the number of good lightbulbs.

7. In the Number_pop box, type the number of items in the population.

For this example, that's 16, the total number of lightbulbs.

The answer appears in the dialog box. The answer is 0.37 and some additional decimal places.

8. Click OK to put the answer into the selected cell.

As I do with the binomial, I use HYPERGEOMDIST to calculate $pr(0)$ through $pr(6)$ for this example. Then I use the Chart Wizard (see Chapter 3) to graph the results. Figure 16-9 shows the data and the graph. My objective is to help you visualize and understand the hypergeometric distribution.

Figure 16-9:
The hypergeometric distribution for x successes in a six-item sample from a population that consists of seven successes and nine failures.

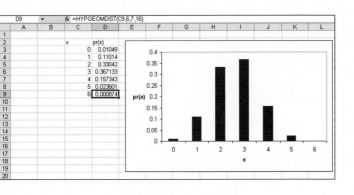

Chapter 17

More on Probability

In This Chapter

▶ The beta version

▶ Pursuing Poisson

▶ Grappling with gamma

▶ Exponentially speaking

....

*I*n the previous chapter, I delve into probability in a semiformal way and introduce distributions of random variables. The binomial distribution is the starting point. In this chapter, I examine additional distributions.

One of the symbols on the pages of this book (and other books in the For Dummies series) lets you know that "Technical Stuff" follows. It might have been a good idea to hang that symbol above this chapter's title. So here's a small note of caution: Some mathematics follows. I put the math in to help you understand what you're doing when you work with the dialog boxes of the Excel functions I describe.

Are these functions on the esoteric side? Well . . . yes. Will you ever have occasion to use them? Well . . . you just might.

Beta

I begin with an esoteric one because it connects with the binomial distribution, which I discuss in Chapter 16. The beta distribution (not to be confused with "beta," the probability of a Type 2 error) is a sort of chameleon in the world of distributions. It takes on a wide variety of appearances, depending on the circumstances. I won't give you all the mathematics behind the beta distribution because the full treatment involves calculus.

The connection with the binomial is this: In the binomial, the random variable x is the number of successes in N trials with p as the probability of a success. N and p are constants. In the beta distribution, the random variable x is the probability of a success, with N and the number of successes as constants.

Why is this useful? In the real world, you usually don't know the value of p, and you're trying to find it. Typically, you conduct a study, find the number of successes in a set of trials, and then you have to estimate p. Beta shows you the likelihood of possible values of p for the number of trials and successes in your study.

Some of the math is complicated, but I can at least show you the rule that generates the density function for N trials with r successes, when N and r are whole numbers:

$$f\left(x\middle|r,N\right) = \frac{(N-1)!}{(r-1)!(N-r-1)!}\,x^{r-1}(1-x)^{N-r-1}$$

The vertical bar in the parentheses on the left means "given that." So this density function is for specific values of N and r. Calculus enters the picture when N and r aren't whole numbers.

To give you an idea of what this function looks like, I use Excel to generate and graph the density function for four successes in ten trials. Figure 17-1 shows the data and the graph. Each value on the x-axis is a possible value for the probability of a success. The curve shows probability density. As I point out in Chapter 16, probability density is what makes the area under the curve correspond to probability. The curve's maximum point is at $x = .4$, which is what you would expect for four successes in ten trials.

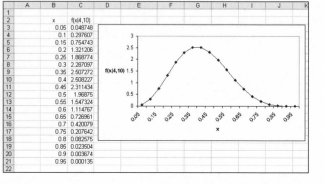

Figure 17-1:
The Beta Density function for four successes in ten trials.

Suppose I toss a die (one of a pair of dice), and I define a success as any toss that results in a 3. I assume I'm tossing a fair die, so I assume that $p = pr(3) = 1/6$. Suppose I toss a die ten times and get four 3s. How good does that fair-die assumption look?

The graph in Figure 17-1 gives you a hint: The area to the left of .16667 (the decimal equivalent of 1/6) is a pretty small proportion of the total area, meaning that the probability that p is 1/6 or less is pretty low.

Now, if you have to go through all the trouble of creating a graph, and then guesstimate proportions of area to come up with an answer like "pretty low," you're doing a whole lot of work for very little return. Fortunately, Excel has a better way.

BETADIST

BETADIST eliminates the need for all the graphing and guesstimating. This function enables you to work with the cumulative beta distribution to determine the probability that p is less than or equal to some value. Considering the complexity of beta, BETADIST is surprisingly easy to work with.

In the BETADIST Function Arguments dialog box, and in the BETADIST help file, you see "Alpha" and "Beta." The dialog box tells you each one is a "parameter *to* the distribution" and the help file tells you each is "a parameter *of* the distribution." Aside from altering the preposition, neither one is much help — at least, not in any way that helps you apply Alpha and Beta.

So here are the nuts and bolts: For the example I'm working through, Alpha is the number of successes and Beta is the number of failures.

When you put the density function in terms of Alpha (α) and Beta (β), it's

$$f(x) = \frac{(\alpha + \beta - 1)!}{(\alpha - 1)!(\beta - 1)!} x^{\alpha - 1}(1 - x)^{\beta - 1}$$

Again, this only applies when α and β are both whole numbers. If that's not the case, you need calculus to compute $f(x)$.

The steps are:

1. **Select a cell for BETADIST's answer.**

2. **Click the Insert Function button to open the Insert Function dialog box.**

3. **In the Insert Function dialog box, select BETADIST and click OK to open its Function Arguments dialog box (see Figure 17-2).**

4. **In the X box, type the probability of a success.**

   For this example, the probability of a success is 1/6.

5. **In the Alpha box, type the number of successes.**

   Excel refers to Alpha and Beta (in the next step) as "parameters to the distribution." I treat them as "number of successes" and "number of failures." So I entered 4 for this example.

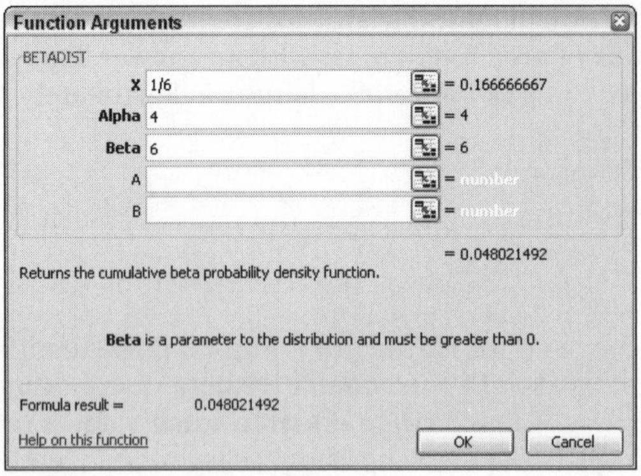

Function Arguments

BETADIST

X	1/6	= 0.166666667
Alpha	4	= 4
Beta	6	= 6
A		= number
B		= number

= 0.048021492

Returns the cumulative beta probability density function.

Beta is a parameter to the distribution and must be greater than 0.

Formula result = 0.048021492

Help on this function   [OK]   [Cancel]

Figure 17-2:
The
Function
Arguments
dialog
box for
BETADIST.

6. In the Beta box, type the number of failures (NOT the number of trials).

I entered 6.

The A box and the B box are evaluation limits for the value in the X box. These aren't relevant for this type of example. I left them blank, which by default sets A = 0 and B = 1.

The answer, .048021492, appears in the dialog box. "Pretty low" indeed. With four successes in ten tosses, you'd intuitively expect that p is greater than 1/6.

7. Click OK to put the answer into the selected cell.

The beta distribution has wider applicability than I showed you here. Consequently, you can put all kinds of numbers (within certain restrictions) into the various boxes. For example, the value you put into the X box can be greater than 1.00, and you can type values that aren't whole numbers into the Alpha box and the Beta box.

BETAINV

This one is the inverse of BETADIST. If you enter a probability and values for successes and failures, it returns a value for p. For example, if you supply it with .048021492, four successes, and six failures, it returns 0.1666667 — the decimal equivalent of 1/6.

BETAINV has a more helpful application. You can use it to find the confidence limits for the probability of a success.

Suppose you've found r successes in N trials, and you're interested in the 95 percent confidence limits for the probability of a success. The lower limit is:

BETAINV(.025, r, N-r)

The upper limit is:

BETAINV(.975, *r*, *N-r*)

1. **Select a cell for BETAINV's answer.**

2. **Click the Insert Function button to open the Insert Function dialog box.**

3. **In the Insert Function dialog box, select BETAINV and click OK to open its Function Arguments dialog box (see Figure 17-3).**

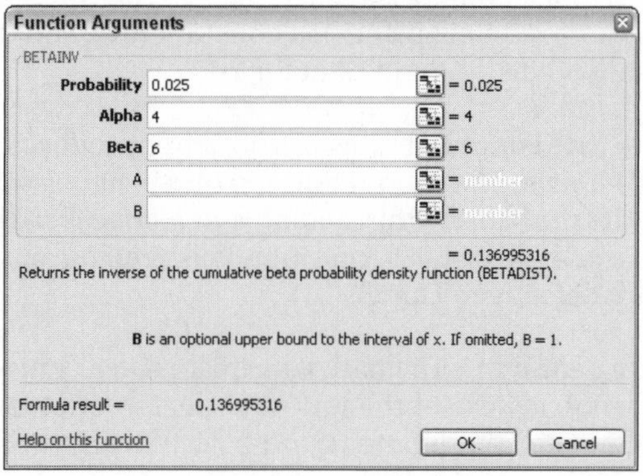

Figure 17-3:
The
Function
Arguments
dialog
box for
BETAINV.

4. **In the Probability box, type a cumulative probability.**

   For the lower bound of the 95 percent confidence limits, the probability is .025.

5. **In the Alpha box, type the number of successes.**

   I entered 4.

6. **In the Beta box, type the number of failures (NOT the number of trials).**

   I entered 6.

   The A box and the B box are evaluation limits for the value in the X box. Again, these aren't relevant for this type of example. I left them blank, which by default sets A = 0 and B = 1.

   The answer, .13699536, appears in the dialog box.

7. **Click OK to put the answer into the selected cell.**

Entering .975 in Step 4 gives .700704575 as the result. So the 95 percent confidence limits for the probability of a success are .137 and .701 (rounded off) if you have four successes in ten trials.

With more trials, of course, the confidence limit narrows. For 40 successes in 100 trials, the confidence limits are .307 and .497.

Poisson

If you have the kind of process that produces a binomial distribution, and you have an extremely large number of trials and a very small number of successes, the *Poisson distribution* approximates the binomial. The equation of the Poisson is

$$pr(x) = \frac{\mu^x e^{-\mu}}{x!}$$

In the numerator, μ is the mean number of successes in the trials, and e is 2.71828 (and infinitely more decimal places), a constant near and dear to the hearts of mathematicians. (I discuss e in Chapter 20.)

Here's an example: The FarKlempt Robotics Corporation produces a universal joint for its robots' elbows. The production process is under strict computer control, so that the probability that a joint is defective is .001. What's the probability that in a sample of 1,000, one joint is defective? What's the probability that two are defective? Three?

Named after nineteenth-century mathematician Siméon-Denis Poisson, this distribution is computationally easier than the binomial — or at least it was when mathematicians had no computational aids. With Excel, you can easily use BINOMDIST to do the binomial calculations.

First, I apply the Poisson distribution to the FarKlempt example. If π = .001 and N = 1000, the mean is

$\mu = N\pi = (1000)(.001) = 1$

(See Chapter 16 for an explanation of $\mu = N\pi$.)

Now for the Poisson. The probability that one joint in a sample of 1,000 is defective is:

$$pr(1) = \frac{\mu^x e^{-\mu}}{x!} = \frac{1^1 (2.71828)^{-1}}{1!} = .368$$

For two defective joints in 1,000, it's

$$pr(2) = \frac{\mu^x e^{-\mu}}{x!} = \frac{1^2 (2.71828)^{-2}}{2!} = .184$$

And for three defective joints in 1,000:

$$pr(3) = \frac{\mu^x e^{-\mu}}{x!} = \frac{1^3 (2.71828)^{-1}}{3!} = .061$$

As you read through this, it might seem odd that I refer to a defective item as a success. Remember, that's just a way of labeling a specific event.

POISSON

Here are the steps for using Excel's POISSON for the preceding example:

1. **Select a cell for POISSON's answer.**

2. **Click the Insert Function button to open the Insert Function dialog box.**

3. **In the Insert Function dialog box, select POISSON and click OK to open its Function Arguments dialog box (see Figure 17-4).**

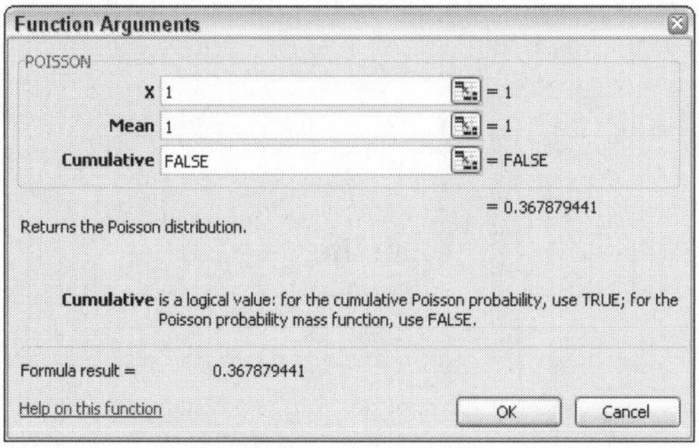

Figure 17-4:
The Function Arguments dialog box for POISSON.

Function Arguments	
POISSON	
X 1	= 1
Mean 1	= 1
Cumulative FALSE	= FALSE
	= 0.367879441

Returns the Poisson distribution.

Cumulative is a logical value: for the cumulative Poisson probability, use TRUE; for the Poisson probability mass function, use FALSE.

Formula result = 0.367879441

Help on this function   OK   Cancel

4. **In the X box, type the number of events for which you're determining the probability.**

   I'm looking for $pr(1)$, so I entered 1.

5. **In the Mean box, type the mean of the process.**

   That's $N\pi$, which for this example is 1.

6. **In the Cumulative box, type TRUE for the cumulative probability; type FALSE for just the probability of the number of events.**

   I entered FALSE.

   The answer, .367879441, appears in the dialog box.

7. **Click OK to put the answer into the selected cell.**

In the example, I showed you the probability for two defective joints in 1,000 and the probability for three. To follow through with the calculations, in Step 4, type 2 into the X box to calculate $pr(2)$, and then type 3 to find $pr(3)$.

As I said before, in the twenty-first century it's pretty easy to calculate the binomial probabilities directly. Figure 17-5 shows you the Poisson and the binomial probabilities for the numbers in Column B and the conditions of the example. I graphed the probabilities so you can see how close the two really are. I selected Cell D3 so the Formula Bar shows you how I used BINOMDIST to calculate the binomial probabilities.

Although the Poisson's usefulness as an approximation is outdated, it has taken on a life of its own. Phenomena as widely disparate as reaction time data in psychology experiments, degeneration of radioactive substances, and scores in professional hockey games seem to fit Poisson distributions. This is why business analysts and scientific researchers like to base models on this distribution. ("Base models on?" What does *that* mean? I tell you all about it in Chapter 18.)

Figure 17-5:
Poisson probabilities and binomial probabilities.

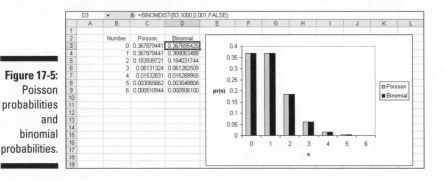

Gamma

The gamma distribution is related to the Poisson distribution in the same way the negative binomial distribution is related to the binomial. The negative binomial tells you the number of trials until a specified number of successes in a binomial distribution. The gamma distribution tells you how many samples you go through to find a specified number of successes in a Poisson distribution. Each sample can be a set of objects (as in the FarKlempt Robotics universal joint example), a physical area, or a time interval.

The probability density function for the gamma distribution is:

$$f(x) = \frac{1}{\beta^{\alpha}(\alpha - 1)!} x^{\alpha - 1} e^{\frac{-x}{\beta}}$$

Again, this works when α is a whole number. If it's not, you guessed it — calculus. (By the way, when this function has only whole-number values of α, it's called the *Erlang distribution,* just in case anybody ever asks you.) The letter *e,* once again, is the constant 2.7818 I told you about earlier.

Don't worry about the exotic-looking math. As long as you understand what each symbol means, you're in business. Excel does the heavy lifting for you.

So here's what the symbols mean. For the FarKlempt Robotics example, α is the number of successes and β corresponds to μ the Poisson distribution. The variable x tracks the number of samples. So if x is 3, α is 2, and β is 1, you're talking about the probability density associated with finding the second success in the third sample if the average number of successes per sample (of 1,000) is 1. (Where does 1 come from, again? That's 1,000 universal joints per sample multiplied by .001, the probability of producing a defective one.)

To determine probability, you have to work with area under the density function. This brings me to the Excel worksheet function designed for gamma.

GAMMADIST

GAMMADIST gives you a couple of options. You can use it to calculate the probability density, and you can use it to calculate probability. Figure 17-6 shows how I used the first option to create a graph of the probability density so you can see what the function looks like. Working within the context of the example I just laid out, I set Alpha to 2, Beta to 1, and calculated the density for the values of x in Column D.

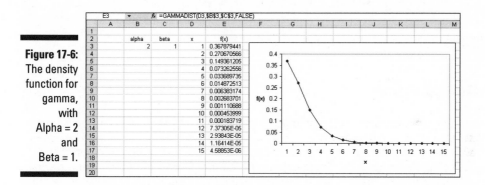

Figure 17-6: The density function for gamma, with Alpha = 2 and Beta = 1.

The values in Column E show the probability densities associated with finding the second defective universal joint in the indicated number of samples of 1,000. For example, Cell E5 holds the probability density for finding the second defective joint in the third sample.

In real life, you work with probabilities rather than densities. Next, I show you how to use GAMMADIST to determine the probability of finding the second defective joint in the third sample. Here it is:

1. **Select a cell for GAMMADIST's answer.**

2. **Click the Insert Function button to open the Insert Function dialog box.**

3. **In the Insert Function dialog box, select GAMMADIST and click OK to open its Function Arguments dialog box (see Figure 17-7).**

4. **In the X box, type the number of samples for which you're determining the probability.**

   I'm looking for *pr*(3), so I entered 3.

5. **In the Alpha box, type the number of successes.**

   I want to find the second success in the third sample, so I entered 2.

6. **In the Beta box, type the average number of successes that occur within a sample.**

   For this example, that's 1.

7. **In the Cumulative box, type** TRUE **for the cumulative distribution; type** FALSE **to find the probability density.**

   I want to find the probability, not the density, so I entered TRUE.

   The answer, .800851727, appears in the dialog box.

8. **Click OK to put the answer into the selected cell.**

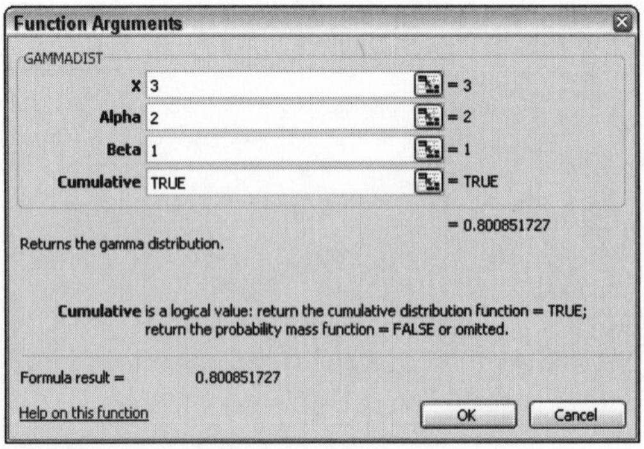

Figure 17-7: The Function Arguments dialog box for GAMMADIST.

GAMMAINV

If you want to know, at a certain level of probability, how many samples it takes to observe a specified number of successes, this is the function for you.

GAMMAINV is the inverse of GAMMADIST. Enter a probability along with Alpha and Beta and it returns the number of samples. To access GAMMAINV's Function Arguments dialog box, follow the first two steps from the preceding section, and then in Step 3 select GAMMAINV instead of GAMMADIST from the Insert Function dialog box.

The GAMMAINV Function Arguments dialog box has a Probability box, an Alpha box, and a Beta box. Figure 17-8 shows that if you enter the answer for the preceding section into the Probability box and the same numbers for Alpha and Beta, the answer is 3. (Well, actually, a tiny bit more than 3.)

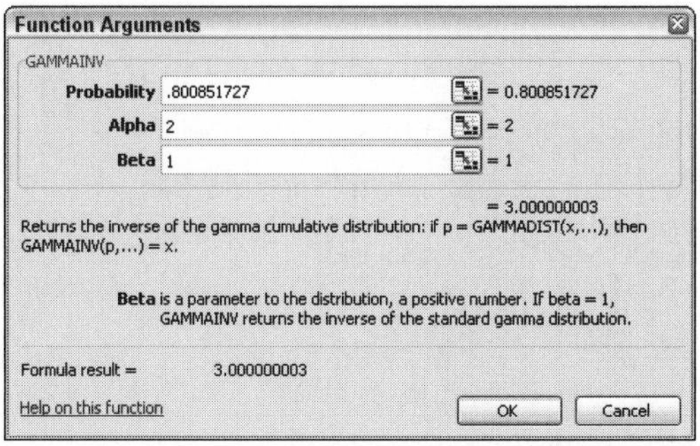

Figure 17-8:
The
Function
Arguments
dialog
box for
GAM-
MAINV.

Exponential

If you're dealing with the gamma distribution and you have Alpha = 1, you have the exponential distribution. This gives the probability that it takes a specified number of samples to get to the first success.

What does the density function look like? Excuse me . . . I'm about to go mathematical on you for a moment. Here, once again, is the density function for gamma:

$$f(x) = \frac{1}{\beta^{\alpha}(\alpha - 1)!} x^{\alpha - 1} e^{\frac{-x}{\beta}}$$

If $\alpha = 1$, it looks like this:

$$f(x) = \frac{1}{\beta} e^{\frac{-x}{\beta}}$$

Statisticians like substituting λ (the Greek letter *lambda*) for $\frac{1}{\beta}$, so here's the final version:

$$f(x) = \lambda e^{-\lambda x}$$

I bring this up because Excel's EXPONDIST Function Arguments dialog box has a box for LAMBDA, and I want you to know what it means.

EXPONDIST

Use EXPONDIST to determine the probability that it takes a specified number of samples to get to the first success in a Poisson distribution. Here, I work once again with the universal joint example. I show you how to find the probability that you'll see the first success in the third sample.

1. **Select a cell for EXPONDIST's answer.**

2. **Click the Insert Function button to open the Insert Function dialog box.**

3. **In the Insert Function dialog box, select EXPONDIST and click OK to open its Function Arguments dialog box (see Figure 17-9).**

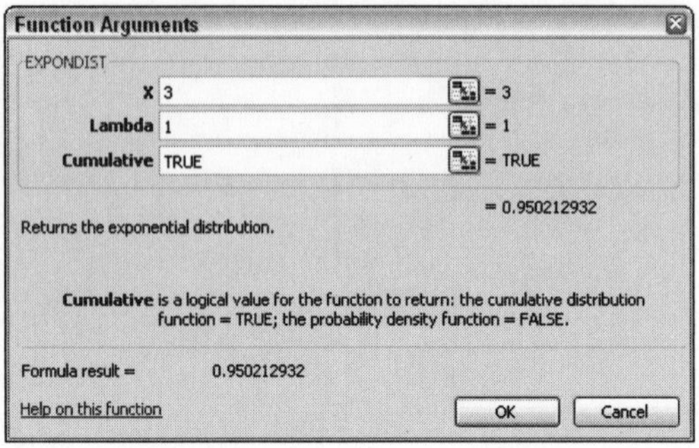

Figure 17-9:
The
Function
Arguments
dialog
box for
EXPONDIST.

Function Arguments

EXPONDIST

X   3   = 3
Lambda 1   = 1
Cumulative TRUE   = TRUE

                               = 0.950212932

Returns the exponential distribution.

Cumulative is a logical value for the function to return: the cumulative distribution function = TRUE; the probability density function = FALSE.

Formula result =   0.950212932

Help on this function   OK   Cancel

4. **In the X box, type the number of samples for which you're determining the probability.**

   I'm looking for *pr*(3), so I entered 3.

5. **In the Lambda box, type the average number of successes per sample.**

   This goes back to the numbers I gave you in the example — the probability of a success (.001) times the number of universal joints in each sample (1,000). I entered 1.

6. **In the Cumulative box, type** TRUE **for the cumulative distribution; type** FALSE **to find the probability density.**

   I want to find the probability, not the density, so I entered TRUE.

   The answer, .950212932, appears in the dialog box.

7. **Click OK to put the answer into the selected cell.**

Chapter 18

A Career in Modeling

Model is a term that gets thrown around a lot these days. Simply put, a *model* is something you know and can work with that helps you understand something you know little about. A model is supposed to mimic, in some way, the thing it's modeling. A globe, for example, is a model of the earth. A street map is a model of a neighborhood. A blueprint is a model of a house.

Researchers use models to help them understand natural processes and phenomena. Business analysts use models to help them understand business processes. The models these people use might include concepts from mathematics and statistics — concepts that are so well known they can shed light on the unknown. The idea is to create a model that consists of concepts you understand, put the model through its paces, and see if the results look like real-world results.

In this chapter, I discuss modeling. My goal is to show how you can harness Excel's statistical capabilities to help you understand processes in your world.

Modeling a Distribution

In one approach to modeling, you gather data and group them into a distribution. Next, you try to figure out a process that results in that kind of a distribution. Restate that process in statistical terms so that it can generate a distribution, and then see how well the generated distribution matches up to the real one. This "process you figure out and restate in statistical terms" is the model.

If the distribution you generate matches up well with the real data, does this mean your model is right? Does it mean the process you guessed is the process that produces the data?

Unfortunately, no. The logic doesn't work that way. You can show that a model is wrong, but you can't prove that it's right.

Plunging into the Poisson distribution

In this section, I go through an example of modeling with the Poisson distribution. I introduced this distribution in Chapter 17, and I told you it seems to characterize an array of processes in the real world. By characterize a process I mean that a distribution of real-world data looks a lot like a Poisson distribution. When this happens, it's possible that the kind of process that produces a Poisson distribution is also responsible for producing the data.

What is that process? Start with a random variable x that tracks the number of occurrences of a specific event in an interval. In Chapter 17, the interval is a sample of 1,000 universal joints, and the specific event is defective joint. Poisson distributions are also appropriate for events occurring in intervals of time, and the event can be something like arrival at a toll booth. Next, I outline the conditions for a *Poisson process,* and use both defective joints and toll booth arrivals to illustrate:

- ✔ The numbers of occurrences of the event in two nonoverlapping intervals are independent.

  The number of defective joints in one sample is independent of the number of defective joints in another. The number of arrivals at a toll booth during one hour is independent of the number of arrivals during another.

- ✔ The probability of an occurrence of the event is proportional to the size of the interval.

  The chance that you'll find a defective joint is larger in a sample of 10,000 than it is in a sample of 1,000. The chance of an arrival at a toll booth is greater for one hour than it is for a half-hour.

- ✔ The probability of more than one occurrence of the event in a small interval is 0 or close to 0.

  In a sample of 1,000 universal joints, you have an extremely low probability of finding two defective ones right next to one another. At any time, two vehicles don't arrive at a toll booth simultaneously.

As I show you in the preceding chapter, the formula for the Poisson distribution is

$$pr(x) = \frac{\mu^x e^{-\mu}}{x!}$$

In this equation, μ represents the average number of occurrences of the event in the interval you're looking at, and e is the constant 2.781828 (followed by infinitely many more decimal places).

It's time to use the Poisson in a model. At the FarBlonJet Corporation, Web designers track the number of hits per hour on the intranet home page. They monitor the page for 200 consecutive hours, and group the data as in Table 18-1.

Table 18-1	Hits Per Hour on the FarBlonJet Intranet Home Page	
Hits/Hour	*Observed Hours*	*Hits/Hour X Observed Hours*
0	10	0
1	30	30
2	44	88
3	44	132
4	36	144
5	18	90
6	10	60
7	8	56
Total	200	600

The first column shows the variable Hits/Hour. The second column, Observed Hours, shows the number of hours in which each value of Hits/Hour occurred. In the 200 hours observed, ten of those hours went by with no hits, 30 hours had one hit, 44 had two hits, and so forth. These data lead the Web designers to use a Poisson distribution to model Hits/Hour. Another way to say this: They believe a Poisson process produces the number of hits per hour on the Web page.

Multiplying the first column by the second column results in the third column. Summing the third column shows that in the 200 observed hours the intranet page received 600 hits. So the average number of hits/hour is 3.00.

Applying the Poisson distribution to this example,

$$pr(x) = \frac{\mu^x e^{-\mu}}{x!} = \frac{3^x e^{-3}}{x!}$$

From here on, I pick it up in Excel.

Using POISSON

Figure 18-1 shows each value of x (hits/hour), the probability of each x if the average number of hits per hour is 3, the predicted number of hours, and the observed number of hours (taken from the second column in Table 18-1). I selected cell C3 so that the Formula Bar shows how I used the POISSON worksheet function. I autofilled Column C down to cell C10. (For the details on using POISSON, see Chapter 17.)

To get the predicted number of hours, I multiplied each probability in Column C by 200 (the total number of observed hours). I used the chart wizard to show you how close the predicted hours are to the observed hours. They look pretty close, don't they?

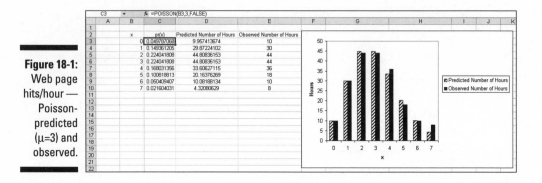

Figure 18-1: Web page hits/hour — Poisson-predicted (μ=3) and observed.

Testing the model's fit

Well, "looking pretty close" isn't enough for a statistician. A statistical test is a necessity. As is the case with all statistical tests, this one starts with a null hypothesis and an alternative hypothesis. Here they are:

H_0: The distribution of observed hits/hour follows a Poisson distribution.

H_1: Not H_0

The appropriate statistical test involves an extension of the binomial distribution. It's called the *multinomial distribution* — *multi* because it encompasses more categories than just *success* and *failure*. It's difficult to work with, and Excel has no worksheet function to handle the computations.

Fortunately, pioneering statistician Karl Pearson (inventor of the correlation coefficient) noticed that χ^2 (chi-square), a distribution I show you in Chapter 10, approximates the multinomial. Originally intended for one-sample hypothesis tests about variances, χ^2 has become much better known for applications like the one I'm about to show you.

Pearson's big idea was this: If you want to know how well a hypothesized distribution (like the Poisson) fits a sample (like the observed hours), use the distribution to generate a hypothesized sample (our predicted hours, for example), and work with this formula:

$$\chi^2 = \sum \frac{(\text{Observed} - \text{Predicted})^2}{\text{Predicted}}$$

Usually, this is written with *Expected* rather than *Predicted,* and both Observed and Expected are abbreviated. The usual form of this formula is:

$$\chi^2 = \sum \frac{(O - E)^2}{E}$$

For this example

$$\chi^2 = \sum \frac{(O-E)^2}{E} = \frac{(10 - 9.9574)^2}{9.9574} + \frac{(30 - 29.8722)^2}{29.8722} + \ldots + \frac{(87 - 4.3208)^2}{4.3208}$$

What does that total up to? Excel figures it out for us. Figure 18-2 shows the same columns as before, with Column F holding the values for $(O-E)^2/E$. The Formula Bar shows I used this formula

=((E3-D3)^2)/D3

to calculate the value in F3. I autofilled up to F10.

The sum of the values in column F is in cell F11, and that's χ^2. If you're trying to show that the Poisson distribution is a good fit to the data, you're looking for a low value of χ^2.

Figure 18-2:
Web page
hits/hour —
Poisson-
predicted
(μ=3) and
observed,
along
with the
calculations
needed to
compute χ^2.

	A	B	C	D	E	F
1						
2		x	pr(x)	Predicted Number of Hours	Observed Number of Hours	((O-E)*2)/E
3		0	0.049787068	9.957413674	10	0.000182135
4		1	0.149361205	29.87224102	30	0.000546405
5		2	0.224041808	44.80836153	44	0.014583179
6		3	0.224041808	44.80836153	44	0.014583179
7		4	0.168031356	33.60627115	36	0.170502041
8		5	0.100818813	20.16376269	18	0.232192228
9		6	0.050409407	10.08188134	10	0.000665010
10		7	0.021604031	4.32080629	8	3.132856565
11					SUM=	3.566110742
12						

OK. Now what? Is 3.5661 (the value in cell F11) high or is it low?

To find out, you evaluate the calculated value of χ^2 against the χ^2 distribution. The goal is to find the probability of getting a value at least as high as the calculated value, 3.5661. The trick is to know how many degrees of freedom (df) you have. When you use χ^2 to determine how well a model fits a set of data

df = k - m - 1

where k = the number of categories and m = the number of parameters estimated from the data. The number of categories is 8 (0 Hits/Hour through 7 Hits/Hour). The number of parameters? I used the observed hours to estimate the parameter μ, so m in this example is 1. That means df = 8-1-1= 6.

Use the worksheet function CHIDIST on the value in F11, with 6 df. CHIDIST returns .73515, the probability of getting a χ^2 of at least 3.5661 if H_0 is true. (See Chapter 10 for more on CHIDIST.) Figure 18-3 shows the χ^2 distribution with 6 df and the area to the right of 3.5661.

If α = .05, the decision is to not reject H_0 — meaning you can't reject the hypothesis that the observed data come from a Poisson distribution.

This is one of those infrequent times when it's beneficial to not reject H_0 — if you want to make the case that a Poisson process is producing the data. If the probability had been just a little greater than .05, not rejecting H_0 would look suspicious. The large probability, however, makes nonrejection of H_0 — and an underlying Poisson process — seem more reasonable. (For more on this see the sidebar "A Point to Ponder" at the end of Chapter 10.)

Figure 18-3:
The χ^2 distribution with df = 6. The darkened area is the probability of getting a χ^2 of at least 3.5661 if H_0 is true.

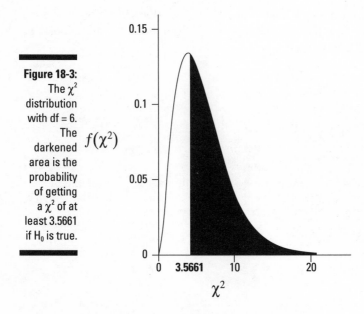

A word about CHITEST

Excel provides CHITEST, a worksheet function that on first look appears to carry out the test I showed you with about one-tenth the work I did on the worksheet. Its Function Arguments dialog box provides one box for the observed values and another for the expected values.

The problem is that CHITEST does not return a value for χ^2. It skips that step and returns the probability that you'll get a χ^2 at least as high as the one you calculate from the observed values and the predicted values.

Why is that a problem? Because CHITEST's degrees of freedom are wrong for this case. CHITEST goes ahead and assumes that df = k-1 (7) rather than k-m-1 (6). You lose a degree of freedom because you estimate μ from the data. In other kinds of modeling, you lose more than one degree of freedom. Suppose, for example, you believe that a normal distribution characterizes the underlying process. In that case, you estimate μ and σ from the data, and you lose two degrees of freedom.

By basing its answer on less than the correct df, CHITEST gives you an inappropriately large (and misleading) value for the probability.

CHITEST would be perfect if it had an option for entering df, or if it returned a value for χ^2 (which you could then evaluate via CHIDIST and the correct df).

When you don't lose any degrees of freedom, CHITEST works as advertised. Does that ever happen? In the next section, it does.

Playing ball with a model

Baseball is a game that generates huge amounts of statistics — and many study these statistics closely. SABR, the Society for American Baseball Research, has sprung from the efforts of a band of dedicated fan-statisticians (fantasticians?) who delve into the statistical nooks and crannies of the Great American Pastime. They call their work *sabremetrics*. (I made up "fantasticians." They call themselves "sabremetricians.")

The reason I mention this is that sabremetrics supplies a nice example of modeling. It's based on the obvious idea that during a game a baseball team's objective is to score runs, and to keep its opponent from scoring runs. The better a team does at both, the more games it wins. Bill James, the founding father of sabremetrics, discovered a neat relationship between the amount of runs a team scores, the amount of runs the team allows, and its winning percentage. He calls it the *Pythagorean percentage:*

$$\text{Pythagorean Percentage} = \frac{(\text{Runs Scored})^2}{(\text{Runs Scored})^2 + (\text{Runs Allowed})^2}$$

Think of it as a model for predicting games won. Calculate this percentage and multiply it by the number of games a team plays. Then, compare the answer to the team's wins. How well does the model predict the number of games each team won during the 2004 season?

To find out, I found all the relevant data for every major-league team for 2004. I put the data into the worksheet in Figure 18-4.

	A	B	C	D	E	F
1	Team	Runs Scored	Runs Allowed	Pythagorean	Predicted Wins	Wins
2	Anaheim	836	734	0.564695084	91	92
3	Arizona	615	899	0.318793587	52	51
4	Atlanta	803	668	0.591007788	96	96
5	Baltimore	842	830	0.507176664	82	78
6	Boston	949	768	0.604257846	98	98
7	Chicago Cubs	789	765	0.515440333	84	89
8	Chicago White Sox	865	831	0.520039116	84	83
9	Cincinnati	750	907	0.406093496	66	76
10	Cleveland	858	857	0.50058309	81	80
11	Colorado	833	923	0.448881434	73	68
12	Detroit	827	844	0.489827504	79	72
13	Florida	718	700	0.51269189	83	83
14	Houston	803	698	0.569612716	92	92
15	Kansas City	720	905	0.38761052	63	58
16	Los Angeles	761	684	0.553136315	90	93
17	Milwaukee	634	757	0.41226045	67	67
18	Minnesota	780	715	0.543396226	88	92
19	Montreal	635	769	0.405419944	66	67
20	New York Mets	684	731	0.466821058	76	71
21	New York Yankees	897	808	0.552057568	89	101
22	Oakland	793	742	0.53318812	86	91
23	Philadelphia	840	781	0.536349132	87	86
24	Pittsburgh	680	744	0.455146781	74	72
25	San Diego	768	705	0.542691763	88	87
26	San Francisco	850	770	0.549262582	89	91
27	Seattle	698	823	0.418368565	68	63
28	St Louis	855	659	0.627324497	102	105
29	Tampa Bay	714	842	0.418290722	68	70
30	Texas	860	794	0.539839829	87	89
31	Toronto	719	823	0.432860528	70	67

Figure 18-4:
Runs scored, runs allowed, predicted wins, and wins for each Major League Baseball team in 2004.

I used this formula to calculate the Pythagorean percentage in Cell D2:

$$=B2^2/((B2^2)+(C2^2))$$

Then, I autofilled the remaining cells in Column D.

Finally, I multiplied each Pythagorean percentage in Column D by the number of games each team played (26 teams played 162 games, four played 161) to get the predicted wins in Column E. Because the number of wins can only be a whole number, I used the ROUND function to round off the predicted wins. For example, the formula that supplies the value in E2 is:

$$=ROUND(D2*162,0)$$

The zero in the parentheses indicates that I wanted no decimal places.

How well does the model fit with reality? This time, CHITEST can supply the answer. I don't lose any degrees of freedom here: I didn't use the Wins data in Column F to estimate any parameters, like a mean or a variance, and then apply those parameters to calculate Predicted Wins. Instead, the predictions came from other data — the Runs Scored and the Runs Allowed. For this reason, $df = k - m - 1 = 30 - 0 - 1 = 29$.

Here's how to use CHITEST (when it's appropriate!):

1. **With the data entered, select a cell for CHITEST's answer.**

2. **Click the Insert Function button to open the Insert Function dialog box.**

3. **In the Insert Function dialog box, select CHITEST and click OK to open the Function Arguments dialog box for CHITEST. (See Figure 18-5.)**

4. **In the Actual_range box, type the cell range that holds the scores for the observed values.**

   For this example, that's F2:F32, the Wins.

5. **In the Expected_range box, type the cell range that holds the predicted values.**

   For this example, it's E2:E32, the Predicted Wins. When you get to this step, the dialog box mentions a product of row totals and column totals. Don't let that confuse you. That has to do with a slightly different application of this function (which I show you in Chapter 20).

   The answer, .999993876, appears in the dialog box. This means that with 29 degrees of freedom you have a huge chance of finding a value of χ^2 at least as high as the one you'd calculate from these observed values and these predicted values. Bottom line: The model is a great fit to the data.

6. **Click OK to put the answer into the selected cell.**

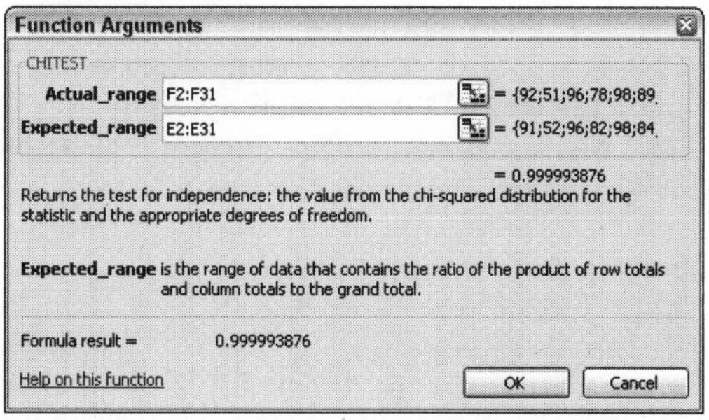

Figure 18-5:
The
Function
Arguments
dialog
box for
CHITEST.

A Simulating Discussion

Another approach to modeling is to simulate a process. The idea is to define as much as you can about what a process does and then somehow use numbers to represent that process and carry it out. It's a great way to find out what a process does in case other methods of analysis are very complex.

Taking a chance: The Monte Carlo method

Many processes contain an element of randomness. You just can't predict the outcome with certainty. To simulate this type of process, you have to have some way of simulating the randomness. Simulation methods that incorporate randomness are called *Monte Carlo* simulations. The name comes from the city in Monaco whose main attraction is gambling casinos.

In the next sections, I show you a couple of examples. These examples aren't so complex that you can't analyze them. I use them for just that reason: You can check the results against analysis.

Loading the dice

In Chapter 16, I talk about a die (one member of a pair of dice) that's biased to come up according to the numbers on its faces: A 6 is six times as likely as a 1, a 5 is five times as likely, and so on. On any toss, the probability of getting a number n is $n/21$.

Suppose you have a pair of dice loaded this way. What would the outcomes of 200 tosses of these dice look like? What would be the average of those 200 tosses? What would be the variance and the standard deviation? You can use Excel to set up Monte Carlo simulations and answer these questions.

To start, I use Excel to calculate the probability of each outcome. Figure 18-6 shows how I do it. Column A holds all the possible outcomes of tossing a pair of dice (2-12). Columns C through N hold the possible ways of getting each outcome. Columns C, E, G, I, K, and M show the possible outcomes on the first die. Columns D, F, H, J, L, and N show the possible outcomes on the second die. Column B gives the probability of each outcome, based on the numbers in Columns C-M. I highlighted B7 so the Formula Bar shows I used this formula to have Excel calculate the probability of a 7:

=((C7*D7)+(E7*F7)+(G7*H7)+(I7*J7)+(K7*L7)+(M7*N7))/21^2

I autofilled the remaining cells in Column B.

The sum in B14 confirms that I considered every possibility.

Figure 18-6: Outcomes and probabilities for a pair of loaded dice.

	B7	▼	f_x	=((C7*D7)+(E7*F7)+(G7*H7)+(I7*J7)+(K7*L7)+(M7*N7))/21^2										
	A	B	C	D	E	F	G	H	I	J	K	L	M	N
	x	pr(x)	1st	2nd	1st	2nd	1st	2nd	1st	2nd	1st	2nd	1st	2nd
1														
2	2	0.0022676	1	1										
3	3	0.0090703	2	1	1	2								
4	4	0.0226757	3	1	2	2	3	1						
5	5	0.0453515	4	1	3	2	2	3	4	1				
6	6	0.0793651	5	1	4	2	3	3	2	4	1	5		
7	7	0.1269841	6	1	5	2	4	3	3	4	2	5	1	6
8	8	0.1587302	6	2	5	3	4	4	3	5	2	6		
9	9	0.1723356	6	3	5	4	4	5	3	6				
10	10	0.1655329	6	4	5	5	4	6						
11	11	0.1360544	6	5	5	6								
12	12	0.0816327	6	6										
13														
14	Sum =	1.0000000												

Next, it's time to simulate the process of tossing the dice. Each toss, in effect, generates a value of the random variable x according to the probability distribution defined by Column A and Column N. How do you simulate these tosses?

Data analysis tool: Random Number Generation

Excel's Random Number Generation tool is tailor-made for this kind of simulation. Tell it how many values you want to generate, give it a probability distribution to work with, and it randomly generates numbers according to the parameters of the distribution. Each randomly generated number corresponds to a toss of the dice.

Here's how to use the Random Number Generation tool:

1. **From the Tools menu, select Data Analysis to open the Data Analysis dialog box.**

2. **In the Data Analysis dialog box, scroll down the Analysis Tools list and select Random Number Generation. Click OK to open the Random Number Generation dialog box.**

   Figure 18-7 shows the Random Number Generation dialog box.

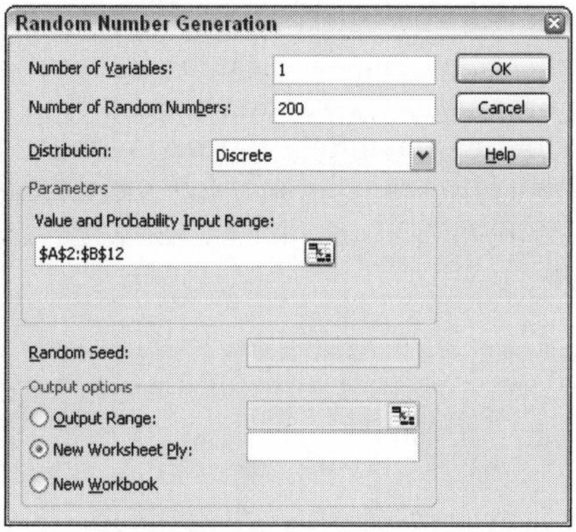

Figure 18-7:
The Random Number Generation dialog box.

3. **In the Number of Variables box, type the number of variables you want to create random numbers for.**

   I know, I know . . . Don't end a sentence with a preposition. As Winston Churchill once said: "That's the kind of nonsense up with which I will not put." Hey but seriously, I entered 1 for this example. I'm only interested in the outcomes of tossing a pair of dice.

4. **In the Number of Random Numbers box, type the number of numbers to generate.**

   I entered 200 to simulate 200 tosses of the loaded dice.

5. **In the Distribution box, click the down arrow to select the type of distribution.**

You have seven options here. The choice you make determines what appears in the Parameters area of the dialog box, because different types of distributions have different types (and numbers) of parameters. You're dealing with a discrete random variable here, so the appropriate choice is Discrete.

6. **Type the array of cells that holds the values of the variable and the associated probabilities in the Value and Probability Input Range box.**

   Choosing Discrete in Step 5 causes the Value and Probability Input Range box to appear in the Parameters section. The possible outcomes of the tosses of the die are in A2:A12, and the probabilities are in B2:B12, so the range is A2:B12. Excel fills in the $-signs for absolute referencing.

7. **In the Output options section, select an option to indicate where you want the results.**

   I selected the New Worksheet Ply option to put the results on a new page in the worksheet.

8. **Click OK.**

Because I selected New Worksheet Ply, a newly created page opens with the results. Figure 18-8 shows the new page. The randomly generated numbers are in Column A. The 200 rows of random numbers are too long to show you. I could have cut and pasted them into ten columns of 20 cells, but then you'd just be looking at 200 random numbers.

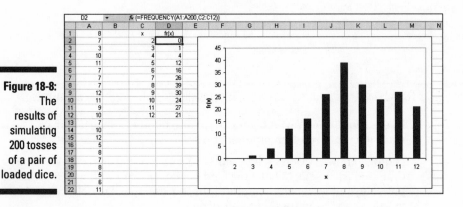

Figure 18-8:
The results of simulating 200 tosses of a pair of loaded dice.

Instead, I use FREQUENCY to group the numbers into frequencies in Columns C and D and then use the Chart Wizard to create a graph of the results. I select D2 so the Formula Bar shows how I used FREQUENCY for that cell.

What about the statistics for these simulated tosses? AVERAGE(A1:A200) tells you the mean is 8.640. VAR(A1:A200) returns 4.573 as the estimate of the variance, and SQRT applied to the variance returns 2.139 as the estimate of the standard deviation.

How do these values match up with the parameters of the random variable? This is what I meant before by checking against analysis. In Chapter 16, I show how to calculate the expected value (the mean), the variance, and the standard deviation for a discrete random variable.

The expected value is:

$$E(x) = \sum x \big(pr(x) \big)$$

In the worksheet in Figure 18-6, I use the SUMPRODUCT worksheet function to calculate $E(x)$. The formula is:

=SUMPRODUCT(A2:A12,B2:B12)

The expected value is 8.667.

The variance is:

$$V(x) = \sum x^2 \big(pr(x) \big) - \big[E(x) \big]^2$$

With $E(x)$ stored in B16, I use this formula

=SUMPRODUCT(A2:A12,A2:A12,B2:B12)-B16^2

Note the use of A2:A12 twice in SUMPRODUCT. That gives you the sum of x^2.

The formula returns 4.444 as the variance. SQRT applied to that number gives 2.108 as the standard deviation.

Table 18-2 shows how closely the results from the simulation match the parameters of the random variable.

Table 18-2   Statistics From the Loaded Dice-Tossing Simulation and the Parameters of the Discrete Distribution

	Simulation Statistic	Distribution Parameter
Mean	8.640	8.667
Variance	4.573	4.444
Standard Deviation	2.139	2.108

Simulating the Central Limit Theorem

This might surprise you, but statisticians often use simulations to make determinations about some of their statistics. They do this when mathematical analysis becomes very difficult.

For example, some statistical tests depend on normally distributed populations. If the populations aren't normal, what happens to those tests? Do they still do what they're supposed to? To answer that question, statisticians might create non-normally distributed populations of numbers, simulate experiments with them, and apply the statistical tests to the simulated results.

In this section, I use simulation to examine an important statistical item — the Central Limit Theorem. In Chapter 9, I introduce the Central Limit Theorem in connection with the sampling distribution of the mean. In fact, I simulate sampling from a population with only three possible values to show you that even with a small sample size the sampling distribution starts to look normally distributed.

Here, I use the Random Number Generation tool to set up a normally distributed population and draw 40 samples. I calculate the mean of each sample, and then set up a distribution of those means. The idea is to see how that distribution matches up with the Central Limit Theorem.

The distribution for this example has the parameters of the population of scores on the IQ test, a distribution I use for examples in several chapters. It's a normal distribution with $\mu=100$ and $\sigma=16$. According to the Central Limit Theorem, the mean of the distribution of means should be 100, and the standard deviation (the standard error of the mean) should be 4.

For a normal distribution, the Random Number Generation dialog box looks like Figure 18-9. The first two entries cause Excel to generate 16 random numbers for a single variable. Choosing Normal in the Distribution box causes the Mean box and the Standard Deviation box to appear under Parameters. As the figure shows, I entered 100 for the Mean and 16 for the Standard Deviation. Under Output Options, I selected Output Range and entered a column of 16 cells. This puts the randomly generated numbers into the indicated column on the current page.

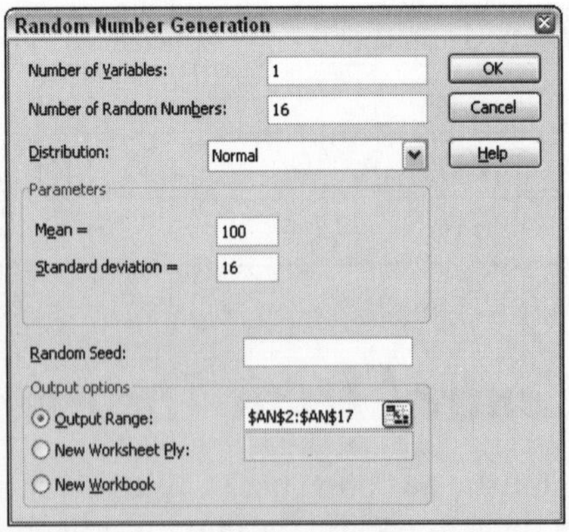

Figure 18-9:
The Random Number Generation dialog box for a normal distribution.

Random Number Generation		
Number of Variables:	1	OK
Number of Random Numbers:	16	Cancel
Distribution:	Normal	Help
Parameters		
Mean =	100	
Standard deviation =	16	
Random Seed:		
Output options		
⊙ Output Range:	$AN$2:$AN$17	
○ New Worksheet Ply:		
○ New Workbook		

I used this dialog box 40 times to generate 40 simulated samples of 16 scores each from a normal population and put the results in adjoining columns. Then I used AVERAGE to calculate the mean for each column.

Next, I copied the 40 means to another worksheet so I could show you how they're distributed. I calculated their mean and the standard deviation. I used FREQUENCY to group the means into a frequency distribution and the Chart Wizard to graph the distribution. Figure 18-10 shows the results.

The mean of the means, 99.671, is close to the Central Limit Theorem's predicted value of 100. The standard deviation of the means, 3.885, is close to the Central Limit's predicted value of 4 for the standard error of the mean. The graph shows the makings of a normal distribution, although it's slightly skewed. In general, the simulation matches up well with the Central Limit Theorem.

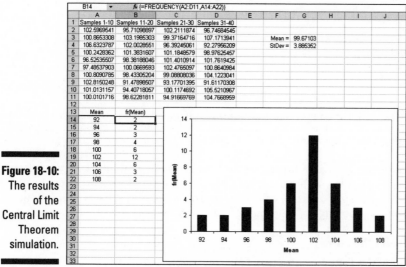

Figure 18-10: The results of the Central Limit Theorem simulation.

A couple of paragraphs ago, I said "I copied the 40 means to another worksheet." That's not quite as straightforward as it sounds. When you try to paste a cell into another worksheet, and that cell holds a formula, Excel usually balks and gives you an ugly-looking error message when you paste. This happens when the formula refers to cell locations that don't hold any values in the new worksheet.

To get around that, you have to do a little trick on the cell you want to copy. You have to convert its contents from a formula into the value that the formula calculates. The steps are:

1. **Select the cell or cell array you want to copy.**

2. **From the Edit menu, select Copy (or press Ctrl + C).**

3. **From the Edit menu again, select Paste Special to open the Paste Special dialog box. (See Figure 18-11.)**

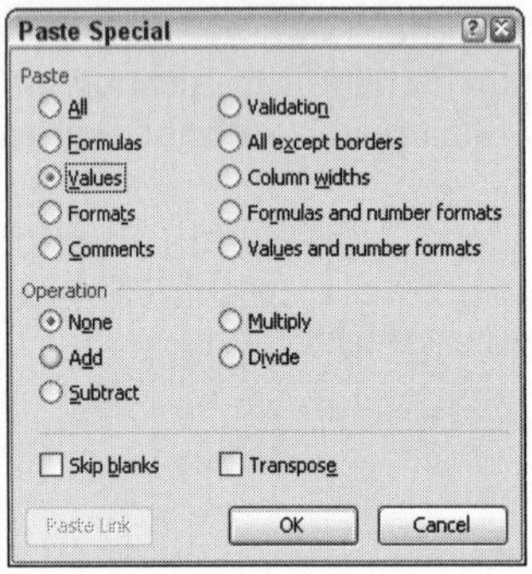

Figure 18-11:
The Paste
Special
Dialog Box.

Paste Special	? ☒
Paste	
○ All	○ Validation
○ Formulas	○ All except borders
⊙ Values	○ Column widths
○ Formats	○ Formulas and number formats
○ Comments	○ Values and number formats
Operation	
⊙ None	○ Multiply
○ Add	○ Divide
○ Subtract	
☐ Skip blanks	☐ Transpose
Paste Link	OK Cancel

4. **In the dialog box, select the Values option.**

5. **Click OK to complete the conversion.**

The Paste Special dialog box offers another helpful capability. Every so often in statistical work, you have to take a row of values and relocate them into a column, or vice versa. Excel calls this *transposition*. In the steps that follow, I describe transposing a row into a column, but it works the other way, too:

1. **Select the row.**

2. **From the Edit menu, select Copy (or press Ctrl + C).**

3. **Select the cell that begins the column you want to put the values into.**

4. **From the Edit menu, select Paste Special to open the Paste Special dialog box. (See Figure 18-11.)**

5. **Select the Transpose option. (It's in the lower-right corner.)**

6. **Click OK to complete the transposition.**

Part V
The Part of Tens

The 5th Wave   By Rich Tennant

SNOW GLOBE DATA STORAGE

Okay let's shake this thing and see what we come up with.

In this part . . .

Y ou've come to the famous Part of Tens. I put two chapters into this part. The first one covers statistical traps and helpful tips — from problems with hypothesis testing to advice on graphs, from pitfalls in regression to advice on graphing variability.

The second chapter goes over a number of Excel features I just couldn't fit anywhere else. This part covers forecasting, graphing, testing for independence, and more. It ends with a look at Excel functions based on logarithms — do you see what I mean about not fitting anywhere else?

Excel provides some database functionality, so you can use a worksheet for record-keeping. The appendix deals with that functionality, some of which is statistical in nature. You can find averages, variances, and standard deviations of numerical data, and set criteria for the calculations.

Another capability I cover is the pivot table — a cross-tabulation technique that enables you to look at your data in many ways. Excel has a wizard for this, and it's a helpful tool. To end this part, I show you how to create a chart that visualizes the pivot table.

Chapter 19

Ten Statistical and Graphical Tips and Traps

. . . .

In This Chapter

▶ Beware of significance

▶ Be wary of graphs

▶ Be cautious with regression

▶ Be careful with concepts

. . . .

*T*he world of statistics is full of pitfalls, but it's also full of opportunities. Whether you're a user of statistics or someone who has to interpret them, it's possible to fall into the pitfalls. It's also possible to walk around them. Here are ten tips and traps from the areas of hypothesis testing, regression, correlation, and graphs.

Significant Doesn't Always Mean Important

As I said earlier in the book, "significance" is, in many ways, a poorly chosen term. When a statistical test yields a significant result, and the decision is to reject H_0, that doesn't guarantee that the study behind the data is an important one. Statistics can only help decision-making about numbers and inferences about the processes that produced them. They can't make those processes important or earth shattering. Importance is something you have to judge for yourself — and no statistical test can do that for you.

Trying to Not Reject a Null Hypothesis Has a Number of Implications

Let me tell you a story: Some years ago, an industrial firm was trying to show it was finally in compliance with environmental cleanup laws. The firm's engineers took numerous measurements of the pollution in the body of water surrounding their factory, compared the measurements with a null hypothesis-generated set of expectations, and found they couldn't reject H_0 with α = .05. The measurements didn't differ significantly (there's that word again) from "clean" water.

This, the company claimed, was evidence that they had cleaned up their act. Closer inspection revealed that the data approached significance, but the pollution wasn't quite of a high enough magnitude to reject H_0. Does this mean the company is not polluting?

Not at all. In striving to "prove" a null hypothesis, they had stacked the deck in favor of themselves. They set a high barrier to get over, didn't clear it, and then patted itself on the back.

Every so often, it's appropriate to try to not reject H_0. When you set out on that path, be sure to set a high value of α (about .20-.30), so that small divergences from H_0 cause rejection of H_0. (I discuss this in Chapter 10 in the sidebar "A point to ponder," and I mention it in other parts of the book, but I think it's important enough to mention again here.)

Regression Isn't Always Linear

When trying to fit a regression model to a scatterplot, the temptation is to immediately use a line. This is the best understood regression model, and once you get the hang of it, slopes and intercepts aren't all that daunting.

But linear regression isn't the only kind of regression. It's possible to fit a curve through a scatterplot. I won't kid you: The statistical concepts behind curvilinear regression are more difficult to understand than the concepts behind linear regression.

It's worth taking the time to learn those concepts, however. Sometimes, a curve is a much better fit than a line. (This is partly a preview of Chapter 20, where I take you through curvilinear regression — and some of the concepts behind it).

Extrapolating Beyond a Sample Scatterplot Is a Bad Idea

Whether you're working with linear regression or curvilinear regression, keep in mind that it's inappropriate to generalize beyond the boundaries of the scatterplot.

Suppose you've established a solid predictive relationship between a test of mathematics aptitude and performance in mathematics courses, and your scatterplot only covers a narrow range of mathematics aptitude. You have no way of knowing whether the relationship holds up beyond that range. Predictions outside that range aren't valid.

Your best bet is to expand the scatterplot by testing more people. You might find that the original relationship only tells part of the story.

Examine the Variability Around a Regression Line

Careful analysis of residuals (the differences between observed and predicted values) can tell you a lot about how well the line fits the data. A foundational assumption is that variability around a regression line is the same up and down the line. If it isn't, the model might not be as predictive as you think. If the variability is systematic (greater variability at one end than at the other), curvilinear regression might be more appropriate than linear. The standard error of estimate won't always be the indicator.

A Sample Can Be Too Large

Believe it or not. This sometimes happens with correlation coefficients. A very large sample can make a small correlation coefficient statistically significant. For example, with 100 degrees of freedom and $\alpha = .05$, a correlation coefficient of .195 is cause for rejecting the null hypothesis that the population correlation coefficient is equal to zero.

But what does that correlation coefficient really mean? The coefficient of determination — r^2 — is just .038, meaning that the $SS_{Regression}$ is less than 4 percent of the SS_{Total} (see Chapter 16.) That's a very small association.

Bottom line: When looking at a correlation coefficient, be aware of the sample size. If it's large enough, it can make a trivial association turn out statistically significant. (Hmmm . . . significance . . . there it is again!)

Consumers: Know Your Axes

When you look at a graph, make sure you know what's on each axis. Make sure you understand the units of measure. Do you understand the independent variable? Do you understand the dependent variable? Can you describe each one in your own words? If the answer to any of those questions is "No," you don't understand the graph you're looking at.

When looking at a graph in a TV ad, be very wary if it disappears too quickly, before you can see what's on the axes. The advertiser may be trying to create a lingering false impression about a bogus relationship inside the graph. The graphed relationship might be as valid as that other staple of TV advertising — scientific proof via animated cartoon: Tiny animated scrub brushes cleaning cartoon teeth might not necessarily guarantee whiter teeth for you if you buy the product. (I know that's off-topic, but I had to get it in.)

Graphing a Categorical Variable as Though It's a Quantitative Variable Is Just Wrong

So you're just about ready to compete in the Rock-Paper-Scissors World Series. In preparation for this international tournament, you've tallied all your matches from the past ten years, listing the percentage of times you won when you played each role.

To summarize all the outcomes, you're about to use Excel's Chart Wizard to create a graph. One thing's sure: Whatever your preference rock-paper-scissors-wise, the graph absolutely, positively had better NOT look like Figure 19-1.

So many people create these kinds of graphs — people who should know better. The line in the graph implies continuity from one point to another. With these data, of course, that's impossible. What's between Rock and Paper? Why are they equal units apart? Why are the three categories in that order? (Can you tell this is my pet peeve?)

Simply put, a line graph is not the proper graph when at least one of your variables is a set of categories. Instead, create a column graph. A pie chart works here, too, because the data are percentages and you have just a few slices. (See Chapter 3 for Yogi Berra's pie-slice guidelines.)

Percentage of Victories Playing Rock Paper Scissors

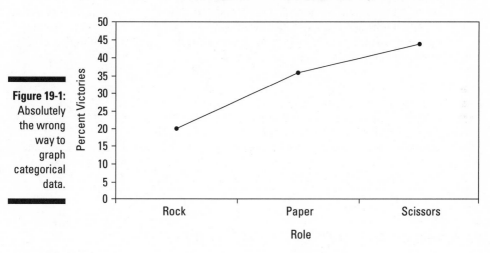

Figure 19-1: Absolutely the wrong way to graph categorical data.

Whenever Appropriate, Include Variability in Your Graph

When the points in your graph represent means, make sure the graph includes the standard error of each mean. This gives the viewer an idea of the variability in the data — which is an important aspect of the data. Here's another plug: In Chapter 20, I show you how to do that in Excel.

Means by themselves don't always tell you the whole story. Take every opportunity to examine variances and standard deviations. You might find some hidden nuggets. Systematic variation — high values of variance associated with large means, for example — might be a clue about a relationship you didn't see before.

Be Careful When Relating Statistics-Book Concepts to Excel

If you're serious about doing statistical work, you'll probably have occasion to look into a statistics text or two. Bear in mind that the symbols in some areas of statistics aren't standard: For example, some texts use M rather than $\bar{x}$ to represent the sample mean, and some represent a deviation from the mean with just x.

Connecting textbook concepts to Excel's statistical functions can be a challenge, because of the texts and because of Excel. Messages on dialog boxes and in help files might contain symbols other than the ones you read about, or they might use the same symbols but in a different way. The discrepancy might lead you to make an incorrect entry into a parameter in a dialog box, resulting in an error that's hard to trace.

Chapter 20

Ten (Or So) Things That Didn't Fit in Any Other Chapter

I wrote this book to show you all of Excel's statistical capabilities. My intent was to tell you about them in the context of the world of statistics, and I had a definite path in mind.

Some of the capabilities didn't fit into that path. I still want you to be aware of them, however, so here they are.

Some Forecasting

Here are a couple of useful techniques to help you come up with some forecasts. Although they don't quite fit into the regression chapter, and they really don't go into the descriptive statistics chapters, they deserve a section of their own.

A moving experience

In many contexts, it makes sense to gather data over periods of time. When you do this, you have a *time series*.

Investors often have to base their decisions on time series — like stock prices — and the numbers in a time series typically show numerous ups and downs. A mean that takes all the peaks and valleys into account might obscure the big picture of the overall trend.

One way to smooth out the bumps and see the big picture is to calculate a *moving average.* This is an average calculated from the most recent scores in the time series. It moves because you keep calculating it over the time series. As you add a score to the front end, you delete one from the back end.

Suppose you have daily stock prices for the last 20 days of a particular stock, and you decide to keep a moving average for the most recent 5 days. Start with the average from the first five of those twenty days. Then, average the prices from days 2-6. Next, average days 3-7, and so on, until you average the final five days of the time series.

Excel's Moving Average data analysis tool does the work for you. Figure 20-1 shows a fictional company's stock prices for 20 days, and the dialog box for the Moving Average tool.

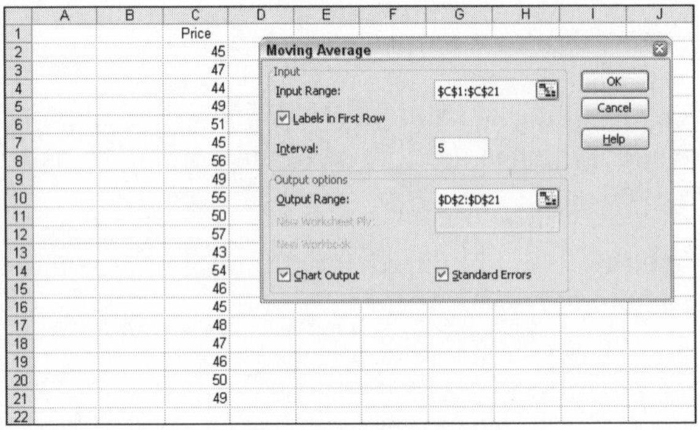

Figure 20-1:
Fictional
stock prices
and the
Moving
Average
dialog box.

The figure shows my entries for Moving Average. The Input Range is cells C1 through C21, the Labels in First Row option is selected, and the Interval is 5. That means each average consists of the most recent five days. Cells D2 through D21 are the Output Range, and I selected the Chart Output and Standard Errors options.

The results are in Figure 20-2. Ignore the ugly-looking #N/A symbols. Each number in Column E is a moving average — a forecast of the price on the basis of the preceding five days.

Each number in Column E is a standard error. In this context, a standard error is the square root of the average of the squared difference between the price and the forecast for the previous five days. So the first standard error in cell E10 is

$$\sqrt{\frac{(51-47.2)^2+(45-47.2)^2+(56-49)^2+(49-50)^2+(55-51.2)^2}{5}} = 4.091943$$

In the graph (which I stretched out from its original appearance), the moving average is in the series labeled Forecast. Sometimes the forecast matches the data, and sometimes it doesn't.

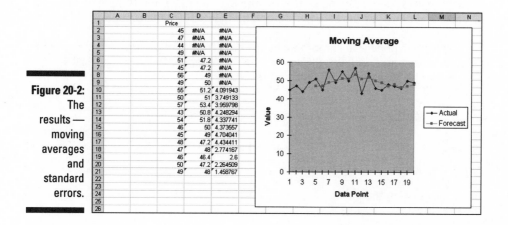

Figure 20-2: The results — moving averages and standard errors.

As the figure shows, the moving average smoothes out the peaks and valleys in the price data. By the way, some analysts say that a stock price that goes below the moving average is a signal to sell, and a price that goes above the moving average is a signal to buy.

In general, how many scores do you include? That's up to you. Include too many and you risk obsolete data influencing your result. Include too few and you risk missing something important.

How to be a smoothie, exponentially

Exponential smoothing is similar to a moving average. It's a technique for forecasting based on prior data. In contrast with the moving average, which works just with a sequence of actual values, exponential smoothing takes its previous prediction into account.

Exponential smoothing operates according to a *damping factor,* a number between zero and one. With α representing the damping factor, the formula is

$$y_t' = (1-\alpha)y_{t-1} + \alpha y'_{t-1}$$

In terms of stock prices from the preceding example, y_t' represents the predicted stock price at a time t. If t is today, $t-1$ is yesterday. So y_{t-1} is yesterday's actual price and y'_{t-1} is yesterday's predicted price. The sequence of predictions begins with the first predicted value as the observed value from the day before.

A larger damping factor gives more weight to yesterday's prediction. A smaller damping factor gives greater weight to yesterday's actual value. A damping factor of 0.5 weighs each one equally.

Figure 20-3 shows the dialog box for the Exponential Smoothing data analysis tool. It's similar to the Moving Average tool, except for the Damping Factor box.

I applied exponential smoothing to the data from the previous example. I did this three times with 0.1, 0.5, and 0.9 as the damping factors. Figure 20-4 shows the graphic output for each result.

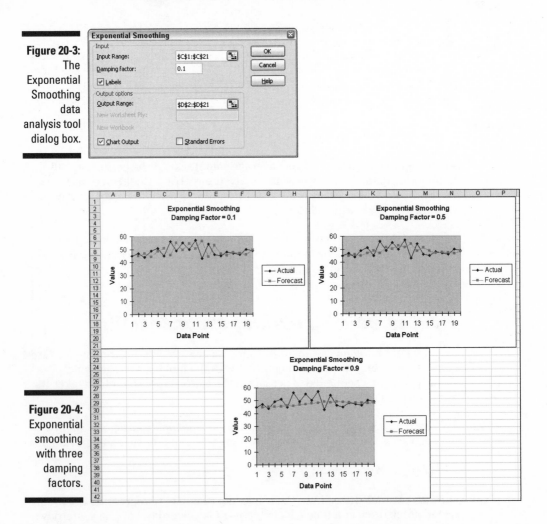

Figure 20-3: The Exponential Smoothing data analysis tool dialog box.

Figure 20-4: Exponential smoothing with three damping factors.

The highest damping factor, 0.9, results in the flattest sequence of predictions. The lowest, 0.1, predicts the most pronounced set of peaks and valleys. How should you set the damping factor? Like the interval in the moving average, that's up to you. Your experience and the specific area of application are the determining factors.

Graphing the Standard Error of the Mean

When you create a graph and your data are means, it's a good idea to include the standard error of each mean in your graph. This gives the viewer an idea of the spread of scores around each mean.

Figure 20-5 gives an example of a situation where this arises. The data are (fictional) scores for four groups of people on a test. Each column header indicates the amount of preparation time for the eight people within the group. I used the Chart Wizard (see Chapter 3) to draw the graph. Because the independent variable is quantitative, a line graph is appropriate. (Sometimes a line graph isn't appropriate: See Chapter 19 for a rant on my biggest peeve.)

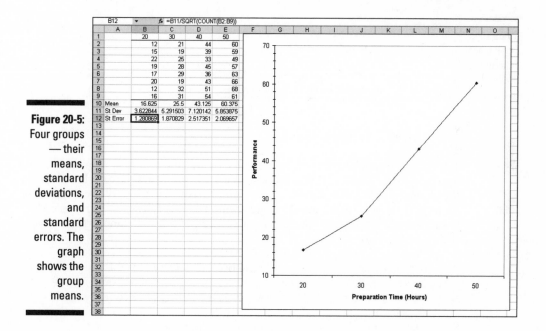

Figure 20-5: Four groups — their means, standard deviations, and standard errors. The graph shows the group means.

For each group I use AVERAGE to calculate the mean and STDEV to calculate the standard deviation. I also calculate the standard error of each mean. I selected cell B12, so the Formula Bar shows you that I calculated the standard error for Column B via this formula:

=B11/SQRT(COUNT(B2:B9))

The trick is to get each standard error into the graph. Start by right-clicking on any data point, which opens a menu whose first choice is Format Data Series. Select that choice to open the Format Data Series dialog box. Click the Y Error Bars tab (see Figure 20-6).

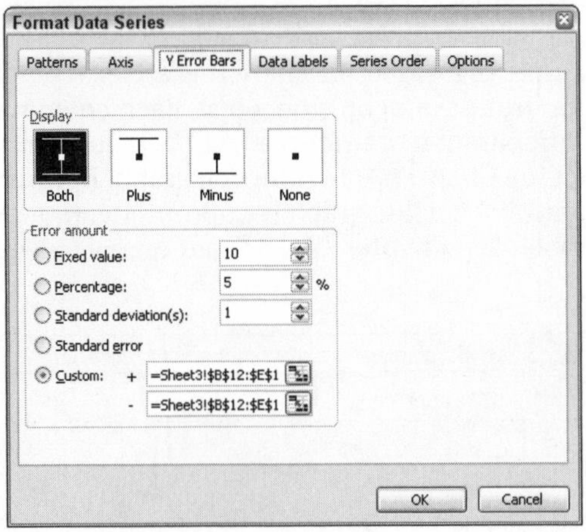

Figure 20-6:
The Format
Data Series
dialog box.

In the Display area, select Both.

In the Error Amount area, you have to be careful. One option is labeled Standard Error. Avoid it. If you think selecting this button tells Excel to put the standard error of each mean on the graph, rest assured that Excel has absolutely no idea what you're talking about. For this selection, Excel calculates the standard error of the set of four means — not the standard error within each group.

Instead, select the Custom option. In the box labeled +, enter the range of cells in the row that holds the calculated standard error for each group. Do the same for the box labeled –. Those cells are B12 through E12 in this example.

Click OK and a graph appears that looks like Figure 20-7.

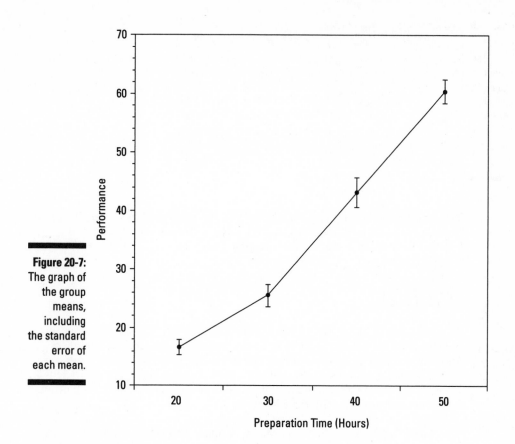

Probabilities and Distributions

Here are some probability-related worksheet functions. A little on the eso-teric side, you might find some use for them.

PROB

If you have a probability distribution of a discrete random variable, and you want to find the probability that the variable takes on a particular value, PROB is for you. Figure 20-8 shows the Function Arguments dialog box for PROB along with a distribution.

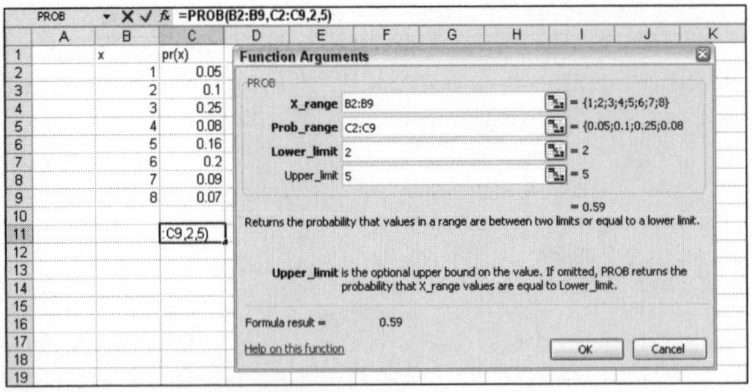

Figure 20-8:
The
Function
Arguments
dialog box
for PROB
and a
probability
distribution.

You supply the random variable (X_Range), the probabilities (Prob_Range), a Lower Limit, and an Upper Limit. PROB returns the probability that the random variable takes on a value between those limits.

WEIBULL

This is a probability density function that's mostly applicable to engineering. It serves as a model for the time until a physical system fails. As engineers know, in some systems the number of failures stays the same over time because shocks to the system cause failure. In others, like some microelectronic components, the number of failures decreases with time. In still others, wear and tear increase failures with time.

The Weibull distribution's two parameters allow it to reflect all these possibilities. One parameter, α, determines how wide or narrow it is. The other, β, determines where it's centered on the x-axis.

The Weibull probability density function is a rather complicated equation. Thanks to Excel, you don't have to worry about it. Figure 20-9 shows WEIBULL's Function Arguments dialog box.

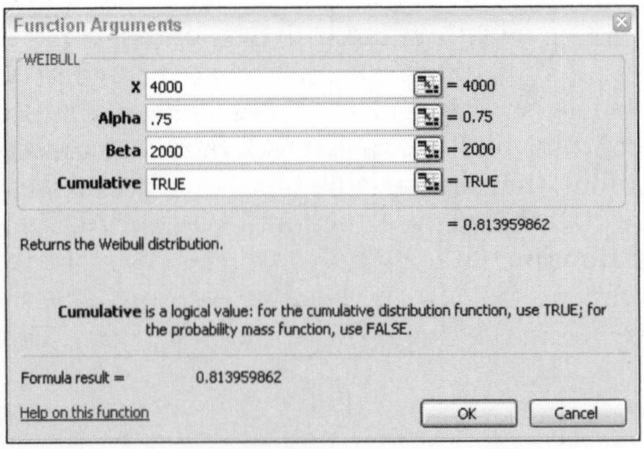

Figure 20-9:
The
Function
Arguments
dialog box
for
WEIBULL.

The dialog box in the figure answers the kind of question a product engineer would ask: Assume the time to failure of a bulb in an LCD projector follows a Weibull distribution with $\alpha = .75$ and $\beta = 2,000$ hours. What's the probability the bulb lasts at most 4,000 hours? The dialog box shows that the answer is .814.

Drawing Samples

Excel's Sampling data analysis tool is helpful for creating samples. You can tailor it in a couple of ways. If you're trying to put a focus group together and you have to select the participants from a pool of people, you can assign each one a number and have the Sampling tool select your group.

One way to select is *periodically,* where you supply *n,* and Excel samples every *n*th number. The other way to select is *randomly.* You supply the number of individuals you want randomly selected and Excel does the rest.

Figure 20-10 presents the Sampling dialog box, three groups I had it sample from, and two columns of output.

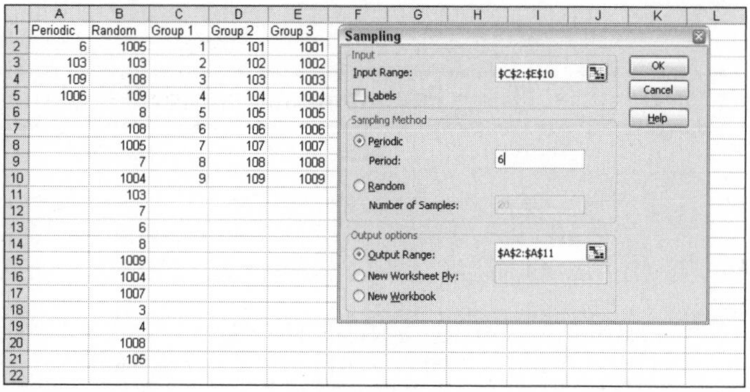

Figure 20-10:
The
Sampling
data
analysis tool
dialog box,
sampled
groups, and
results.

Column A, the first output column, shows the results of periodic sampling with a period of 6. Sampling begins with the sixth score in Group 1. Excel then counts out scores and delivers the sixth, and goes through that process again until it finishes in the last group. The process, as you can see, doesn't recycle. I supplied an output range up to cell A11, but Excel stopped after four numbers.

Column B, the second output column, shows the results of random sampling. I asked for 20 and that's what I got. If you closely examine the numbers in Column B, you see that the process can select a number more than once.

Beware of a little quirk: If you select the Labels option it seems to have no effect. When I specified an input range that includes C1, D1, and E1, and selected the Labels option, I received an error message: "Sampling - Input range contains non-numeric data." Not a show-stopper, but a little annoying.

Testing Independence: The True Use of CHITEST

In Chapter 18, I show you how to use CHITEST to test the goodness of fit of a model to a set of data. I warn you about the pitfalls of using this function in that context, and I mention that it's really intended for something else.

Here's the something else. Imagine you've surveyed 200 people. Each person lives in a rural area, an urban area, or a suburb. Your survey asks them their favorite type of movie — drama, comedy, or animation. You want to know if their movie preference is independent of the environment in which they live.

Table 20-1 shows the results.

Table 20-1	Living Environment and Movie Preference			
	Drama	*Comedy*	*Animation*	*Total*
Rural	40	30	10	80
Urban	20	30	20	70
Suburban	10	20	20	50
Total	70	80	50	200

The number in each cell represents the number of people in the environment indicated in the row who prefer the type of movie indicated in the column.

Do the data show that preference is independent of environment? This calls for a hypothesis test:

H_0: Movie preference is independent of environment

H_1: Not H_0

$\alpha = .05$

To get this done, you have to know what to expect if the two are independent. Then, you can compare the data with the expected numbers and see if they match. If they do, you can't reject H_0. If they don't, you reject H_0.

Concepts from probability help determine the expected data. In Chapter 16, I told you that if two events are independent, you multiply their probabilities to find the probability that they occur together. Here, you can treat the tabled numbers as proportions and the proportions as probabilities.

For example, in your sample, the probability is 80/200 that a person is from a rural environment. The probability is 70/100 that a person prefers drama. What's the probability that a person is in the category "rural and likes drama"? If the environment and preference are independent, that's $(80/200) \times (70/200)$. To turn that probability into an expected number of people, you multiply it by the total number of people in the sample — 200. So the expected number of people is $(80 \times 70)/200$, which is 28.

In general,

$$\text{Expected number in a cell} = \frac{\text{Row Total} \times \text{Column Total}}{\text{Total}}$$

Once you have the expected numbers, you compare them to the observed numbers (the data) through this formula:

$$\chi^2 = \sum \frac{(\text{Observed} - \text{Expected})^2}{\text{Expected}}$$

You test the result against a χ^2 (chi-square) distribution with df = (Number of Rows - 1) $\times$ (Number of Columns - 1), which in this case comes out to 4.

The CHITEST worksheet function performs the test. You supply the observed numbers and the expected numbers, and CHITEST returns the probability that a χ^2 at least as high as the result from the preceding formula could have resulted if the two types of categories are independent. If the probability is small (less than .05), reject H_0. If not, don't reject. CHITEST doesn't return a value of χ^2, it just returns the probability (under a χ^2 distribution with the correct df).

Figure 20-11 shows a worksheet with both the observed data and the expected numbers, along with the Function Arguments dialog box for CHITEST.

The figure shows that I've entered D3:D5 into the Actual_range box and D10:F12 into the Expected_range box. The dialog box shows a very small probability, .00068, so the decision is to reject H_0. The data are consistent with the idea that movie preference is not independent of environment.

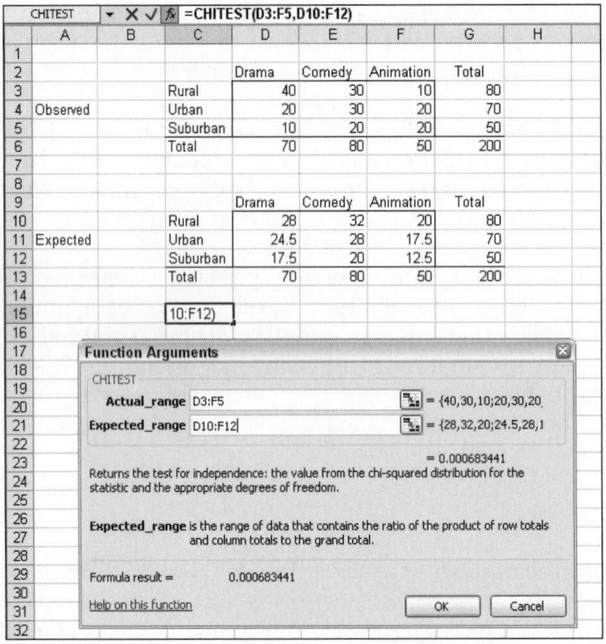

CHITEST	▾	✕ ✓	ƒx	=CHITEST(D3:F5,D10:F12)				
	A	B	C	D	E	F	G	H
1								
2				Drama	Comedy	Animation	Total	
3			Rural	40	30	10	80	
4	Observed		Urban	20	30	20	70	
5			Suburban	10	20	20	50	
6			Total	70	80	50	200	
7								
8								
9				Drama	Comedy	Animation	Total	
10			Rural	28	32	20	80	
11	Expected		Urban	24.5	28	17.5	70	
12			Suburban	17.5	20	12.5	50	
13			Total	70	80	50	200	
14								
15			10:F12)					
16								

Function Arguments   ✕

CHITEST

Actual_range   D3:F5   = {40,30,10;20,30,20

Expected_range   D10:F12   = {28,32,20;24.5,28,1

= 0.000683441

Returns the test for independence: the value from the chi-squared distribution for the statistic and the appropriate degrees of freedom.

Expected_range is the range of data that contains the ratio of the product of row totals and column totals to the grand total.

Formula result =   0.000683441

Help on this function   OK   Cancel

Figure 20-11:
The
Function
Arguments
dialog box
for CHITEST
with
observed
data and
expected
numbers.

Logarithmica Esoterica

The functions in this section are *really* out there. That's why I left them for last. Unless you're a tech-head, you'll probably never use them. I present them for completeness. You might run into them while you're wandering through Excel's statistical functions and wonder what they are.

They're based on what mathematicians call *natural logarithms,* which in turn are based on *e,* that constant I use at various points throughout the book. I begin with a brief discussion of logarithms, and then I turn to *e.*

What is a logarithm?

Plain and simple, a logarithm is an *exponent* — a power to which you raise a number. In the equation

$$10^2 = 100$$

2 is an exponent. Does that mean that 2 is also a logarithm? Well . . . yes. In terms of logarithms,

$$\log_{10}100 = 2$$

That's really just another way of saying $10^2 = 100$. Mathematicians read it as "the logarithm of 100 to the base 10 equals 2." It means that if you want to raise 10 to some power to get 100, that power is 2.

How about 1,000? As you know

$$10^3 = 1,000$$

so

$$\log_{10}1,000 = 3$$

How about 453? Uh . . . Hmmm . . . That's like trying to solve

$$10^x = 453$$

What could that answer possibly be? 10^2 means 10×10 and that gives you 100. 10^3 means $10 \times 10 \times 10$ and that's 1,000. But 453?

Here's where you have to think outside the dialog box. You have to imagine exponents that aren't whole numbers. I know, I know . . . how can you multiply a number by itself a fraction at a time? If you could, somehow, the number in that 453 equation would have to be between 2 (which gets you to 100) and 3 (which gets you to 1,000).

In the sixteenth century, mathematician John Napier showed how to do it and logarithms were born. Why did Napier bother with this? One reason is that it was a great help to astronomers. Astronomers have to deal with numbers that are . . . well . . . astronomical. Logarithms ease computational strain in a couple of ways. One way is to substitute small numbers for large ones: The logarithm of 1,000,000 is 6 and the logarithm of 100,000,000 is 8. Also, working with logarithms opens up a helpful set of computational shortcuts. Before calculators and computers appeared on the scene, this was a very big deal.

Incidentally,

$$10^{2.6560982} = 453$$

meaning that

$$\log_{10}453 = 2.6560982$$

You can use Excel to check that out if you don't believe me. Select a cell and type = LOG(453,10), press Enter, and watch what happens. Then, just to close the loop, reverse the process. If your selected cell is — let's say — D3, select another cell and type

=POWER(10,D3)

or

=10^D3

Either way, the result is 453.

Ten, the number that's raised to the exponent, is called the *base*. Because it's also the base of our number system and we're so familiar with it, logarithms of base 10 are called *common logarithms*.

Does that mean you can have other bases? Absolutely. *Any* number (except 0 or 1 or a negative number) can be a base. For example,

$6.4^2 = 40.96$

so

$\log_{6.4} 40.96 = 2$

If you ever see log without a base, base 10 is understood, so

$\log 100 = 2$

In terms of bases, one number is special . . .

What is e?

Which brings me to *e*, a constant that's all about growth. Before I get back to logarithms, I'll tell you about *e*.

Imagine the princely sum of $1 deposited in a bank account. Suppose the interest rate is 2 percent a year. (Good luck with *that*.) If it's simple interest, the bank adds $.02 every year, and in 50 years you have $2.

If it's compound interest, at the end of 50 years you have $(1 + .02)^{50}$ — which is just a bit more than $2.68, assuming the bank compounds the interest once a year.

Of course, if they compound it twice a year, each payment is $.01, and after 50 years they've compounded it 100 times. That gives you $(1 + .01)^{100}$, or just over $2.70. What about compounding it four times a year? After 50 years — 200 compoundings — you have $(1 + .005)^{200}$, which results in the don't-spend-it-all-in-one-place amount of $2.71 and a tiny bit more.

Focusing on "just a bit more" and "a tiny bit more," and taking it to extremes, after 100,000 compoundings you have $2.718268. After 100 million, you have $2.718282.

If you could get the bank to compound many more times in those 50 years, your sum of money approaches a *limit* — an amount it gets ever so close to, but never quite reaches. That limit is *e*.

The way I set up the example, the rule for calculating the amount is

$(1 + (1/n))^n$

where *n* represents the number of payments. Two cents is 1/50th of a dollar and I specified 50 years — 50 payments. Then I specified two payments a year (and each year's payments have to add up to 2 percent), so that in 50 years you have 100 payments of 1/100th of a dollar, and so on.

To see this in action, enter numbers into a column of a spreadsheet as I have in Figure 20-12. In cells C2 through C20, I have the numbers 1 through 10 and then selected steps through 100 million. In cell D2, I put this formula

=(1+(1/C2))^C2

and then autofilled to D20. The entry in D20 is very close to *e*.

	n	f(n)
1		
2	1	2
3	2	2.25
4	3	2.37037037
5	4	2.44140625
6	5	2.48832
7	6	2.521626372
8	7	2.546499697
9	8	2.565784514
10	9	2.581174792
11	10	2.59374246
12	25	2.665836331
13	50	2.691588029
14	100	2.704813829
15	200	2.711517123
16	400	2.714891744
17	800	2.716584847
18	1000	2.716923932
19	100000	2.718268237
20	100000000	2.718281786
21		

D2 =(1+(1/C2))^C2

Figure 20-12:
Getting to *e*.

Mathematicians can tell you another way to get to *e*:

$$e = 1 + \frac{1}{1!} + \frac{1}{2!} + \frac{1}{3!} + \frac{1}{4!} + \dots$$

Those exclamation points signify *factorial*. 1! = 1, 2! = 2 × 1, 3! = 3 × 2 × 1. (For more on factorials, see Chapter 16.)

Excel helps visualize this one, too. Figure 20-13 lays out a spreadsheet with selected numbers up to 170 in Column C. In D2, I put this formula:

=1+ 1/FACT(C2)

and, as the Formula Bar in the figure shows, in D3 I put this one:

=D2 +1/ FACT(C3)

Then I autofilled up to D17. The entry in D17 is very close to *e*. In fact, from D11 on, you see no change, even if you increase the amount of decimal places.

Figure 20-13:
Another
path to *e*.

D3	▼	*fx* =D2 +1/ FACT(C3)	
	A	B	C	D
1			n	f(n)
2			1	2
3			2	2.5
4			3	2.6666666667
5			4	2.7083333333
6			5	2.7166666667
7			6	2.7180555556
8			7	2.7182539683
9			8	2.7182787698
10			9	2.7182815256
11			10	2.7182818011
12			25	2.7182818011
13			50	2.7182818011
14			100	2.7182818011
15			150	2.7182818011
16			160	2.7182818011
17			170	2.7182818011
18				

Why did I stop at 170? Because that takes Excel to the max. At 171, you get an error message.

So *e* is associated with growth. Its value is 2.781828 . . . The three dots mean you never quite get to the exact value (like π, the constant that enables you to find the area of a circle).

This number pops up in all kinds of places. It's in the formula for the normal distribution (see Chapter 8), and it's in distributions I discuss in Chapter 17. Many natural phenomena are related to *e*.

It's so important that scientists, mathematicians, and business analysts use it as the base for logarithms. Logarithms to the base *e* are called *natural logarithms*. A natural logarithm is typically abbreviated as *ln*. (In Chapter 15, in the section "Do two correlation coefficients differ?," I use the notation log_e. It's the same as *ln*.)

Table 20-2 presents some comparisons (rounded to three decimal places) between common logarithms and natural logarithms:

Table 20-2	Some Common Logarithms (Log) and Natural Logarithms (Ln)	
Number	**Log**	**Ln**
e	0.434	1.000
10	1.000	2.303
50	1.699	3.912
100	2.000	4.605
453	2.656	6.116
1000	3.000	6.908

One more thing. In many formulas and equations, it's often necessary to raise *e* to a power. Sometimes the power is a fairly complicated mathematical expression. Because superscripts are usually printed in small font, it can be a strain to have to constantly read them. To ease the eyestrain, mathematicians have invented a special notation: *exp*. Whenever you see *exp* followed by something in parentheses, it means to raise *e* to the power of whatever's in the parentheses. For example,

$$exp(1.6) = e^{1.6} = 4.953$$

Excel's EXP function does that calculation for you.

Speaking of raising *e*, when Google, Inc. filed their IPO they said they wanted to raise $2,718,281,828, which is *e* times a billion dollars rounded to the nearest dollar.

On to the Excel functions.

LOGNORMDIST

A random variable is said to be *lognormally* distributed if its natural logarithm is normally distributed. Maybe the name is a little misleading because I just said *log* means "common logarithm" and *ln* means "natural logarithm."

Unlike the normal distribution, the lognormal can't have a negative number as a possible value for the variable. Also unlike the normal, the lognormal is not symmetric — it's skewed to the right.

Like the Weibull distribution, engineers use it to model the breakdown of physical systems — particularly of the wear-and-tear variety. Here's where the large numbers-to-small numbers property of logarithms comes into play. When huge numbers of hours figure into a system's life cycle, it's easier to think about the distribution of logarithms than the distribution of the hours.

Excel's LOGNORMDIST works with the lognormal distribution. You specify a value, a mean, and a standard deviation for the lognormal. LOGNORMDIST returns the probability that the variable is at most that value.

For example, the FarKlempt Robotics Corporation has gathered extensive hours-to-failure data on a universal joint component that goes into their robots. They find that hours-to-failure is lognormally distributed with a mean of 10 and a standard deviation of 2.5. What is the probability that this component fails in at most 10,000 hours?

Figure 20-14 shows the LOGNORMDIST Function Arguments dialog box for this example. In the X box, I entered ln(10000). I entered 10 into the Mean box and 2.5 into the Standard_dev box. The dialog box shows the answer, .000929 (and some more decimals).

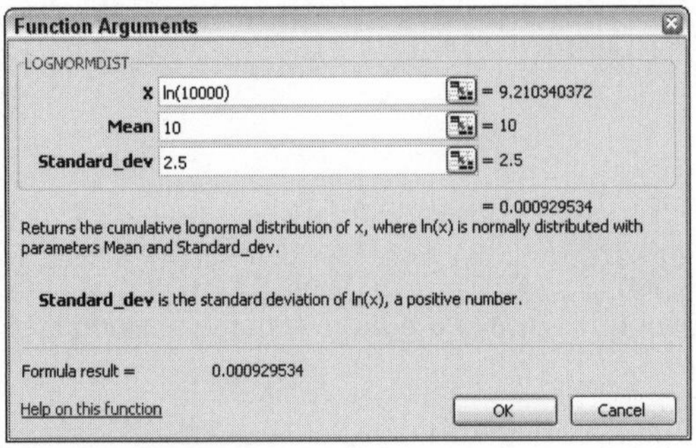

Figure 20-14:
The
Function
Arguments
dialog
box for
LOGNORM-
DIST.

LOGINV

LOGINV turns LOGNORMDIST around. You supply a probability, a mean, and a standard deviation for a lognormal distribution. LOGINV gives you the value of the random variable that cuts off that probability.

To find the value that cuts off .001 in the preceding example's distribution, I use the Function Arguments dialog box for LOGINV (see Figure 20-15). With the indicated entries, the dialog box shows that the value is 9.722 (and more decimals).

By the way, in terms of hours that's 16,685 — just for .001.

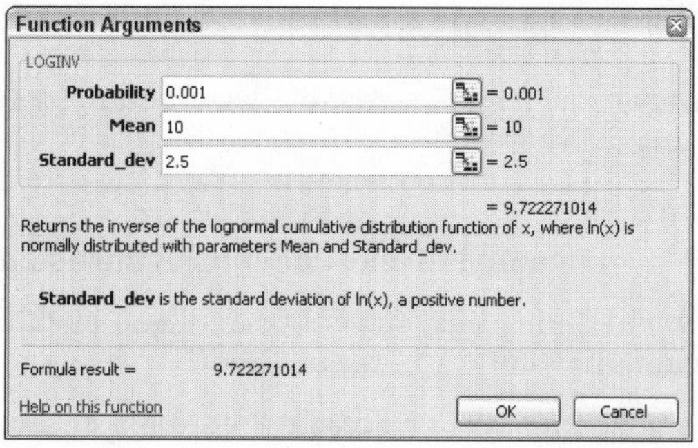

Figure 20-15:
The Function Arguments dialog box for LOGINV.

Array function: LOGEST

In Chapter 14, I tell you all about linear regression. It's possible to have a relationship between two variables that's curvilinear rather than linear.

The equation for a line that fits a scatterplot is

$$y' = a + bx$$

One way to fit a curve through a scatterplot is with this equation:

$$y' = ae^{bx}$$

LOGEST estimates a and b for this curvilinear equation. Figure 20-16 shows the LOGEST Function Arguments dialog box and the data for this example. It also shows an array for the results.

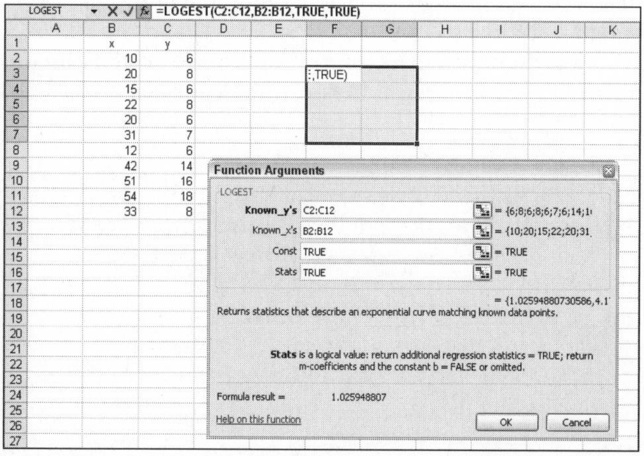

Figure 20-16:
The Function Arguments dialog box for LOGEST along with data and the selected array for the results.

Array functions can be a bit tricky, so I'll go through all the steps for you:

1. **With the data entered, select a five-row-by-two-column array of cells for LOGEST's results.**

   I selected F3:G7.

2. **Click the Insert Function button to open the Insert Function dialog box.**

3. **In the Insert Function dialog box, select LOGEST and click OK to open the Function Arguments dialog box for LOGEST.**

4. **In the Known_y's box, enter the cell range that holds the scores for the y-variable.**

   For this example, that's C2:C12.

5. **In the Known_x's box, enter the cell range that holds the scores for the x-variable.**

   For this example, it's B2:B12.

6. **In the Const box, type TRUE (or leave it blank) to calculate the value of a in the curvilinear equation I showed you; type FALSE to set a to 1.**

   I entered TRUE. The dialog box uses b where I use a. No set of symbols is standard.

7. **In the Stats box, type TRUE to return the regression statistics in addition to a and b; type FALSE (or leave it blank) to return just a and b.**

   Again, the dialog box uses b where I use a, and *m-coefficient* where I use b.

8. **IMPORTANT: Do NOT click OK. Because this is an array function, press Ctrl+Shift+Enter to put LOGEST's answers into the selected array.**

Figure 20-17 shows LOGEST's results. They're not labeled in any way, so I added the labels for you in the worksheet. The left column gives you $exp(b)$ (more on that in a moment), standard error of b, R Square, F, and the $SS_{regression}$. The right column provides a, standard error of a, standard error of estimate, degrees of freedom, and $SS_{residual}$. For more on these statistics, see Chapters 14 and 15.

Figure 20-17:
LOGEST's
results in
the selected
array.

	A	B	C	D	E	F	G	H
1		x	y					
2		10	6					
3		20	8		exp(slope)	1.025949	4.171775	intercept
4		15	6		st error of slope	0.003109	0.098679	st error of intercept
5		22	8		R-Square	0.882981	0.150622	st error of estimate
6		20	6		F	67.9108	9	df
7		31	7		SSregression	1.540693	0.204183	SSresidual
8		12	6					
9		42	14					
10		51	16					
11		54	18					
12		33	8					

About *exp(b)*. LOGEST, unfortunately, doesn't return the value of *b* — the exponent for the curvilinear equation. To find the exponent, you have to calculate the natural logarithm of what it does return. Applying Excel's LN worksheet function here gives 0.0256 as the value of the exponent.

So the curvilinear regression equation for the sample data is:

$$y' = 4.1718e^{0.0256x}$$

or in that *exp* notation I told you about,

$$y' = 4.1718exp(0.0256x)$$

A good way to help yourself understand all this is to use the Chart Wizard to create a scatterplot (see Chapter 3). Then, right-click on a data point in the plot and select Add Trendline from the menu. That opens the Add Trendline dialog box (see Figure 20-18). On the Type tab, select Exponential, as I've done in the figure.

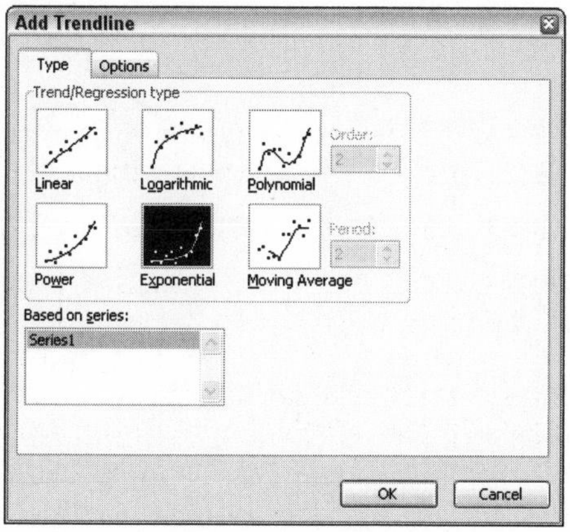

Figure 20-18:
The Type
tab on
the Add
Trendline
dialog box.

Next, click the Options tab. Select the Display Equation on Chart option (see Figure 20-19).

Click OK, and you have a scatterplot complete with curve and equation. I reformatted mine in several ways to make it look clearer on the printed page. Figure 20-20 shows the result.

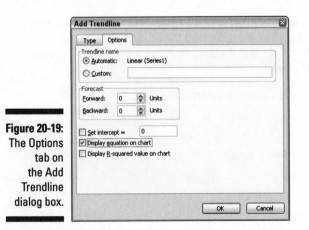

Figure 20-19:
The Options
tab on
the Add
Trendline
dialog box.

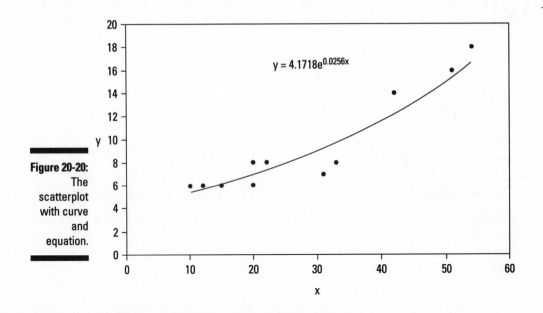

Figure 20-20:
The
scatterplot
with curve
and
equation.

Array function: GROWTH

GROWTH is curvilinear regression's answer to TREND (see Chapter 14). You can use this function two ways — to predict a set of *y*-values for the *x*-values in your sample, or to predict a set of *y*-values for a new set of *x*-values.

Predicting y's for the x's in your sample

Figure 20-21 shows GROWTH set up to calculate y's for the x's I already have. I include the Formula Bar in this screen shot so you can see what the formula looks like for this use of GROWTH.

Figure 20-21:
The
Function
Arguments
dialog
box for
GROWTH
along with
the sample
data.
GROWTH is
set up to
predict x's
for the
sample y's.

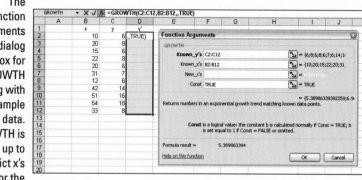

1. **With the data entered, select a cell range for GROWTH's answers.**

   I selected D2:D12 to put the predicted y's right next to the sample y's.

2. **Click the Insert Function button to open the Insert Function dialog box.**

3. **In the Insert Function dialog box, select GROWTH and click OK to open the Function Arguments dialog box for GROWTH.**

4. **In the Known_y's box, enter the cell range that holds the scores for the y-variable.**

   For this example, that's C2:C12.

5. **In the Known_x's box, enter the cell range that holds the scores for the x-variable.**

   For this example, it's B2:B12.

6. **Leave the New_x's box blank.**

7. **In the Const box, type TRUE (or leave it blank) to calculate a; type FALSE to set a to 1.**

   I entered TRUE. (I really don't know why you'd enter FALSE.) Again, the dialog uses b where I use a.

8. **IMPORTANT: Do NOT click OK. Because this is an array function, press Ctrl+Shift+Enter to put GROWTH's answers into the selected column.**

Figure 20-22 shows the answers in D2:D12.

	A	B	C	D	E
1		x	y	y'	
2		10	6	5.389863	
3		20	8	6.963614	
4		15	6	6.126412	
5		22	8	7.329697	
6		20	6	6.963614	
7		31	7	9.230331	
8		12	6	5.673214	
9		42	14	12.23488	
10		51	16	15.40746	
11		54	18	16.63827	
12		33	8	9.715578	

Predicting a new set of y's for a new set of x's

Here, I use GROWTH to predict y's for a new set of *x*'s. Figure 20-23 shows GROWTH set up for this. The figure also shows the selected array of cells for the results. Again, I include the Formula Bar to show you the formula for this use of the function.

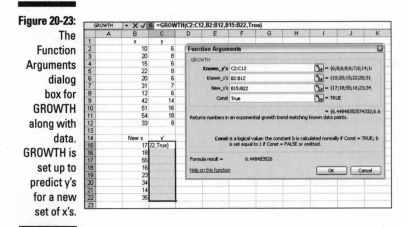

1. **With the data entered, select a cell range for GROWTH's answers.**

   I selected C15:C22.

2. **Click the Insert Function button to open the Insert Function dialog box.**

3. **In the Insert Function dialog box, select GROWTH and click OK to open the Function Arguments dialog box for GROWTH.**

4. **In the Known_y's box, enter the cell range that holds the scores for the *y*-variable.**

   For this example, that's C2:C12.

5. **In the Known_x's box, enter the cell range that holds the scores for the *x*-variable.**

   For this example, it's B2:B12.

6. **In the New_x's box enter the cell range that holds the new scores for the *x*-variable.**

   That's B15:B22.

7. **In the Const box, type TRUE (or leave it blank) to calculate *a*; type FALSE to set *a* to 1.**

   I typed TRUE. (Again, I really don't know why you'd enter FALSE.)

8. **IMPORTANT: Do NOT click OK. Because this is an array function, press Ctrl+Shift+Enter to put GROWTH's answers into the selected column.**

Figure 20-24 shows the answers in C15:C22.

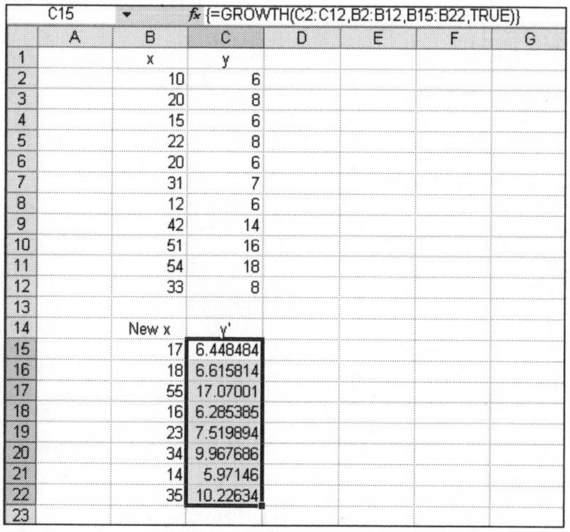

Figure 20-24:
The results
of GROWTH:
Predicted
y's for a new
set of x's.

	C15	▼	*fx* {=GROWTH(C2:C12,B2:B12,B15:B22,TRUE)}				
	A	B	C	D	E	F	G
1		x	y				
2		10	6				
3		20	8				
4		15	6				
5		22	8				
6		20	6				
7		31	7				
8		12	6				
9		42	14				
10		51	16				
11		54	18				
12		33	8				
13							
14		New x	y'				
15		17	6.448484				
16		18	6.615814				
17		55	17.07001				
18		16	6.285385				
19		23	7.519894				
20		34	9.967686				
21		14	5.97146				
22		35	10.22634				
23							

When Your Worksheet Is a Database

· ·

In This Appendix

▶ Databases in Excel

▶ Naming arrays

▶ Statistics in databases

▶ Pivot tables

· ·

*E*xcel's main function in life is to perform calculations. Many of those calculations revolve around built-in statistical capabilities.

You can also set up a worksheet to store information in something like a database, although Excel is not as sophisticated as a dedicated database package. Excel offers database functions that are much like its statistical functions, so I thought I'd familiarize you with them.

Introducing Excel Databases

Strictly speaking, Excel provides a *data list*. This is an array of worksheet cells into which you enter related data in a uniform format. You organize the data in columns, and you put a name at the top of each column. In database terminology, each named column is a *field*. Each row is a separate *record*.

This type of structure is useful for keeping inventories, as long as they're not overly huge. You wouldn't use an Excel database for record-keeping in a warehouse or a large corporation, for example. For a small business, however, it might fit the bill.

The Satellites database

Figure A-1 shows an example. This is an inventory of the classic satellites in our solar system. By "classic," I mean that astronomers discovered most of them before the twentieth century via conventional telescopes. The three twentieth-century entries are so dim that astronomers discovered them by examining photographic plates. Today's super telescopes and space probes have revealed many more satellites that I don't include.

	A	B	C	D	E	F	G	H
1		Name	Planet	Orbital Period (Days)	Average Distance (X 1000 km)	Year Discovered	Discoverer	
2		*io	Saturn	1.26	>150	>1877	Galileo	
3			>20				Cassini	
4								
5								
6								
7								
8								
9								
10		Name	Planet	Orbital Period (Days)	Average Distance (X 1000 km)	Year Discovered	Discoverer	
11		Amalthea	Jupiter	0.50	181.30	1892	Barnard	
12		Ariel	Uranus	2.52	191.24	1851	Lassell	
13		Callisto	Jupiter	16.69	1883.00	1610	Galileo	
14		Charon	Pluto	6.39	19.64	1978	Christy	
15		Deimos	Mars	1.26	23.46	1877	Hall	
16		Dione	Saturn	2.74	377.40	1684	Cassini	
17		Enceladus	Saturn	1.37	238.02	1789	Herschel	
18		Europa	Jupiter	3.55	670.90	1610	Galileo	
19		Ganymede	Jupiter	7.15	1070.00	1610	Galileo	
20		Hyperion	Saturn	21.28	1481.00	1848	Bond	
21		Iapetus	Saturn	79.33	3561.30	1671	Cassini	
22		Io	Jupiter	1.77	421.60	1610	Galileo	
23		Mimas	Saturn	9.42	185.52	1789	Herschel	
24		Miranda	Uranus	1.41	129.78	1948	Kuiper	
25		Moon	Earth	27.32	384.40	N/A	N/A	
26		Nereid	Neptune	360.14	5513.40	1949	Kuiper	
27		Oberon	Uranus	13.46	582.50	1787	Herschel	
28		Phobos	Mars	0.32	9.38	1877	Hall	
29		Phoebe	Saturn	-550.48	12952.00	1898	Pickering	
30		Rhea	Saturn	4.52	527.04	1672	Cassini	
31		Tethys	Saturn	1.89	294.66	1684	Cassini	
32		Titan	Saturn	15.94	1221.85	1655	Huygens	
33		Titania	Uranus	8.71	435.84	1787	Herschel	
34		Triton	Neptune	-5.88	354.80	1846	Lassell	
35		Umbriel	Uranus	4.14	265.97	1851	Lassell	
36								

Figure A-1:
The Satellites database.

The database is in cells B10:G35. The Name field provides the name of the satellite; the Planet field indicates the planet around which the satellite revolves.

Orbital Period (Days) shows how long it takes for a satellite to make a complete revolution around its planet. Our Moon, for example, takes a little over 27 days. A couple of records have negative values for this. That means they revolve around the planet in a direction opposite to the planet's rotation.

Average Distance (×1000 km) is the average distance from the planet to the satellite in thousands of kilometers. The last two fields provide the year of discovery and the astronomer who discovered the satellite. For our Moon, of course, those two are unknown.

The criteria range

I copied the column headers — excuse me, field names — into the top row. I also put some information into nearby cells. This area is for *criteria* — a part and parcel of each database function. It's not necessary to have this at the top of the worksheet. You can designate any range in the worksheet to hold the criteria.

When you use a function it's in this format:

=FUNCTION(Database, Field, Criteria)

The function operates on the specified database, in the designated field, according to the indicated criteria. (Criteria is a plural. The singular form is criterion.)

For example, if you want to know how many satellites revolve around Saturn, you select a cell and type

=DCOUNT(B1:G35,E10,C1:C2)

Here's what this formula means: In the database (B1:G35), DCOUNT tallies the amount of number-containing cells in the Average Distance ($\times$ 1000 km) field, constrained by the criterion specified in the cell range C1:C2. That criterion is equivalent to Planet = Saturn. Note that a criterion has to include at least one column header . . . uh . . . field-name, and at least one row.

When you include more than one row, you're saying "or." For example, if your criterion happens to be G1:G3, you're specifying satellites discovered by Galileo or Cassini.

When you include more than one column in a criterion, you're saying "and." If your criterion is E1:F2, you're specifying satellites farther than 150,000 km from their planets and discovered after 1877.

The game of the name

Database function formulas are more informative if you can use names instead of cell arrays. Here's an example:

=DCOUNT(Satellites,Average_Distance__X_1000_km,E1:E2)

Excel allows you to make this happen.

According to Form

Excel provides a Data Form to help you work with databases. Highlight the entire cell range of the database, including the column headers. Select the Data menu and then select Form.

The figure shows the appearance of the Data Form when you open it with the entire database selected. Excel fills in the field names automatically, and the fields populate with the values from the first record. You can use the form to navigate through the database, and you can use it to add a record. You can start with one record and use the add-a-record capability to enter all the rest, but for me it's easier to just type each record.

Sheet1		
Name :	Amalthea	1 of 25
Planet :	Jupiter	New
Orbital Period (Days):	0.5	Delete
Average Distance (X 1000 km):	181.3	Restore
Year Discovered :	1892	Find Prev
Discoverer:	Barnard	Find Next
		Criteria
		Close

Whenever you add records (and whichever way you add them), be sure to go back to the Define Name dialog box (see Figure A-2) and increase the cell range attached to the database name.

You can attach a name to an array of cells and then use that name whenever you refer to that array. You can attach a name to a single cell and use that name to refer to the cell. Excel refers to this as *inserting* a name.

Databases are a natural home for this capability. To give the name Satellites to the database, follow these steps:

1. **Select the range you want to name.**

   For this example, that's B10:G35.

2. **From the Insert menu, select Name.**

   This opens a menu, from which you select Define to open the dialog box in Figure A-2.

3. **In the Names in Workbook box, type the name you want to attach to the selected range.**

   I typed Satellites.

4. **Click OK.**

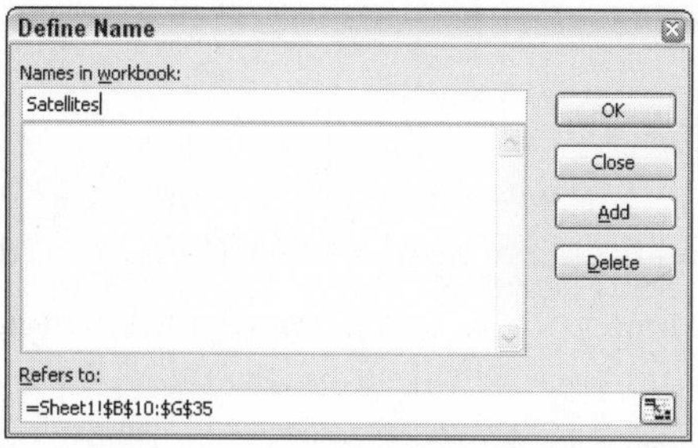

Figure A-2:
The Define
Name
dialog box.

From here on, I refer to the database as Satellites.

To attach a name to a field name, select only the cell that holds the name. Then, go through the four steps I just outlined.

The format of a database function

The formula I just showed you

=DCOUNT(Satellites,Average_Distance__X_1000_km,E1:E2)

is accessible through a Function Arguments dialog box, as is the case for all the other worksheet functions in Excel. Figure A-3 shows the equivalent dialog box for the preceding formula set against the backdrop of the database and the criteria range.

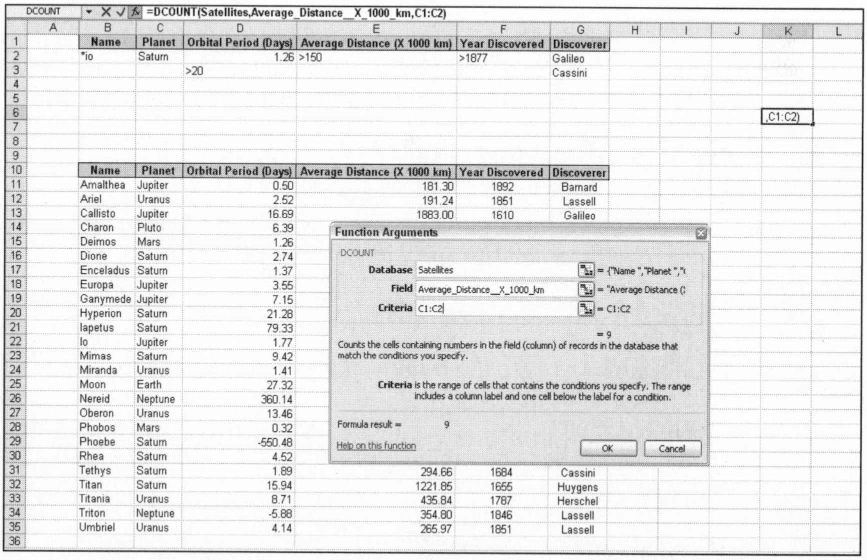

Figure A-3:
The
Function
Arguments
dialog
box for
DCOUNT.

To open the dialog box:

1. **Select a worksheet cell.**

   I selected K6.

2. **Click the Insert Function button to open the Insert Function dialog box.**

3. **In the Insert Function dialog box, select a function to open its Function Arguments dialog box.**

   From the Database category, I selected DCOUNT, the dialog box in Figure A-3.

4. **In the Function Arguments dialog box, put an entry into each box.**

   For the Database, I highlighted the entire cell range (B10:G3) to make the database name, Satellites, appear in the Database box. For the Field, I clicked in cell E10 to put its name in the Field box. For the Range, I highlighted C1:C2.

   The answer, 9, appears in the dialog box.

5. **Click OK to put the answer into the selected cell.**

All the database functions follow the same format. You access them the same way, and you fill in the same type of information in their dialog boxes. So, I'm going to skip over that sequence of steps as I describe each function, and just discuss the equivalent worksheet formula.

Counting and Retrieving

One essential database capability is to let you know how many records meet a particular criterion. Another is to retrieve records. Here are the Excel versions.

DCOUNT and DCOUNTA

DCOUNT counts records. The restriction is that the field you specify has to contain numbers. If it doesn't, the answer is zero, as in

=DCOUNT(Satellites,Name,C1:C2)

because no records in the Name field contain numbers.

DCOUNTA counts records in a different way. This one works with any field. It counts the number of nonblank records in the field that satisfy the criterion. So this formula returns 9:

=DCOUNTA(Satellites,Name,C1:C2)

Getting to "or"

Here's a tally that involves "or":

=DCOUNTA(Satellites,Name,D1:D3)

The criterion D1:D3 specifies satellites whose orbital period is 1.26 days or greater than 20 days. Multiple rows mean "or." Five satellites meet that criterion: Deimos, Hyperion, Iapetus, our Moon, and Nereid.

Wildcards

Look closely at Figure A-1 and you see the cryptic entry *io in cell B2. I included this entry to let you know that Excel database functions can deal with wildcard characters. The formula

=DCOUNTA(Satellites,Name,B1:B2)

returns 3, the number of satellites with the letter string *io* anywhere in their names (Dione, Io, and Hyperion).

DGET

DGET retrieves exactly one record. If the criteria you specify result in more than one record (or in no records), DGET returns an error message.

This formula

=DGET(Satellites,Name,D1:D2)

retrieves Deimos, the name of the satellite whose orbital period is 1.26 days.

This one

=DGET(Satellites,Name,E1:E2)

results in an error message because the criterion specifies more than one record.

Arithmetic

Excel wouldn't be Excel without calculation capabilities. Here are the ones it offers for its databases.

DMAX and DMIN

As their names suggest, these provide the maximum value and the minimum value according to your specifications. The formula

=DMAX(Satellites,Orbital_Period__Days,E1:E2)

returns 360.14. This is the maximum orbital period for any satellite that's farther than 150,000 km from its planet.

For the minimum value that meets this criterion,

=DMIN(Satellites,Orbital_Period__Days,E1:E2)

gives you -550.48. That's Phoebe, a satellite that revolves in the opposite direction to its planet's rotation.

DSUM

This one adds the values in a field. To add all the orbital periods in the satellites discovered by Galileo or by Cassini, use this formula:

=DSUM(Satellites,Orbital_Period__Days,G1:G3)

That sum is 117.64.

Want to total up all the orbital periods? (I know . . . =SUM(B11:B35). Just work with me here.)

This formula gets it done:

=DSUM(Satellites,Orbital_Period__Days,C1:C2)

Why? It's all in the criterion. E1:E3 means that Planet = Saturn or . . . anything else, because C2 is empty. The sum, by the way, is 35.457. Bottom line: Be careful whenever you include an empty cell in your criteria.

DPRODUCT

Here's a function that's probably here only because Excel's designers could create it. You specify the data values, and DPRODUCT multiplies them.

The formula

=DPRODUCT(Satellites,Orbital_Period__Days,G1:G2)

returns the product (749.832) of the orbital periods of the satellites Galileo discovered — a calculation I'm pretty sure Galileo never thought about.

Statistics

Now for the statistical database functions. These work just like the similarly named worksheet functions.

DAVERAGE

Here's the formula for the average of the orbital periods of satellites discovered after 1887:

=DAVERAGE(Satellites,Orbital_Period__Days,F1:F2)

The average is negative (-36.4086) because the specification includes those two satellites with the negative orbital periods.

DVAR and DVARP

DVAR is the database counterpart of VAR, which divides the sum of N squared deviations by N-1. This is called *sample variance*.

DVARP is the database counterpart of VARP, which divides the sum of N squared deviations by N. This is the *population variance*. (For details on VAR and VARP, sample variance and population variance, and the implications of N and N-1, see Chapter 5.)

Here's the sample variance for the orbital period of satellites farther than 150,000 km from their planets and discovered after 1877:

=DVAR(Satellites,Orbital_Period__Days,E1:F2)

That turns out to be 210,358.1.

The population variance for that same subset of satellites is

=DVARP(Satellites,Orbital_Period__Days,E1:F2)

which is 140,238.7.

Once again, if you have multiple columns in the criteria, you're dealing with "and."

DSTDEV and DSTDEVP

These two return standard deviations. The standard deviation is the square root of the variance (see Chapter 5). DSTDEV returns the sample standard deviation, which is the square root of DVAR's returned value. DSTDEVP returns the population standard deviation, the square root of DVARP's returned value.

For the specifications in the preceding example, the sample standard deviation is

=DSTDEV(Satellites,Orbital_Period__Days,E1:F2)

which is 458.6481.

The population standard deviation is

=DSTDEVP(Satellites,Orbital_Period__Days,E1:F2)

This result is 374.4846.

Pivot Tables

A *pivot table* is a cross-tabulation — another way of looking at the data. You can reorganize the database, and turn it (literally) on its side and inside out. And you can do it in any number of ways.

For example, you can set up a pivot table that has the satellites in the rows and a planet in each column, and has the data for orbital period inside the cells. Figure A-4 shows what I mean.

Figure A-4:
A pivot table of the Satellites data showing satellites, planets, and orbital period.

	A	B	C	D	E	F	G	H	I	J
1										
2										
3	Sum of Orbital Period (Days)	Planet								
4	Name	Earth	Jupiter	Mars	Neptune	Pluto	Saturn	Uranus	Grand Total	
5	Amalthea		0.5						0.5	
6	Ariel							2.52	2.52	
7	Callisto		16.69						16.69	
8	Charon					6.387			6.387	
9	Deimos			1.26					1.26	
10	Dione						2.74		2.74	
11	Enceladus						1.37		1.37	
12	Europa		3.55						3.55	
13	Ganymede		7.15						7.15	
14	Hyperion						21.28		21.28	
15	Iapetus						79.33		79.33	
16	Io		1.77						1.77	
17	Mimas						9.42		9.42	
18	Miranda							1.41	1.41	
19	Moon	27.32							27.32	
20	Nereid				360.14				360.14	
21	Oberon							13.46	13.46	
22	Phobos			0.32					0.32	
23	Phoebe						-550.48		-550.48	
24	Rhea						4.52		4.52	
25	Tethys						1.89		1.89	
26	Titan						15.94		15.94	
27	Titania							8.71	8.71	
28	Triton				-5.88				-5.88	
29	Umbriel							4.14	4.14	
30	Grand Total	27.32	29.66	1.58	354.26	6.387	-413.99	30.24	35.457	
31										

Figure A-5 shows a pivot table that presents another view of the data. This one takes the spotlight off the individual satellites and puts it on the planets. Each planet's row is divided into two rows — one for the Orbital Period and one for the Average Distance. The numbers are the sums across each planet's satellites. Adding the Orbital Period for all of Jupiter's satellites gives you 29.66, for example.

Figure A-5: Another pivot table of the Satellites data, showing planets, orbital period, and average distance.

	A	B	C	D
1				
2				
3	Planet	Data	Total	
4	Earth	Sum of Orbital Period (Days)	27.32	
5		Sum of Average Distance (X 1000 km)	384.4	
6	Jupiter	Sum of Orbital Period (Days)	29.66	
7		Sum of Average Distance (X 1000 km)	4226.8	
8	Mars	Sum of Orbital Period (Days)	1.58	
9		Sum of Average Distance (X 1000 km)	32.84	
10	Neptune	Sum of Orbital Period (Days)	354.26	
11		Sum of Average Distance (X 1000 km)	5868.2	
12	Pluto	Sum of Orbital Period (Days)	6.387	
13		Sum of Average Distance (X 1000 km)	19.64	
14	Saturn	Sum of Orbital Period (Days)	-413.99	
15		Sum of Average Distance (X 1000 km)	20838.79	
16	Uranus	Sum of Orbital Period (Days)	30.24	
17		Sum of Average Distance (X 1000 km)	1605.33	
18	Total Sum of Orbital Period (Days)		35.457	
19	Total Sum of Average Distance (X 1000 km)		32976	
20				

Excel provides a wizard to help you create this kind of table. Here's how to use the wizard. This example focuses on creating the pivot table in Figure A-4.

1. **Select the Data menu and choose PivotTable and PivotChart Report to open the PivotTable and PivotChart Report Wizard (see Figure A-6).**

   In this dialog box you tell the wizard the kind of data you want to analyze and the type of report you want to create.

Figure A-6: The PivotTable and PivotChart Report Wizard's first dialog box.

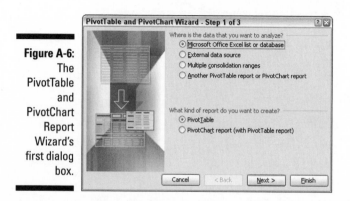

PivotTable and PivotChart Wizard - Step 1 of 3

Where is the data that you want to analyze?
- ⦿ Microsoft Office Excel list or database
- ○ External data source
- ○ Multiple consolidation ranges
- ○ Another PivotTable report or PivotChart report

What kind of report do you want to create?
- ⦿ PivotTable
- ○ PivotChart report (with PivotTable report)

Cancel | < Back | Next > | Finish

2. Make your selections.

For this example, the default settings work just fine — the Microsoft Office Excel List or Database option in the upper area, and the PivotTable option in the lower area.

3. Click Next to continue (see Figure A-7).

Figure A-7:
The PivotTable and PivotChart Report Wizard's second dialog box.

4. Type the source of the data in the Range box.

This is where naming the database comes in handy. All I had to do was type Satellites.

5. Click Next to continue.

In the final dialog box (see Figure A-8), you indicate where you want the report. I chose the New Worksheet option to create a new page for the report.

Figure A-8:
The PivotTable and PivotChart Report Wizard's last dialog box.

6. Click Finish.

Because I chose New Worksheet in Step 5, a new page opens with the layout of the Pivot Table, along with the PivotTable toolbar and the PivotTable Field List (see Figure A-9).

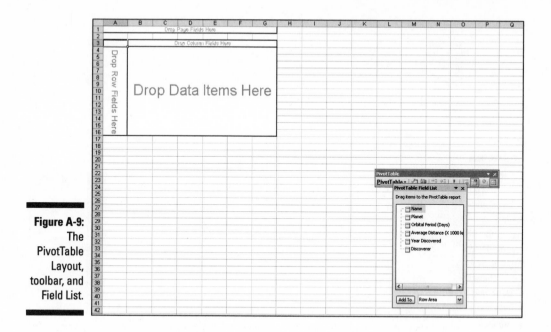

Figure A-9:
The
PivotTable
Layout,
toolbar, and
Field List.

7. Fill in the PivotTable Layout.

To populate the PivotTable Layout, select a field from the PivotTable Field List, drag it to the Layout, and drop it. Where you drop it, of course, determines its appearance in the PivotTable.

I begin with the rows. To make the names of the satellites appear in the rows, I selected Name and dragged it into the area labeled Drop Row Fields Here. Figure A-10 shows the result. In addition to the satellite names in the rows, Name becomes bolded in the Field List to indicate it's in the table.

Next, I dragged Planet to the area labeled Drop Column Fields Here (see Figure A-11).

Dragging Orbital Period (Days) from the Field List and dropping it into the Drop Data Items Here area results in the table shown in Figure A-4.

You can make the table more elaborate by dropping another set of data items into the Drop Data Items Here area. Figure A-12 shows what happens after I dropped Average Distance (× 1000 km) into that area.

Notice that each row in Figure A-12 begins with the words "Sum of," indicating that each cell entry is a sum. From the Satellites database, you know that each entry is an individual number, not a sum. What's the story? In this database, each satellite-planet combination is unique. That's why the numbers in this pivot table aren't really sums. By contrast, the pivot table shown in Figure A-5 is an example of summed data. In looking at just the planets, that table adds up the data for each planet's satellites.

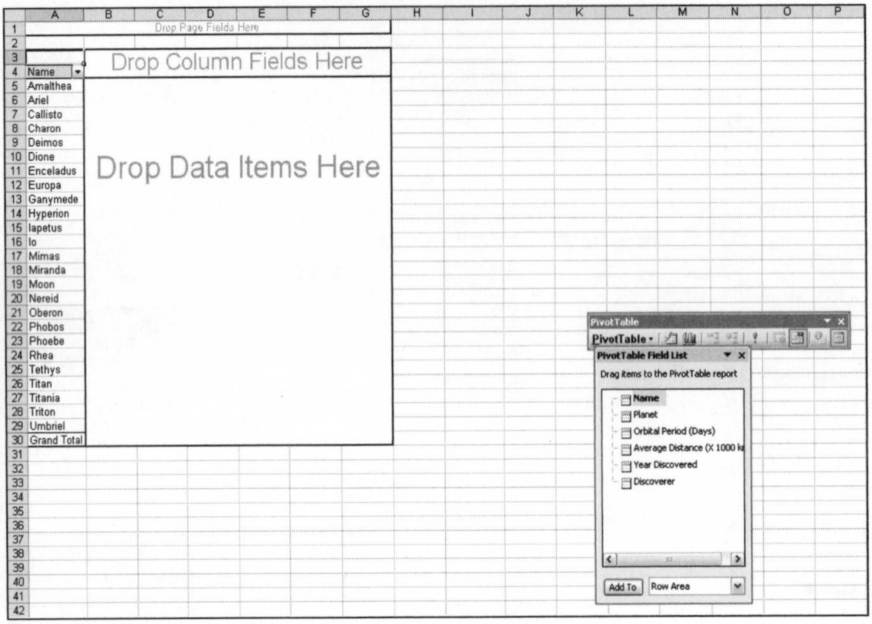

Figure A-10:
The
PivotTable
Layout after
dragging the
Name Field
into the area
labeled
Drop Row
Fields Here.

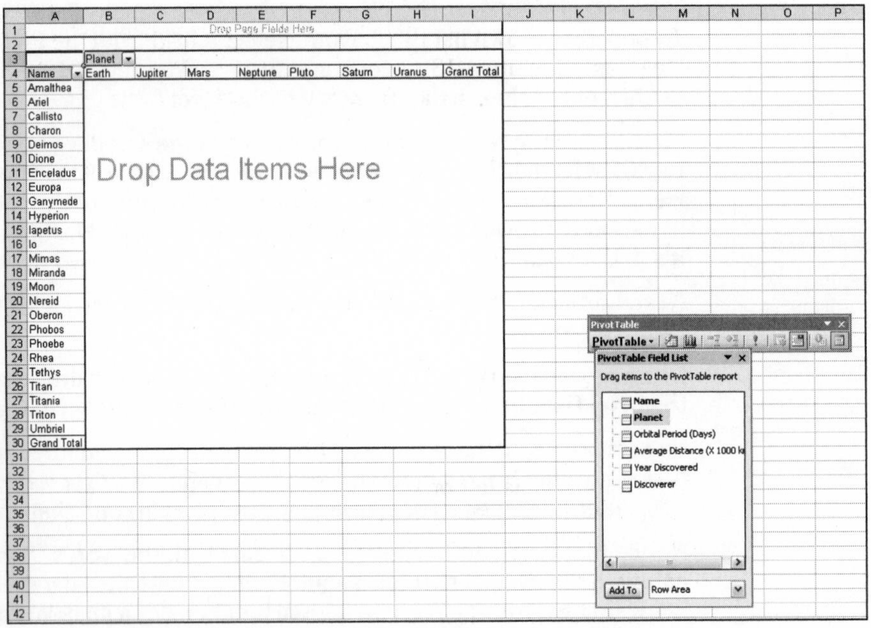

Figure A-11:
The
PivotTable
Layout after
dragging the
Planet Field
into the area
labeled
Drop
Column
Fields Here.

	A	B	C	D	E	F	G	H	I	J	K	L
1				Drop Page Fields Here								
2												
3			Planet ▼									
4	Name ▼	Data ▼	Earth	Jupiter	Mars	Neptune	Pluto	Saturn	Uranus	Grand Total		
5	Amalthea	Sum of Orbital Period (Days)		0.5						0.5		
6		Sum of Average Distance (X 1000 km)		181.3						181.3		
7	Ariel	Sum of Orbital Period (Days)							2.52	2.52		
8		Sum of Average Distance (X 1000 km)							191.24	191.24		
9	Callisto	Sum of Orbital Period (Days)		16.69						16.69		
10		Sum of Average Distance (X 1000 km)		1883						1883		
11	Charon	Sum of Orbital Period (Days)					6.387			6.387		
12		Sum of Average Distance (X 1000 km)					19.64			19.64		
13	Deimos	Sum of Orbital Period (Days)			1.26					1.26		
14		Sum of Average Distance (X 1000 km)			23.46					23.46		
15	Dione	Sum of Orbital Period (Days)						2.74		2.74		
16		Sum of Average Distance (X 1000 km)						377.4		377.4		
17	Enceladus	Sum of Orbital Period (Days)						1.37		1.37		
18		Sum of Average Distance (X 1000 km)						238.02		238.02		
19	Europa	Sum of Orbital Period (Days)		3.55						3.55		
20		Sum of Average Distance (X 1000 km)		670.9						670.9		
21	Ganymede	Sum of Orbital Period (Days)		7.15						7.15		
22		Sum of Average Distance (X 1000 km)		1070								
23	Hyperion	Sum of Orbital Period (Days)										
24		Sum of Average Distance (X 1000 km)										
25	Iapetus	Sum of Orbital Period (Days)										
26		Sum of Average Distance (X 1000 km)										
27	Io	Sum of Orbital Period (Days)		1.77								
28		Sum of Average Distance (X 1000 km)		421.6								
29	Mimas	Sum of Orbital Period (Days)										
30		Sum of Average Distance (X 1000 km)										
31	Miranda	Sum of Orbital Period (Days)										
32		Sum of Average Distance (X 1000 km)										
33	Moon	Sum of Orbital Period (Days)	27.32									
34		Sum of Average Distance (X 1000 km)	384.4									
35	Nereid	Sum of Orbital Period (Days)				360.14						
36		Sum of Average Distance (X 1000 km)				5513.4						
37	Oberon	Sum of Orbital Period (Days)										
38		Sum of Average Distance (X 1000 km)										
39	Phobos	Sum of Orbital Period (Days)		0.32						0.32		
40		Sum of Average Distance (X 1000 km)		9.38						9.38		
41	Phoebe	Sum of Orbital Period (Days)								0.48		
42		Sum of Average Distance (X 1000 km)								12952		

Figure A-12: The PivotTable Layout after dragging Average Distance (× 1000 km) into the area labeled Drop Column Fields Here.

Note the down arrow next to each field name. Clicking the down arrow opens a drop-down list that allows you to display some or all of that field's data. Figure A-13 shows the drop-down list for Planet. Deselecting a check box removes that planet from the table.

Figure A-13: The Planet drop-down list.

One more area is left on the layout — Drop Page Fields Here, at the top of the page. Dropping a field into that area turns the table into a kind of multipage catalog. Each page shows just the data for an individual in that field.

I dropped Discoverer into that area. This creates a drop-down list for that field, and I use it to page through each astronomer's discoveries. Figure A-14 shows the four satellites Herschel discovered.

	A	B	C	D	E
1	Discoverer	Herschel ▼			
2					
3			Planet ▼		
4	Name ▼	Data ▼	Saturn	Uranus	Grand Total
5	Enceladus	Sum of Orbital Period (Days)	1.37		1.37
6		Sum of Average Distance (X 1000 km)	238.02		238.02
7	Mimas	Sum of Orbital Period (Days)	9.42		9.42
8		Sum of Average Distance (X 1000 km)	185.52		185.52
9	Oberon	Sum of Orbital Period (Days)		13.46	13.46
10		Sum of Average Distance (X 1000 km)		582.5	582.5
11	Titania	Sum of Orbital Period (Days)		8.71	8.71
12		Sum of Average Distance (X 1000 km)		435.84	435.84
13	Total Sum of Orbital Period (Days)		10.79	22.17	32.96
14	Total Sum of Average Distance (X 1000 km)		423.54	1018.34	1441.88
15					

Figure A-14:
Selecting
from the
Page field
(Discoverer).

The PivotTable toolbar (see Figure A-15) provides a number of tools that allow you to craft the table's appearance. Table Options, for example, opens a dialog box you can use to remove the row totals and column totals. Formulas enables you to create new fields by performing calculations on existing fields.

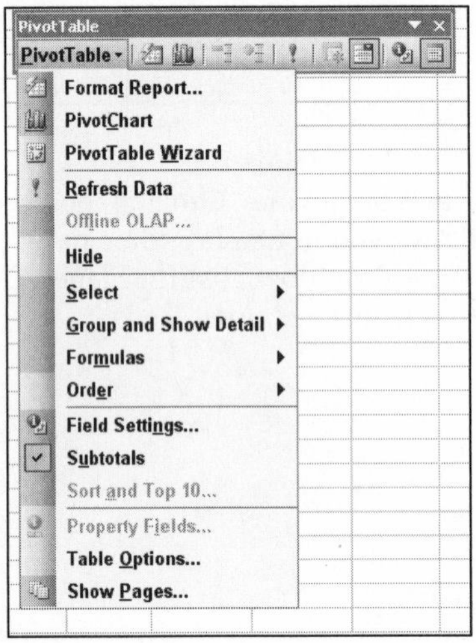

Figure A-15:
The Pivot
Table
toolbar.

If you select PivotChart, Excel creates a column graph of the table (see Figure A-16). You can reformat this chart the same way you reformat any other (see Chapter 3). You can even change the chart type, although I advise against it. In fact, Excel has some chart types it doesn't allow you to use for PivotCharts.

The importance of pivot tables is that they allow you to get your hands dirty with the data. By dropping fields into and out of the table, you might see relationships and carry out analyses that might not occur to you if you just look at the original database.

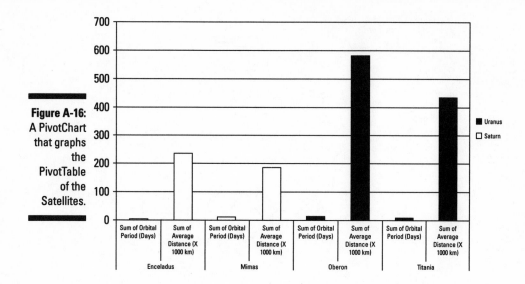

Figure A-16:
A PivotChart that graphs the PivotTable of the Satellites.

Index